Energy Management and Efficiency: Principles and Applications

Energy Management and Efficiency: Principles and Applications

Edited by Marian Lucas

SYRAWOOD
PUBLISHING HOUSE

New York

Published by Syrawood Publishing House,
750 Third Avenue, 9th Floor,
New York, NY 10017, USA
www.syrawoodpublishinghouse.com

Energy Management and Efficiency: Principles and Applications
Edited by Marian Lucas

International Standard Book Number: 978-1-68286-736-5 (Hardback)

Cataloging-in-Publication Data

Energy management and efficiency : principles and applications / edited by Marian Lucas.
    p. cm.
Includes bibliographical references and index.
ISBN 978-1-68286-736-5
1. Power resources--Management. 2. Energy consumption. 3. Energy conservation.
4. Energy development. I. Lucas, Marian.
TJ163.2 .E54 2019
333.79--dc23

TABLE OF CONTENTS

PREFACE

The planning and operation of the consumption and production of energy is referred to as energy management. It strives to achieve climate protection, resource conservation and cost savings. Energy assessment is one of the significant initial stages in developing an effective cost control energy program. Facility management, logistics, production, energy procurement, maintenance, etc. are the operational functions in which energy management is required. Formulating energy strategies for industries involves the considerations of the use of renewable energies, yield expectations, energy investments, etc. Potential energy strategies fall under the classification of passive strategy, maximum strategy and strategy aimed at short-term or long-term profit maximization. Energy efficiency is achieved when there is optimum reduction in the amount of energy expenditure required for providing services. This book elucidates the concepts and innovative models around prospective developments with respect to energy management and efficiency. Also included in this book is a detailed explanation of the various principles and applications of energy management. It is a resource guide for experts as well as students.

Various studies have approached the subject by analyzing it with a single perspective, but the present book provides diverse methodologies and techniques to address this field. This book contains theories and applications needed for understanding the subject from different perspectives. The aim is to keep the readers informed about the progresses in the field; therefore, the contributions were carefully examined to compile novel researches by specialists from across the globe.

Indeed, the job of the editor is the most crucial and challenging in compiling all chapters into a single book. In the end, I would extend my sincere thanks to the chapter authors for their profound work. I am also thankful for the support provided by my family and colleagues during the compilation of this book.

Editor

The Effect of Distributed Parameters on Conducted EMI from DC-Fed Motor Drive Systems in Electric Vehicles

Li Zhai [1,2,*], **Liwen Lin** [1,2], **Xinyu Zhang** [3] **and Chao Song** [1,2]

[1] National Engineering Laboratory for Electric Vehicle, Beijing Institute of Technology, Beijing 100081, China; 15201430735@163.com (L.L.); 18210606506@163.com (C.S.)

[2] Co-Innovation Center of Electric Vehicles in Beijing, Beijing Institute of Technology, Beijing 100081, China

[3] Beijing Institute of Radio Metrology and Measurement, Beijing 100854, China; xinyuzhang203@163.com

* Correspondence: zhaili26@bit.edu.cn

Academic Editor: Joeri Van Mierlo

Abstract: The large dv/dt and di/dt outputs of power devices in DC-fed motor drive systems in electric vehicles (EVs) always introduce conducted electromagnetic interference (EMI) emissions and may lead to motor drive system energy transmission losses. The effect of distributed parameters on conducted EMI from the DC-fed high voltage motor drive systems in EVs is studied. A complete test for conducted EMI from the direct current fed(DC-fed) alternating current (AC) motor drive system in an electric vehicle (EV) under load conditions is set up to measure the conducted EMI of high voltage DC cables and the EMI noise peaks due to resonances in a frequency range of 150 kHz–108 MHz. The distributed parameters of the motor can induce bearing currents under low frequency sine wave operation. However the impedance of the distributed parameters of the motor is very high at resonance frequencies of 500 kHz and 30 MHz, and the effect of the bearing current can be ignored, so the research mainly focuses on the distributed parameters in inverters and cables at 500 kHz and 30 MHz, not the effect of distributed parameters of the motor on resonances. The corresponding equivalent circuits for differential mode (DM) and common mode (CM) EMI at resonance frequencies of 500 kHz and 30 MHz are established to determine the EMI propagation paths and analyze the effect of distributed parameters on conducted EMI. The dominant distributed parameters of elements responsible for the appearing resonances at 500 kHz and 30 MHz are determined. The effect of the dominant distributed parameters on conducted EMI are presented and verified by simulation and experiment. The conduced voltage at frequencies from 150 kHz to 108 MHz can be mitigated to below the limit level-3 of CISPR25 by changing the dominant distributed parameters.

Keywords: electric vehicle; DC-fed; motor drive system; conducted electromagnetic interference (EMI); distributed parameter

1. Introduction

In the face of the worldwide demand for reduction in greenhouse gas emissions and $PM_{2.5}$ production [1], recently many countries have adopted policies, mainly in the form of tax incentives for the purchase, to increase the number of electric vehicles (EVs) and thus reduce pollutant emissions and improve the air quality, especially in urban areas [2–5]. Electromagnetic interference (EMI) considerations in EVs have become increasingly important, as the electromagnetic compatibility (EMC) regulations for EVs (typically defined from 10 kHz to 30 MHz) have become more stringent [6]. The DC-fed motor drive system of EVs, consisting of the electric motor, power inverter, and electronic controller has an essential role in EVs [7]. Large dv/dt and di/dt due to high-speed switching of power devices within a DC-fed voltage-type pulse width modulation (PWM) inverter of high-power-density and high-efficiency motor drive system always introduce unwanted higher-order harmonics currents and high frequency noise currents through parasitic/distributed parameters of the motor system [8–10], and are mainly responsible for the conducted and/or radiated electromagnetic interference (EMI) emissions which will greatly affect the behavior of low voltage supply electronic equipment (such as board bus system, sensors, vehicle control units (VCUs), battery management systems (BMSs), power batteries, and the drive motor in EVs [6,11]. Additionally, the unwanted higher-order harmonics current from the motor drive system due to the switching of insulated gate bipolar transistors (IGBTs) can not only generate common mode (CM) EMI and differential mode (DM) EMI emissions, but also increase the motor losses, which may lead to energy transmission losses and thermal problems in the power inverter system.

1.1. Literature Review

Much valuable work involving the EMI emissions of the motor drive system for conventional industrial applications has been widely conducted by many researchers [12–14]. However, the EMI from the drive motor system under varying load conditions for EVs is different from that of a conventional industrial motor with no load or invariant load conditions. Previous studies on EMI emissions from vehicle components are based on the measurements specified in the EMC standard CISPR25 (International Special Committee on Radio Interference 25) [15,16], which are implemented for low voltage components in EVs and not suitable for the high voltage applications in EVs (e.g., motor system, charging system), so we cannot correctly predict the EMI emissions from the high voltage motor system in EVs due to the fact few EMC laboratories have the dynamometer needed to study EVs under varying loading conditions, so the present study on the EMI mechanism and propagation path of the motor system is much less than that on the total EMC performance of EVs [17,18]. The EMI emissions from the high voltage cables of the AC motor drive system of EVs under load condition have not been considered in previous works.

Various parasitics and distributed parameters exist inside the motor system and they play a very important role in the generation of EMI. The high-frequency leakage currents flowing to the ground could be generated through distributed parameters between the components of the motor drive system (such as the motor, inverter, cables, etc.) and the chassis of the body of the EV at high frequency, and introduce the radiation of power cables, shaft voltage and bearing currents in the motor [19,20]. Additionally the EMI emission peaks due to resonances caused by distributed parameters may cause some energy losses of the motor drive system and decrease the efficiency of the system [21–23], so the distributed parameters at high frequency in the system should not be neglected anymore for EVs. Therefore, the effect of the distributed parameters on EMI emissions is important for identification of EMI propagation paths and the critical distributed parameters of elements responsible for EMI, and mitigation of EMI emissions [24].

Models of the motor drive system are necessary to analyze and predict the EMI sources and propagation inside the motor drive system to find the elements responsible for the EMI. Since the most basic and widely applied full-wave models based on the "black box" approach cannot show the location of the noise source or the propagation path inside the motor power inverter [20,25], some terminal modeling techniques for a two-port network were proposed to predict the conducted EMI [14,21,26]. However, there has never been a theoretical analysis of the parasitic effects of the distributed parameters on EMI noise suppression. An equivalent simulation program with integrated circuit (SPICE)-based model is a better approach to find the parts and elements of the motor inverter system responsible for EMI [12] and analyze the effects of the distributed parameters on the EMI noise. A rather simple measurement-based SPICE model of the motor power inverter has been presented [12] to quickly identify the parts responsible for EMI and help predict resonances between the two ports of the motor power inverter by a straightforward correlation between the system geometry and the parasitic circuit elements [27]. A detailed analysis of current paths and the equivalent circuits at three important resonance frequencies have been presented to determine the EMI propagation path in the motor drive system [6]. A combination of mitigation strategies was designed to mitigate the CM conducted emission by IGBT switching and the radiated emissions of AC cables.

Most of the above work focuses on analyses of CM and DM EMI propagation paths in the system based on a "black box" approach, terminal modeling techniques and SPICE-based models, which have not considered the effect of the distributed parameters of the high voltage motor drive system in EVs on EMI noise, so the previous equivalent circuits of the EMI could not be correctly proposed to accurately and effectively predict the actual source and propagation of the EMI in the system. The effects of the distributed parameters on the conducted EMI noise have not been adequately considered previously because of a lack of the modeling of the conducted EMI from the high voltage motor drive system with suitable parameters and better analysis methods.

1.2. Motivation and Innovation

This study proposes a new method to analyze the effect of distributed parameters on conducted EMI from the DC-fed high voltage motor drive systems in EVs. A complete test for conducted EMI emissions from the AC motor drive system of an EV under load conditions will be set up to measure the conducted EMI of high voltage DC cables and EMI noise peaks due to resonances in a frequency range of 150 kHz–108 MHz. The corresponding equivalent circuits for DM and CM EMI at the resonance frequencies of 500 kHz and 30 MHz are established to determine the EMI propagation paths and analyze the effect of distributed parameters on conducted EMI. The dominant distributed parameters of elements responsible for the resonances appearing at 500 kHz and 30 MHz will be determined. The effect of the dominant distributed parameters on conducted EMI will be verified by simulations and experiments.

1.3. Organization of the Paper

The organization of this study is as follows: Section 2 illustrates the structure of the complete test setup for conducted EMI emissions from the AC motor drive system of an EV. Then, the corresponding current paths and equivalent circuits of DM and CM EMI at resonance frequencies of 500 kHz and 30 MHz, and the effect of distributed parameters on CM EMI will be presented in Section 3. After that, the simulation verification and discussion will be illustrated in Section 4. The experiment verification will be discussed in Section 5. Finally, conclusions are provided in Section 6.

2. System Conducted Emission Measurement

2.1. Conducted EMI Emission Setup

The complete test setup for conducted EMI emissions from the DC-fed AC motor drive system on an EV in an EMI laboratory is shown in Figure 1 and mainly consists of a DC power supply such as a Li-ion battery, DC cables, a DC-fed voltage-type PWM three-phase power inverter, AC cables, and an AC motor. Measurements were performed to comply with the CISPR 25 standard which provides conducted EMI emission limits for vehicle components in a frequency range of 150 kHz to 108 MHz [28]. Two standard line impedance stabilization networks (LISNs) terminated with 50 Ω resistances provide DC power from a battery or DC power supply to the three-phase power inverter using two shielded cables (2 m). The power inverter with 330 V DC input is connected to a 50 kW/100 kW permanent magnet synchronous motor (PMSM) using three shielded cables (1 m). As required by the EMC regulations of CISPR 25, all components are connected to a large copper sheet as ground reference plane, except for the AC motor which is located on an insulated bench covered with ferrite material and connected to an electric dynamometer supplying a mechanical load. The output speed and torque of the AC motor can be measured by a meter between the dynamometer and the insulated output shaft of the AC motor. With this configuration, the total conducted EMI noise voltage signals in DC cables can be picked up by any one of the line impedance stabilization network (LISN) impedances connected to an EMI receiver [29].

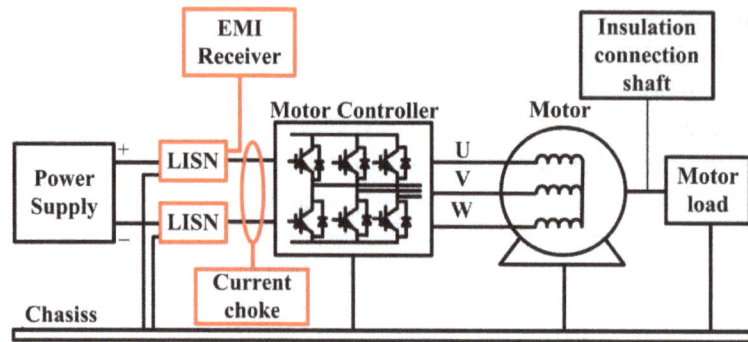

Figure 1. Conducted EMI emission system test setup for the AC motor drive system.

2.2. Conducted EMI Experiment Results

The AC motor is operated continuously at 2000 rpm speed with no-load and 60 N·m loaded torque, as shown in Figure 2. Figure 3 shows the experimental comparison in conducted EMI emission levels between the no-load and load conditions. This experimental result indicates that the conducted EMI noise voltage of the power inverter is dominant in a frequency range of 150 kHz to 108 MHz and is not compliant with CISPR25, as shown in Table 1. Therefore the conducted EMI emission levels in the load condition are more severe and higher than those in the no-load mode. Two noise voltage peaks at frequency around 500 kHz and 30 MHz can be observed and may mainly be caused by PWM switching harmonics or parasitic resonances due to the distributed parameters of the motor system [30]. It is critical to analyze the source and propagation mechanism of EMI to predict the conducted EMI emissions and determine the dominant distributed parameters of the elements in the motor system responsible for the resonances.

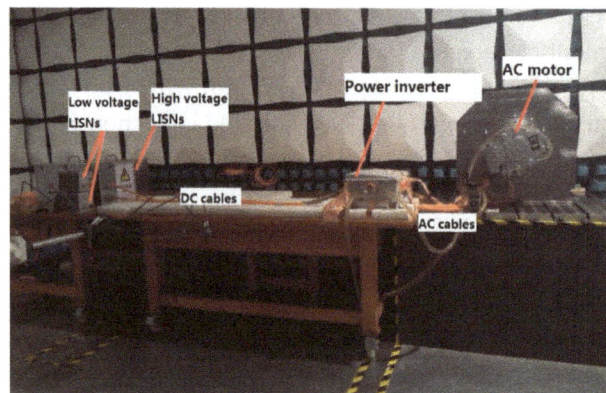

Figure 2. The test platform for conducted EMI emission.

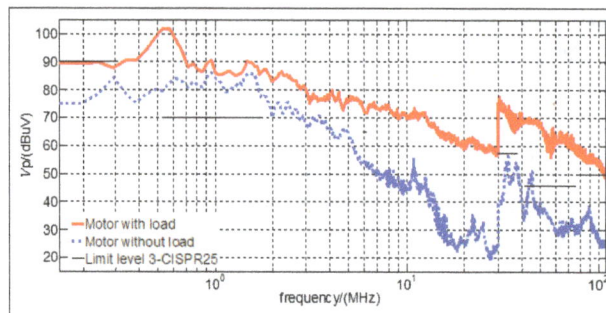

Figure 3. Comparison of measurement with load and measurement without load.

Table 1. CISPR25 class3-peak limits for conducted disturbances.

Service Band	Frequency/MHz	Limit/dB (μv)
Broadcast	0.15–0.30	90
	0.53–1.80	70
	5.9–6.2	65
	41–88	46
	76–108	50
Mobile services	26–28	56
	30–54	56
	68–87	50

3. System Conducted Emission Analysis

3.1. Noise Source

Figure 1 shows the circuit of a full bridge IGBT-based inverter in the motor controller model. The DC-fed PWM power inverter is designed to have a rated 250 V output voltage. The DC bus input voltage is 330 V. The six switches S_1–S_6 in the inverter are 1200 V/600 A full bridge IGBT modules (Infineon) with sinusoidal pulse width-modulation (SPWM) control. Although control methods (like space vector pulse width modulation (SVPWM), direct torque control (DTC), indirect field oriented control (IFOC), etc.) have better characteristics for mitigating harmonics, this is not the case for the EMI noise at high frequency. Therefore, we focus on the effect of characteristics of the trapezoidal wave for PWM on the EMI. We just take SPWM as an example for explaining the principle of the spectrum of the trapezoidal wave. The switching frequency of IGBT was set to 20 kHz and the line frequency for the AC motor was 400 Hz. The SPWM control signals are generated by compared a sinusoidal reference with a 20 kHz triangular carrier signal as illustrated in Figure 4a, which shows the PWM waveforms in a half-period, which have nine duty cycles corresponding to nine pulses with different pulse-widths. The noise source due to the SPWM control is often simplified by assuming a trapezoidal shape for the switching transients [26]. Each PWM pulse can be described as a trapezoidal pulse by an amplitude A, a frequency f, a pulse rise-time τ_r, a pulse fall-time τ_f and a pulse-wide τ. T represents the period of the trapezoidal pulse. The continuous envelope spectrum for a trapezoidal pulse can be given by the following equations [31]:

$$\text{Envelope} = 2A\frac{\tau}{T}\left|\frac{\sin(\pi\tau f)}{\pi\tau f}\right|\left|\frac{\sin(\pi\tau_r f)}{\pi\tau_r f}\right| \tag{1}$$

$$20\log_{10}(\text{envelope}) = 20\log_{10}(2A\tfrac{\tau}{T}) + 20\log_{10}(\tfrac{\sin(\pi\tau f)}{\pi\tau f}) \\ + 20\log_{10}(\tfrac{\sin(\pi\tau_r f)}{\pi\tau_r f}) \tag{2}$$

The τ is smaller under unloaded operation conditions than that under load operation by SPWM control, as shown in Figure 4. Then from (1) and (2) the magnitude of the EMI noise voltage decreases as the value of τ decreases, and is lower under unloaded operation than that under load operation. From (1) and (2), the first break point in the frequency spectral bound is related to τ and is $1/\pi\tau$. The higher the τ, the wider the span related to the DC term, as shown in Figure 4b. The τ is smaller under unloaded operation conditions. Then the magnitude of the EMI noise voltage is lower under unloaded operation than that under load operation, as shown in Figure 3. The second breakpoint in the frequency spectral bound is related to rise/fall time and is $1/\pi\tau_r$. The smaller the rise/fall time, the larger the high-frequency spectral content, as shown in Figure 4c. The frequency band of the EMI noise source due to IGBT switching is from 0 Hz to 1 GHz. Then the resonances could be caused up to 1 GHz by parasitic distributed parameters of the AC motor system and may result in peak voltages exceeding the limit levels specified in the CISPR25 standard, as shown in Table 1.

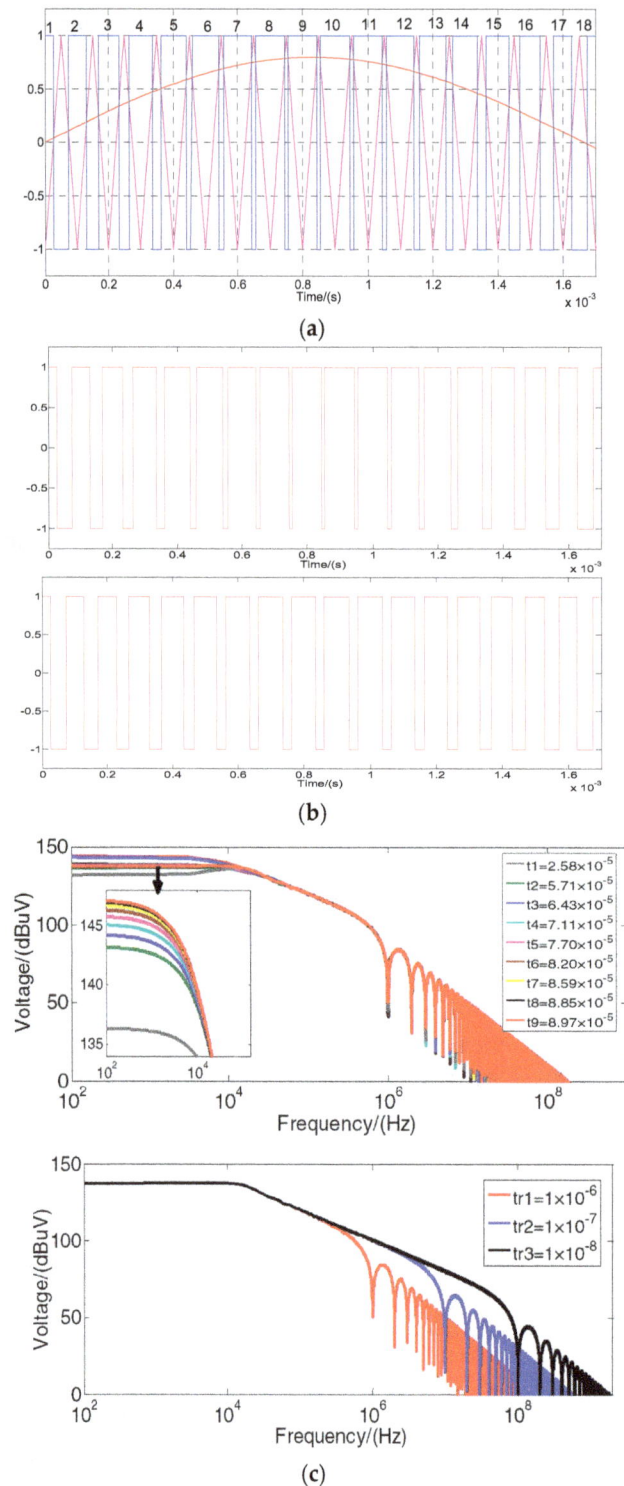

(a)

(b)

(c)

Figure 4. Spectral bounds for a trapezoidal wave: (**a**) SPWM in half-period; (**b**) The effect of pulse-width; (**c**) The effect of rise-time.

3.2. Analysis of the Current Path of Conducted Emissions

The self-inductance and mutual inductance are equivalent to one inductance in order to simplify the equivalent circuit for analyzing EMI propagation path. The motor drive system is constructed in DM and CM situation, as shown in Figure 5 and Figure 7, where, S_1–S_6 represent six IGBTs in the inverter, C_1–C_6 represent the distributed capacitance between the collector and emitter of S_1–S_6,

C_{Y1} and C_{Y2} represent the filter Y capacitors between the positive/negative DC cable and chassis, L_{Y1} and L_{Y2} represent the equivalent series inductances (ESLs) of C_{Y1} and C_{Y2}, C_X represents the filter X capacitor between the DC buses, L_X represents the equivalent series resistance (ESR) of C_X. Two LISNs can be represented by the circuit composed of R_{L1}, C_{L1}, R_{L2}, C_{L2}, C_7, C_8 and C_9 to represent the distributed capacitance from the collector and emitter of the IGBT to the chassis, C_{10} represents the distributed capacitance between the motor and the chassis. L_M represents the inductance of the motor phase winding, $L_{DC \, bus \, bar+}$ and $L_{DC \, bus \, bar-}$ represent the DC bus bars' inductance, which includes self-inductance and mutual inductance between two DC bus bars, so its value is larger than that of the lead stray inductance of IGBT, which is smaller and can be ignored, compared the inductance of the DC bus bars. C_{DC+} and C_{DC-} represent the DC cables' capacitance, L_{DC+} and L_{DC-} represent the DC cables' inductance, R_{DC1} and R_{DC2} represent the DC cables' resistance. The main distributed parameters' values are measured by VNA and shown in Table 2. The EMI noise propagation paths based on distributed parameters and the equivalent circuits of DM and CM noise current are respectively presented as follows.

Table 2. Parameters in the motor drive system.

Parameter	Value	Parameter	Value
L_{Y1}, L_{Y2}	200 nH	C_X	1028 μF
C_7	30 pF	L_X	20 nH
C_8	20 pF	C_1–C_6	20 pF
C_9	20 pF	R_{L1}, R_{L2}	50 Ω
C_{10}	200 pF	C_{L1}, C_{L2}	0.47 μF
C_{Y1}, C_{Y2}	100 pF	L_M	1 mH
L_{DC+}, L_{DC-}	50 nH	C_{DC+}, C_{DC-}	100 nF
R_{DC+}, R_{DC-}	0.0002 Ω	$L_{DC \, bus \, bar+}, L_{DC \, bus \, bar-}$	104 nH

3.2.1. Analysis of the DM Current Path for 500 kHz

The DM EMI emission from the phase node P between the two IGBTs of one phase bridge leg can be equivalent to a DM noise current source I_{DM} between the phase node P and DC bus minus the distributed parameters of the inner elements of the motor system, as shown in Figure 5a. The DM current loop can be illustrated by calculating the impedance of each circuit element ignoring the distributed parameters at 500 kHz, so the DM current flows though the distributed parameters of the motor system is shown in Figure 5b. I_{DM} acts as a driving force to form the following three current loops:

- current loop I: $I_{DM} \to C_4 \to L_{DC \, bus \, bar-} \to R_{DC2} \to L_X \to C_X \to R_{DC1} \to L_{DC \, bus \, bar+} \to C_1 \to I_{DM}$
- current loop II: $I_{DM} \to L_M \to C_6 \to L_{DC \, bus \, bar-} \to R_{DC2} \to L_X \to C_X \to R_{DC1} \to L_{DC \, bus \, bar+} \to C_1 \to I_{DM}$
- current loop III: $I_{DM} \to L_M \to C_2 \to L_{DC \, bus \, bar-} \to R_{DC2} \to L_X \to C_X \to R_{DC1} \to L_{DC \, bus \, bar+} \to C_1 \to I_{DM}$

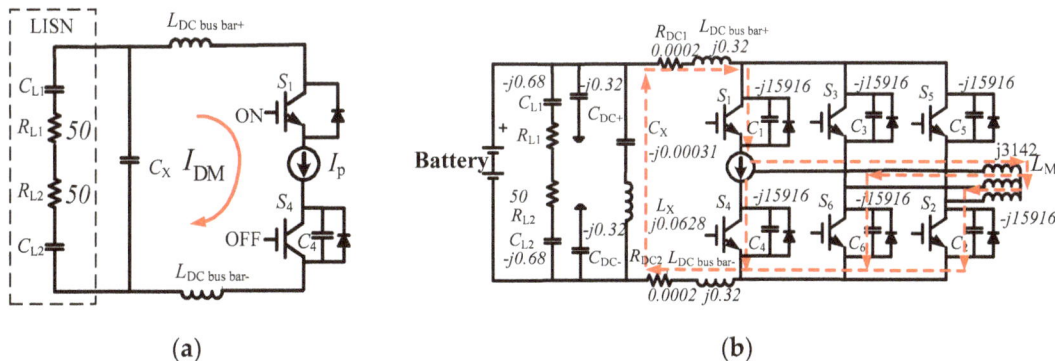

(a)   (b)

Figure 5. (a) Equivalent circuit for DM; (b) DM interference propagation path at 500 kHz.

3.2.2. Analysis of DM Current Path for 30 MHz

The impedance of each circuit element at 30 MHz is calculated as shown in Figure 6. I_{DM} acts as a driving force to form the following three current loops at 30 MHz:

- current loop I: $I_{DM} \rightarrow C_4 \rightarrow L_{DC\ bus\ bar-} \rightarrow R_{DC2} \rightarrow L_X \rightarrow C_X \rightarrow R_{DC1} \rightarrow L_{DC\ bus\ bar+} \rightarrow C_1 \rightarrow I_{DM}$
- current loop II: $I_{DM} \rightarrow C_4 \rightarrow C_6 \rightarrow C_3 \rightarrow C_1 \rightarrow I_{DM}$
- current loop III: $I_{DM} \rightarrow C_4 \rightarrow C_2 \rightarrow C_5 \rightarrow C_1 \rightarrow I_{DM}$

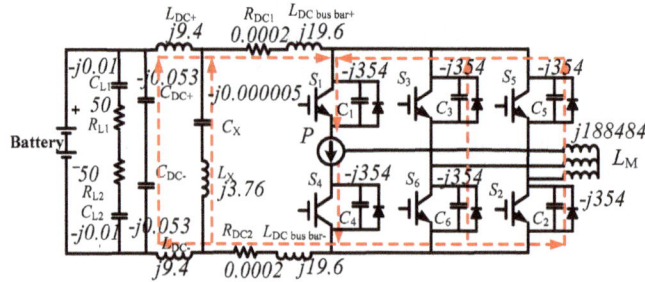

Figure 6. DM interference propagation path at 30 MHz.

The harmonic content of the DM current at 500 kHz and 30 MHz flowing through the 50 Ω impedance of LISN is also very small due to the filtering effect of the X capacitor. Then the resonant peak at about 500 kHz and 30 MHz is not dominated by the DM current.

3.2.3. Analysis of CM Current Path for 500 kHz

The CM noise current is always generated at high frequency and flows through the distributed parameters to the ground [32]. The CM emission from the phase node P of two IGBTs of one phase bridge leg can be equivalent to a CM voltage source U_{DM} between the phase node P and chassis with ignored distributed parameters of inner elements of the motor system, as shown in Figure 7a,b. The CM current loop can be illustrated by calculating the impedance of each circuit element with distributed parameters at 500 kHz. U_{CM} acts as a driving force to form three CM current loops shown in Figure 7c, where current loop1 (red line), current loop2 (blue line) and current loop3 (purple line) are considered in parallel. The CM current flowing paths at 500 kHz are composed of a DC side path and an AC side path. The effective impedance of inductance for one branch of the current loop (red) dominated by $L_{DC\ bus\ bar+}$ or $L_{DC\ bus\ bar-}$ is about j0.32 Ω and the effective impedance of capacitance for one branch of the current loop (red) dominated by the DC cable capacitance is about $-$j0.32 Ω. Therefore the peak voltage at about 500 kHz is mainly caused by the series resonance in the current loop (red).

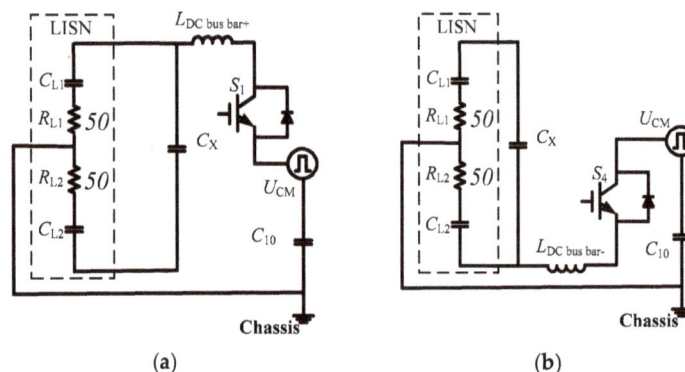

(a)   (b)

Figure 7. *Cont.*

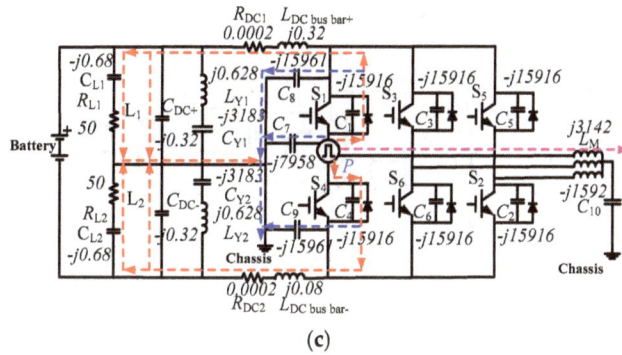

(c)

Figure 7. (a) Equivalent circuit for CM; **(b)** Equivalent circuit for CM; **(c)** CM interference propagation path at 500 KHz.

The elements responsible for the resonance at 500 kHz mainly are the capacitance between the DC cables and the chassis, and the stray inductance of the DC bus bar of IGBT. The resonance peak at about 500 kHz is mainly dominated by the CM current. The effect of the current loop2 (blue line) and current loop3 (purple line) is smaller and can be ignored in the equivalent circuit of CM current at 500 kHz shown in Figure 8.

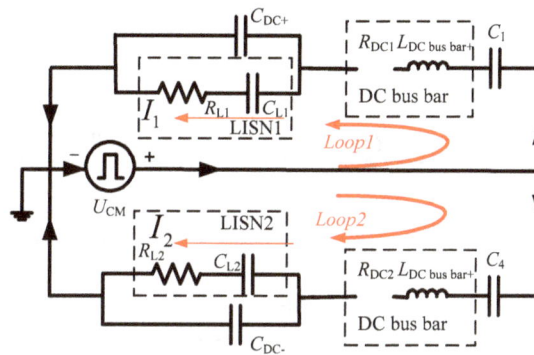

Figure 8. Equivalent circuit of CM interference at 500 KHz.

The CM current at about 500 kHz flowing on LISN can be expressed by the following equation:

$$I_1 = \frac{Z_1 U_{CM}}{(Z_1 + Z_2)Z_3} \lim_{x \to \infty} \tag{3}$$

where Z_1 denotes series-parallel impedance of R_{L1}, C_{L1} and C_{DC+}, Z_2 denotes series impedance of $L_{DC\,bus\,bar+}$, R_{DC1} and C_1, Z_3 denotes the series impedance of R_{L1} and C_{L1}.

3.2.4. Analysis of CM Current Path for 30 MHz

U_{DM} acts as a driving force to form the CM current loops shown in Figure 9. It is difficult to determine the main CM current loop and the elements responsible for the resonance at 30 MHz because of the complexity of CM current equivalent circuit dominated by C_1, C_4, $L_{DC\,bus\,bar+}$, $L_{DC\,bus\,bar-}$, R_{DC2}, R_{DC1}, R_{DC2}, L_{DC+}, L_{DC-}, C_{Y1}, L_{Y1}, C_{Y2}, L_{Y2}, C_7, C_8 and C_9. The model of the CM current equivalent circuit is essential to study the main dominated CM current loop at 30 MHz to determine the distributed parameters responsible for the resonance peak at 30 MHz.

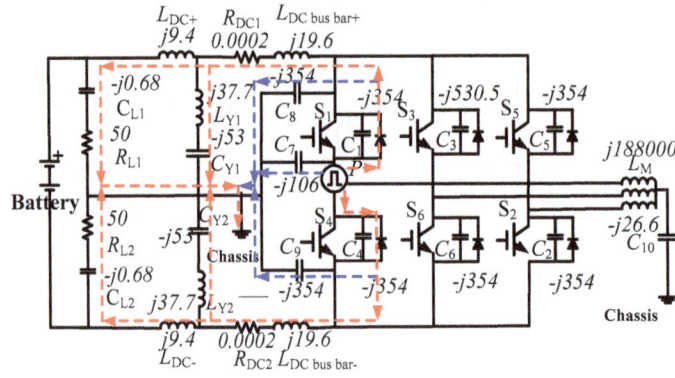

Figure 9. CM interference propagation path at 30 MHz.

The equivalent circuit of CM current at 30 MHz is shown in Figure 10 and the CM current at 30 MHz flowing through the LISN resistor can be expressed by the following equation:

$$I_2 = \frac{Z_1 U_{CM}}{(Z_1 + Z_{C1})(Z_2 + Z_3)Z_4} \tag{4}$$

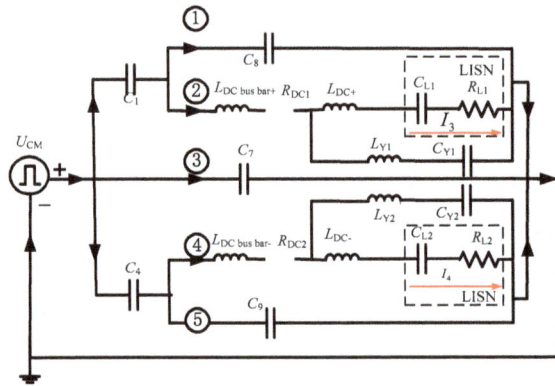

Figure 10. Equivalent circuit of CM interference at 30 MHz.

Z_1 denotes series-parallel impedance of C_8, $L_{DC \, bus \, bar+}$, R_{DC1}, L_{DC+}, C_{Y1}, L_{Y1}, C_{L1} and R_{L1}. Z_2 denotes series impedance of $L_{DC \, bus \, bar+}$ and R_{DC1}. Z_3 denotes series-parallel impedance of L_{DC+}, C_{L1}, R_{L1}, C_{Y1} and L_{Y1}. Z_4 denotes series impedance of L_{DC+}, C_{L1} and R_{L1}.

From Equation (3), C_{DC+}, $L_{DC \, bus \, bar+}$, C_{DC-} and $L_{DC \, bus \, bar-}$ are the dominant distributed parameters causing the series resonance at 500 kHz and the effect on the CM current I_1. Therefore C_1 and C_4 have a very small effect on I_1 and can be ignored. Changing of any parameter among C_{DC+}, $L_{DC \, bus \, bar+}$, C_{DC-} and $L_{DC \, bus \, bar+}$ can reduce the value of I_1 at 500 kHz. From Equation (4), L_{DC+}, C_{Y1}, L_{Y1}, L_{DC-}, C_{Y2}, L_{Y2}, C_8 and C_9 are the effective distributed parameters at 30 MHz and with an effect on I_1.

The effect from distributed parameters on conducted voltage U_R according to above equivalent circuits and calculation at 500 kHz and 30 MHz is shown in Table 3. In a motor drive system, the parameters such as C_1–C_6, C_X, L_X and L_M usually cannot be controlled, and only the distributed parameters $L_{DC \, bus \, bar+}$, $L_{DC \, bus \, bar-}$, C_{DC+}, C_{DC-}, C_8, C_9, L_{Y1}, L_{Y2}, C_{Y1}, C_{Y2} could be changed along with different arrangements, filtering and shielding.

Table 3. The effect of the main distribution parameter on conducted EMI emission.

Changing of Parameters	Current	Voltage
$L_{DC\ bus\ bar+}$ ↑ or ↓	$I_1 \downarrow$	$U_R \downarrow$
$L_{DC\ bus\ bar-}$ ↑ or ↓	$I_1 \downarrow$	$U_R \downarrow$
C_{DC+} ↑ or ↓	$I_1 \downarrow$	$U_R \downarrow$
C_{DC-} ↑ or ↓	$I_1 \downarrow$	$U_R \downarrow$
C_8 ↑	$I_2 \downarrow$	$U_R \downarrow$
C_9 ↑	$I_2 \downarrow$	$U_R \downarrow$
L_{Y1} ↑	$I_2 \downarrow$	$U_R \downarrow$
L_{Y2} ↑	$I_2 \downarrow$	$U_R \downarrow$
C_{Y1} ↑	$I_2 \downarrow$	$U_R \downarrow$
C_{Y2} ↑	$I_2 \downarrow$	$U_R \downarrow$
$L_{DC\ bus\ bar+}$ ↑	$I_2 \downarrow$	$U_R \downarrow$
$L_{DC\ bus\ bar-}$ ↑	$I_2 \downarrow$	$U_R \downarrow$

4. Simulation of the Effect of Distributed Parameters

4.1. System Conducted EMI Modeling

According to the structure of the system and the distributed parameters, the power inverter system was modeled as a simplified single-arm bridge of the power inverter system using the EMC simulation software "Computer Simulation Technology" (CST) that predicts the noise in the entire conducted emissions range from 150 kHz to 108 MHz, as illustrated in Figure 11. The conducted emission (CE) voltage on the resistor of the LISN in the DM and CM network equivalent circuits can be obtained by using time-domain simulation, followed by fast fourier transform(FFT) in the designer platform provided in the CST software. The CM EMI source is equivalent to an ideal trapezoidal shape wave and the EMI voltage measured by a probe P_1 is the positive conducted emission voltage. The EMI voltage simulation result is shown in Figure 12. Table 3 and Figure 12 suggest that the conducted EMI voltage spectrum of the motor drive system in EV can be divided into two different frequency ranges: the low frequency range around 500 kHz that is dominated by DC cables' capacitance and DC bus bars, and the high frequency range around 30 MHz that is related to parasitic resonances due to the distributed parameters of Y capacitors and distributed capacitance from the IGBT phase node to the chassis. The EMI voltage peaks obtained through simulation in Figure 12 correspond to those of measurements at 500 kHz and 30 MHz. There are some larger errors between simulation results and measurement results because the measurement EMI voltage is the total of the EMI noise from the three arms of the power inverter. Conversely, the simulated EMI voltage is obtained from a single-arm. It can be seen that the model is efficient enough to be used to predict the CM current paths and the elements responsible for the EMI.

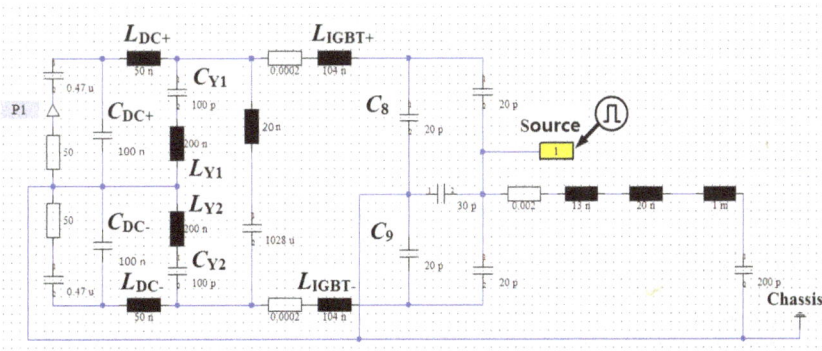

Figure 11. CM circuit model of single-arm bridge of power inverter.

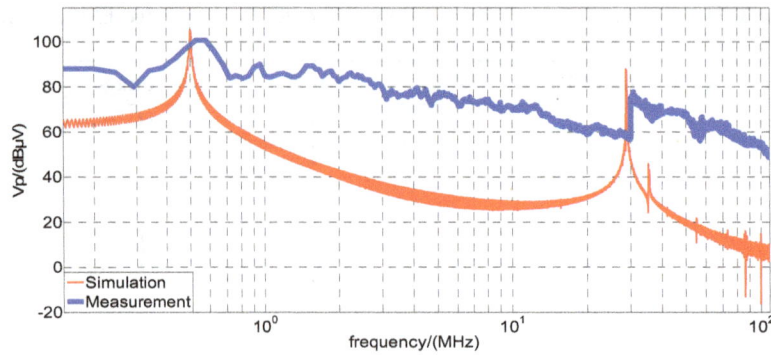

Figure 12. Comparison of measurement and simulation.

4.2. The Effect of Distributed Parameters

4.2.1. The Effect of $L_{DC\ bus\ bar+}$ and $L_{DC\ bus\ bar-}$

According to Table 3, changing the value of $L_{DC\ bus\ bar+}$ and $L_{DC\ bus\ bar-}$ can mitigate the resonance and reduce the conducted emissions at low frequency. Therefore, the EMI voltage peak due to resonance at 500 kHz could be suppressed. The voltage value of conducted emission at 500 kHz can be rapidly decreased by about 50 dB by increasing the value of $L_{DC\ bus\ bar+}$ and $L_{DC\ bus\ bar-}$ from 104 nH to 220 nH to comply with the limit level-3 of CISPR25 regulatory standards, as shown in Figure 13. It suggests that two CM inductors can be placed on the DC bus bar of the power inverter of the AC motor to reduce the CM current and conducted emission at 500 kHz.

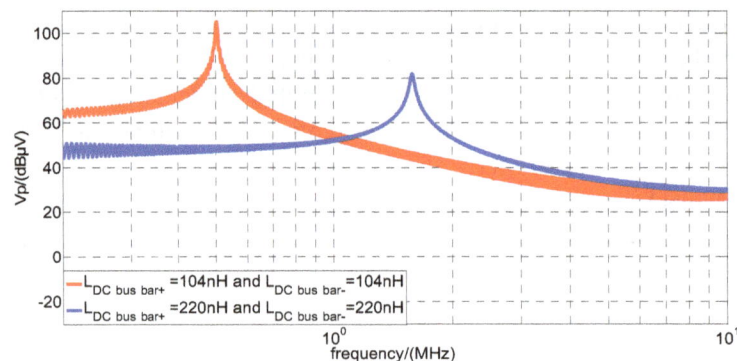

Figure 13. Positive conducted emission(CE) voltage after $L_{DC\ bus\ bar+}$ and $L_{DC\ bus\ bar-}$ were increased.

4.2.2. The Effect of L_{Y1} and L_{Y2}

According to Table 3, L_{Y1} and L_{Y2} can be increased to mitigate the resonance and reduce the conducted emission at high frequency. Therefore, the EMI voltage peak due to resonance at 30 MHz could be suppressed. The voltage value of conducted emission at 30 MHz can be rapidly decreased by about 64 dB by increasing the value of L_{Y1} and L_{Y2} from 200 nH to 300 nH to comply with the level-3 limit of CISPR25, as shown in Figure 14. It suggests that the equivalent series inductance of Y capacitor can affect the conducted EMI at 30 MHz. A better design of the parameters of the Y capacitor can reduce the conducted emissions at high frequency.

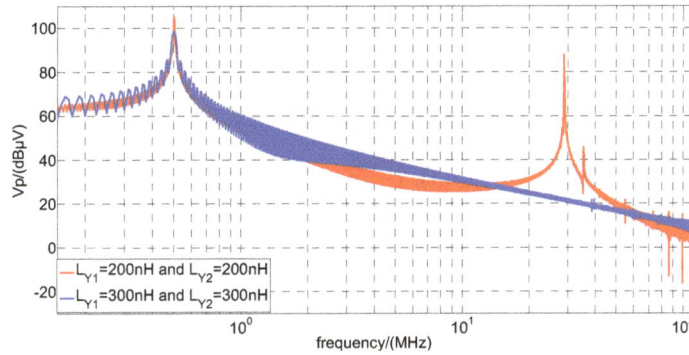

Figure 14. Positive CE voltage after L_{Y1} and L_{Y2} were increased.

4.2.3. The Effect of Combination of $L_{DC\ bus\ bar+}$, $L_{DC\ bus\ bar-}$, L_{Y1} and L_{Y2}

Increasing of the value of $L_{DC\ bus\ bar+}$, $L_{DC\ bus\ bar-}$ (each 220 nH) and L_{Y1}, L_{Y2} (each 300 nH) can decrease the conduced emissions at frequencies from 150 kHz to 108 MHz to below the limit level-3 of CISPR25. Then there are no resonances previously caused by $L_{DC\ bus\ bar+}$ and $L_{DC\ bus\ bar-}$ at 500 kHz, L_{Y1} and L_{Y2} at 30 MHz. Although there still is a resonance point at around 55 MHz, the value of conducted voltage is decreased below the level-3 limit of CISPR25, as shown in Figure 15.

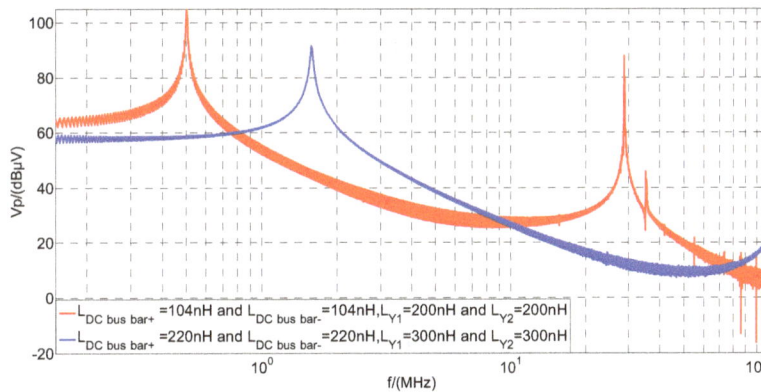

Figure 15. Positive CE voltage after $L_{DC\ bus\ bar+}$, $L_{DC\ bus\ bar-}$, L_{Y1} and L_{Y2} were increased.

4.2.4. The Effect of C_{DC+} and C_{DC-}

From Table 3, the distributed capacitances from DC cables (C_{DC+}, C_{DC-}) can affect the conducted emissions. Changing the value of C_{DC+} and C_{DC-} can mitigate the resonance at 500 kHz and reduce the conducted emission at low frequency. The voltage value of conducted emission at 500 kHz can be decreased by about 25 dB by increasing the value of C_{DC+} and C_{DC-} from 100 nF to 250 nF to comply with the limit level-3 of CISPR25, as shown in Figure 16. It suggests that filtering and shielding of the DC input of the power inverter of the AC motor can be used to change the distributed parameters of DC cables to reduce the CM current at low frequency and suppress the EMI voltage peak due to resonance at 500 kHz. For example, a Y capacitor could be added between the DC cable and chassis to reduce conducted emissions, although the Y capacitor may cause a high ground leakage current.

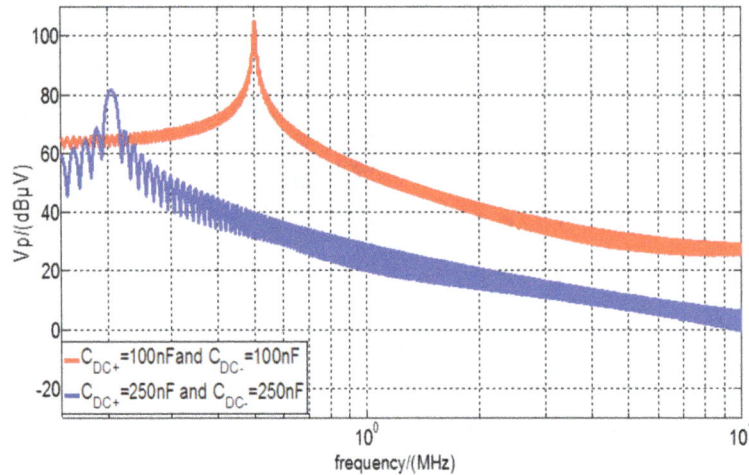

Figure 16. Positive CE voltage after C_{DC+} and C_{DC-} were increased.

4.2.5. The Effect of the Combination of C_{Y1}, C_{Y2}, C_8 and C_9

From Table 3, the capacitances of the Y capacitors (C_{Y1}, C_{Y2}) and the distributed capacitance from the collector and emitter of the IGBT to the chassis (C_8, C_9) can affect the conduced emissions at high frequency. The voltage value of conducted emissions at high frequency can be decreased to comply with CISPR25 by increasing the value of C_{Y1} and C_{Y2} from 100 nF to 500 nF and the value of C_8 and C_9 from 20 pF to 100 pF, as shown in Figure 17. It shows that the peak value of the conducted voltage is reduced by 20 dB at about 30 MHz. Therefore, Y capacitors could be added between the collector and emitter of the IGBT and chassis, and between DC input and chassis to reduce conducted emissions at high frequency.

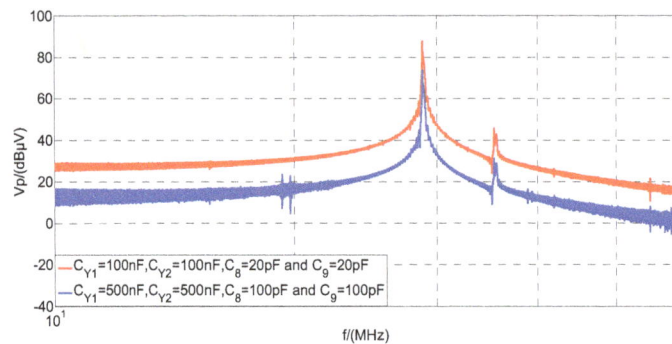

Figure 17. Positive CE voltage after C_{Y1}, C_{Y2}, C_8 and C_9 were increased.

4.2.6. The Effect of the Combination of C_{DC+}, C_{DC-}, C_{Y1}, C_{Y2}, C_8 and C_9

It is a better mitigation method to change the value of the distributed parameters (C_{DC+}, C_{DC-}, C_{Y1}, C_{Y2}, C_8 and C_9) to reduce the conducted emission at frequencies from 150 kHz to 108 MHz defined in CISPR25. The conducted voltage is reduced by increasing the capacitances ($C_{DC+} = C_{DC-} = 250$ nF, $C_{Y1} = C_{Y2} = 500$ nF, $C_8 = C_9 = 100$ pF), as shown in Figure 18.

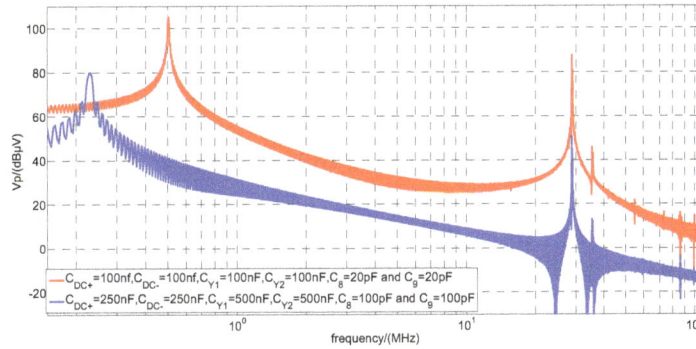

Figure 18. Positive CE voltage after C_{DC+}, C_{DC-}, C_{Y1}, C_{Y2}, C_8 and C_9 were increased.

5. Experimental Verification

According to the above analysis of the effect of distributed parameters on the power inverter system of AC motors for EVs, changing the combination of C_{DC+}, C_{DC-}, C_{Y1}, C_{Y2}, C_7 and C_8 should be a better way to suppress the voltage peak at 500 kHz and 30 MHz to comply with the limit level-3 of CISPR 25. A new pair of Y capacitors are added between the collector and emitter of the IGBT and DC bus bar and the chassis to increase the capacitances between the inverter and the chassis. Experimental verification is conducted and the results are shown in Figure 19. The conducted emission characteristics at around 500 kHz and 30 MHz in Figure 19 are approximately as predicted by the simulation in Figure 18.

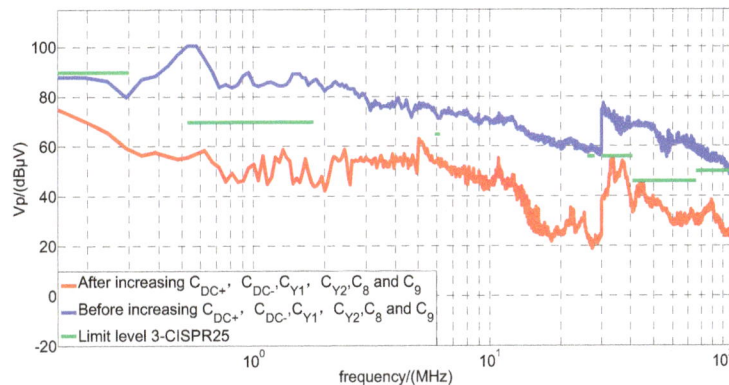

Figure 19. Positive CE voltage after C_{DC+}, C_{DC-}, C_{Y1}, C_{Y2}, C_8 and C_9 were increased.

6. Conclusions

This study proposed a new method to analyze the effects of distributed parameters on conducted EMI from the DC-fed high voltage motor drive systems in EVs. The conducted EMI of high voltage DC cables of the motor drive system in a frequency range of 150 kHz–108 MHz and two EMI noise peaks at resonances frequencies 500 kHz and 30 MHz have been measured by a complete test for conducted EMI emissions from the AC motor drive system of an EV under load conditions. The research mainly focuses on the effects of distributed parameters in the inverter and cables on the resonances at 500 kHz and 30 MHz, nor the distributed parameters of the motor due to the high impedance of the motor model at 500 kHz and 30 MHz. The corresponding equivalent circuits for DM and CM EMI at resonance frequencies of 500 kHz and 30 MHz are established to determine the EMI propagation paths and the dominant distributed parameters of elements responsible for the resonances appearing at 500 kHz and 30 MHz.

The distributed parameters $L_{DC\,bus\,bar+}$, $L_{DC\,bus\,bar-}$, C_{DC+}, C_{DC-}, C_8, C_9, L_{Y1}, L_{Y2}, C_{Y1} and C_{Y2} can affect the EMI emissions from the high voltage motor drive system. The effect of the dominant distributed parameters on conducted voltage is verified by modeling of the CM circuit of a single-arm bridge of the high voltage motor power inverter. Increasing the value of $L_{DC\ bus\ bar+}$, $L_{DC\ bus\ bar-}$ from 104 nH to 220 nH and L_{Y1}, L_{Y2} from 200 nH to 300 nH or increasing the capacitances (C_{DC+}, C_{DC-} from 100 nF to 250 nF, C_{Y1}, C_{Y2} from 100 nF to 500 nF, C_8, C_9 from 20 pF to 100 pF) can mitigate the two resonance peaks at the frequencies of 500 kHz and 30 MHz and decrease the conduced voltage at frequencies from 150 kHz to 108 MHz to below the limit level-3 of CISPR25. The effect of the combination of C_{DC+}, C_{DC-}, C_{Y1}, C_{Y2}, C_8 and C_9 on conducted voltage is verified by experiments. In future work, modeling of a CM circuit of three-arm bridge of the high voltage motor power inverter in EV will be developed. After that the effect of distributed parameters on EMI noise will be further simulated and tested.

Acknowledgments: This study is supported by National Natural Science of Foundation of China and Outstanding Talents Project of Beijing.

Author Contributions: Li Zhai analyzed the system conducted emission; Liwen Lin and Xinyu Zhang performed the modeling; Li Zhai and Chao Song performed the experiments; Li Zhai and Liwen Lin analyzed the data; Li Zhai wrote the paper.

Conflicts of Interest: The authors declare no conflict of interest.

References

1. Sun, F.; Xiong, R.; He, H. A systematic state-of-charge estimation framework for multi-cell battery pack in electric vehicles using bias correction technique. *Appl. Energy* **2016**, *162*, 1399–1409. [CrossRef]
2. Ferrero, E.; Alessandrini, S.; Balanzino, A.; Yan, J. Impact of the electric vehicles on the air pollution from a highway. *Appl. Energy* **2016**, *169*, 450–459. [CrossRef]
3. Hu, X.; Murgovski, N.; Johannesson, L.; Bo, E. Energy efficiency analysis of a series plug-in hybrid electric bus with different energy management strategies and battery sizes. *Appl. Energy* **2013**, *111*, 1001–1009. [CrossRef]
4. Chen, Z.; Xiong, R.; Wang, C.; Cao, J. An on-line predictive energy management strategy for plug-in hybrid electric vehicles to counter the uncertain prediction of the driving cycle. *Appl. Energy* **2016**, *185*, 1663–1672. [CrossRef]
5. Torresa, J.L.; Gonzalezb, R.; Gimeneza, A.; Lopeza, J. Energy management strategy for plug-in hybrid electric vehicles. A comparative study. *Appl. Energy* **2014**, *113*, 816–824. [CrossRef]
6. Zhai, L.; Zhang, X.; Bondarenko, N.; Loken, D.; Doren, T.V.; Beetner, D.G. Mitigation emission strategy based on resonances from a power inverter system in electric vehicles. *Energies* **2016**, *9*, 419. [CrossRef]
7. Ehsani, M.; Gao, Y.; Emadi, A. *Modern Electric, Hybrid Electric, and Fuel Cell Vehicles: Fundamentals, Theory, and Design*; CRC Press: Boca Raton, FL, USA, 2010.
8. Ardon, V.; Aime, J.; Chadebec, O.; Clavel, E. EMC modeling of an industrial variable speed drive with an adapted PEEC method. *IEEE Trans. Magn.* **2010**, *46*, 2892–2898. [CrossRef]
9. Lai, J.S.; Huang, X.; Chen, S.; Nehl, T.W. EMI characterization and simulation with parasitic models for a low-Voltage high-current AC motor drive. *IEEE Trans. Ind. Appl.* **2004**, *40*, 178–185. [CrossRef]
10. Jettanasen, C.; Costa, F.; Vollaire, C. Common-mode emissions measurements and simulation in variable-speed drive systems. *IEEE Trans. Power Electron.* **2009**, *24*, 2456–2464. [CrossRef]
11. Tommasini, R.; Spertino, F. Electric power distribution and environment: Interference of power installation magnetic fields on computer systems. *Appl. Energy* **1999**, *64*, 181–193. [CrossRef]
12. Toure, B.; Schanen, J.-L.; Gerbaud, L.; Meynard, T.; Roudet, J.; Ruelland, R. EMC modeling of drives for aircraft applications: Modeling process, EMI filter optimization and technological choice. *IEEE Trans. Power Electron.* **2013**, *28*, 1145–1156. [CrossRef]
13. Lai, J.S. Resonant snubber-based soft-switching inverters for electric propulsion drives. *IEEE Trans. Ind. Electron.* **1997**, *44*, 71–80.
14. Mutoh, N.; Ogata, M. New methods to control EMI noises generated in motor drive systems. *IEEE Trans. Ind. Appl.* **2004**, *40*, 143–152. [CrossRef]
15. Rebholz, H.M.; Tenbohlen, S.; Kohler, W. Time-domain characterization of RF sources for the design of noise suppression filters. *IEEE Trans. Electromagn. Compat.* **2009**, *51*, 945–952. [CrossRef]

16. Reuter, M.; Friedl, T.; Tenbohlen, S.; Köhler, W. Emulation of conducted emissions of an automotive inverter for filter development in HV networks. In Proceedings of the IEEE International Symposium on Electromagnetic Compatibility (EMC), Denver, CO, USA, 5–9 August 2013; pp. 236–241.

17. Trzynadlowski, A.M.; Wang, Z.; Nagashima, J.; Stancu, C. Comparative investigation of PWM techniques for general motor's new drive for electric vehicles. In Proceedings of the IEEE Industry Applications Conference, Pittsburgh, PA, USA, 13–18 October 2002; pp. 2010–2015.

18. Piazza, M.C.D.; Ragusa, A.; Tine, G.; Vitale, G. A model of electromagnetic radiated emissions for dual Voltage automotive electrical systems. In Proceedings of the IEEE International Symposium on Industrial Electronics, Ajaccio, France, 4–7 May 2004; pp. 317–322.

19. Akagi, H.; Shimizu, T. Attenuation of conducted EMI emissions from an inverter-driven motor. *IEEE Trans. Power Electron.* **2008**, *23*, 282–290. [CrossRef]

20. Wang, S.; Maillet, Y.Y.; Wang, F.; Lai, R.; Luo, F.; Boroyevich, D. Parasitic effects of grounding paths on common-mode EMI filter's performance in power electronics systems. *IEEE Trans. Ind. Electron.* **2010**, *57*, 3050–3059. [CrossRef]

21. Bishnoi, H.; Baisden, A.C.; Mattavelli, P.; Boroyevich, D. Analysis of EMI terminal modeling of switched power converters. *IEEE Trans. Power Electron.* **2012**, *27*, 3924–3933. [CrossRef]

22. Gong, X.; Ferreira, A.J. Comparison and reduction of conducted EMI in SiC JFET and Si IGBT-based motor drives. *IEEE Trans. Power Electron.* **2014**, *29*, 1757–1767. [CrossRef]

23. Revol, B.; Roudet, J.; Schanen, J.L.; Loizelet, P. EMI study of three-phase inverter-fed motor drives. *IEEE Trans. Ind. Appl.* **2011**, *47*, 223–231. [CrossRef]

24. Stevanovic, I.; Skibin, S.; Masti, M.; Laitinen, M. Behavioral modeling of chokes for EMI simulations in power electronics. *IEEE Trans. Power Electron.* **2013**, *28*, 695–705. [CrossRef]

25. Mutoh, N.; Ogata, M.; Gulez, K.; Harashima, F. New methods to suppress EMI noises in motor drive systems. *IEEE Trans. Ind. Electron.* **2002**, *49*, 474–485. [CrossRef]

26. Bishnoi, H.; Mattavelli, P.; Burgos, R.; Boroyevich, D. EMI behavioral models of DC-fed three-phase motor drive systems. *IEEE Trans. Power Electron.* **2014**, *29*, 4633–4645. [CrossRef]

27. Bondarenko, N.; Zhai, L.; Xu, B.; Li, G.; Makharashvili, T.; Loken, D.; Berger, P.; Doren, T.P.V.; Beetner, D.G. A measurement-based model of the electromagnetic emissions from a power inverter. *IEEE Trans. Power Electron.* **2015**, *30*, 5522–5531. [CrossRef]

28. Australia, S. *Vehicle, Boats and Internal Combustion Engines—Limits and Methods of Measurement for the Protection of on-Board Receivers*; CISPR25; International Special Committee on Radio Interference: Geneva, Switzerland, 2012.

29. Mihalic, F.; Kos, D. Reduced conductive EMI in switched-mode DC–DC power converters without EMI filters: PWM versus randomized PWM. *IEEE Trans. Power Electron.* **2006**, *21*, 1783–1794. [CrossRef]

30. Chen, S.; Nehl, T.W.; Lai, J.-S.; Huang, X.; Pepa, E.; de Doncker, R.; Voss, I. Towards EMI prediction of a PM motor drive for automotive applications. In Proceedings of the Applied Power Electronics Conference and Exposition, Miami Beach, FL, USA, 9–13 February 2003; pp. 14–22.

31. Paul, C.R. *Introduction to EMC*; Wiley & Sons. Inc.: New York, NY, USA, 1992; pp. 466–474.

32. Mugur, P.R.; Roudet, J.; Crebier, J.C. Power electronic converter EMC analysis through state variable approach techniques. *IEEE Trans. Electromagn. Compat.* **2001**, *43*, 229–238. [CrossRef]

2

Development of a Nearly Zero Emission Building (nZEB) Life Cycle Cost Assessment Tool for Fast Decision Making in the Early Design Phase

Hae Jin Kang

SAMOO Architects and Engineers, Seoul 138-240, Korea; hjkang@samoo.com

Academic Editor: Monjur Mourshed

Abstract: An economic feasibility optimization method for the life cycle cost (LCC) has been developed to apply energy saving techniques in the early design stages of a building. The method was developed using default data (e.g., operation schedules), energy consumption prediction equations and cost prediction equations utilizing design variables considered in the early design phase. With certain equations developed, an LCC model was constructed using the computational program MATLAB, to create an automated optimization process. To verify the results from the newly developed assessment tool, a case study on an office building was performed to outline the results of the designer's proposed model and the cost optimal model.

Keywords: cost optimal model; designer's proposed model; LCC analysis; nearly Zero Emission Building; investment cost; operation cost

1. Introduction

1.1. Introduction

Korea is the world's fifth largest oil importer and it lacks fossil fuel reserves. Oil imports accounts for a large proportion of overall imports of Korea, and thus, fluctuating oil prices greatly affect the overall balance of payments [1]. The country's continued reliance on depleting fossil fuel energy sources may also undermine its energy security, and as a result the government plans to decrease fossil fuel dependence in the long run. The government encourages buildings to use less fossil fuel by setting energy policies, regulating the minimum U-value and advocating the use of renewable energy sources. The government is also offering incentives and tax benefits. However, building owners and builders manage only to meet the regulations and have no interest to exceed these minimum targets. This is attributed to green premium, the initial cost that rises with use of the technologies necessary to build a nearly Zero Emission Building (nZEB), which occur in the construction stage. Decision makers simply link the green premium and the subsequent rise of the initial cost to a reduced economic benefit.

In fact, energy performance of a building which has been improved by green premium and the economic benefit do not have a conflicting but a causal relationship This is due to the fact that the time and money needed to operate a building is quite substantial and investing in green premium can considerably reduce the operational costs in the operating stage. This is particularly true when the oil price is expected to increase, and with the additional cost of CO_2 emissions, the green premium effect is expected to further increase [2].

Decision makers have yet to fully recognize the life cycle concept incorporating the green premium and the operating costs comprehensively. Insufficient technologies and experts to evaluate their economic effect also limit the information that can be accessed. Though the outline of the initial investment is decided at the initial design stage, there is no system that supports decision making at that stage and this causes hesitation in people to invest in the green premium.

Therefore, providing information on the economic advantages coming from the green premium can be a solution in reducing the obstacles to realize a nZEB. As it is complicated to consider all of the energy consumed and the economic benefits generated by a building during its life cycle, an assessment tool must to be developed that would help general decision makers easily grasp the effects [3].

The current study attempted to develop an assessment model which can be used to calculate the LCC (initial investment cost and energy cost) and thereby assist in decision making during the early design phase. The assessment tool which was developed should operate together with a program that can draw an optimized design. The energy saving design for a building envelope typically takes into account different design variables, such as heat insulation, natural lighting and natural ventilation in the early design stage. The energy saving technologies trade off their heating and cooling load, and requires many cases to be designed and evaluated. This process takes time and repeated trials and is therefore omitted in the early design phase most of the time. A tool is required which can be connected to an auto–optimized design process that helps to generate the cost-optimal design without going through a large number of iterations. Therefore, a supportive decision making tool was developed in the present study.

1.2. Research Scope

A baseline building was selected for the present study to develop a cost optimization model and 153 office buildings in Seoul were surveyed to validate the model. The scope for applying each design variable was set, according to current legal requirements for minimum performance and the latest technology available in the market for the maximum performance.

The study developed a simple prediction equation that allowed calculation of energy consumption and LCC based on the baseline building. This was derived to be easily operated by any architect, as it comprised design variables considered in the early design stage. To develop the energy prediction equation, an orthogonal array was used to drastically reduce the number of experiments to 81 from 3^{24} (=282,429,536,481), the number required for energy simulation. EnergyPlus was used for the computer simulation and the relative importance of each design variable was identified using Analysis of Variance (ANOVA). The ANOVA result was used as multiple regression analysis data to develop the prediction equation.

To calculate initial investment in the LCC model, a prediction equation was also developed. To develop the prediction equation, several cases were established where technological elements designed to improve energy performance, were designed for the baseline model before entering quotation data. A regression formula was then generated through statistical analysis of the calculated data. The equation parameters were made to correspond to the energy prediction equation as far as possible and important factors were included to the LCC which would not affect the energy performance.

The LCC model was constructed by using the energy performance prediction equation and the initial investment cost equation developed above. Based on this LCC model, an optimization model was developed by applying a multi-objective optimization method. Finally, a program was constructed and verified.

2. Literature Research

2.1. Definition and Research of Nearly Zero Emission Building

The concept of "Zero Emission" was first proposed by Pauli (United Nations University) in 1994, and it was meant to describe the process where a minimum amount of waste would be generated, eventually creating zero waste. "Zero Emission Building" in architecture means a building with the objective of creating zero CO_2 emissions during the life cycle of the building. When it is explained that "Zero Emission Building" widened the scope of CO_2 emitted by the building into three that are operation energy, embodied energy and transportation energy [4], scope 1 or the operation energy used by the building accounts for 70% of all CO_2 emissions and has the greatest importance [5].

Buildings in South Korea emit more than 25% of all greenhouse gases. With the idea that there plenty of room to reduce the greenhouse gases by constructing green buildings, a great deal of effort has been made to reduce the greenhouse gas emissions from buildings. Besides, the Ministry of Land, Infrastructure and Transport and the Ministry of Trade, Industry and Energy have designated zero energy as a new industry and made joint efforts to realize the goal through cooperation as one of the key projects to create a future growth engine [6]. "Zero Energy Building" basically means a building where the sum of used energy and produced energy is 0 (Net Zero) [7]. Given the current technologies and economic feasibility, however, the policy defined it as a building that consumes the least energy (Nearly Zero, 90% reduction) [8], with maximized heat insulation performance, minimized building energy load (passive), use of renewables, such as photovoltaic (active) and minimum energy needs for proper functioning [9].

Therefore, studies carried out in the country mainly focus on realizing nearly zero emission by analyzing and combining individual elementary technologies. Song [10] analyzed the energy saving effect of air conditioning technology when applied using simulation while Gang [11] studied the effect of ventilation, eave and lighting control. Though the final goal of the above studies was to make a zero energy building, policy studies were also conducted to overcome the gap between the goal and reality. Policy wise, Shin [12] analyzed the energy demand and consumption depending on combination of technologies to bring the zero energy housing into reality. In the meantime, Jo studied the policies of South Korea and other countries to build zero energy buildings and examined the local policies [13].

2.2. Decision Making for Realization of Zero Emission Building

As the early design stage decides 80% of the initial investment and most of the design decided in this stage influences energy consumption during the operating stage [14], it can be said that the operating cost is decided here as well. Many decisions are made in the early design phase. A high-performance envelope designed with high performance insulation materials can help to reduce the heating costs as well as the operating costs. Building exterior shading can also decrease the cost of the cooling system and also the energy consumption for cooling. Exterior shading can however increase the heating cost as it is likely to block solar heat during winter, which raises the LCC. Accordingly, design variables should be analyzed comprehensively as they have varying impacts on the initial investment, cooling/heating costs and lighting costs.

Financial problem is one of the major obstacles in the realization of nearly Zero Emission Buildings (nZEBs). Some investors believe that increasing initial investment to enhance building performance results in an economic loss. However, it is reasonable to calculate the building cost with the life cycle cost (LCC) that includes both the initial investment cost and the operational cost, as investment to improve the energy performance should eventually lead to a lower energy cost during operation. Thus, the return on investment calculation should be related not only to the investment costs, but should also take into account the operational and maintenance costs of buildings [15]. Reaping the most benefit out of the least investment cost is important to consider. In this concept, the "lowest LCC" option, which pertains to the entire life cycle of a building is the most economical one [16]. Recently, the most economical condition is also called cost optimal, with investors wanting their investment to be cost optimal.

Calculating the optimal LCC at the early design phase requires calculation of the initial investment and operational costs and this also calls for a significant investment of time and money in a database. This suggests that the most early design phases end with only a small portion of the design space examined and a large number of design solutions left untouched. There is therefore no guarantee that the final design will be an optimal one. Besides, it is almost impossible in Korea to make the best decision based on such analysis in the early design phase as a quick decision is usually required. If the optimization includes an automatic process of design creation, simulation and evaluation, building design optimization will become beneficial as it will help to create a design quickly which can realize the performance that the building seeks for and provide an overall knowledge of the whole design space [17]. The optimal design created by an automatic process facilitates the designer's decision making that considers the reality of nZEB [18]. It also acts to relieve the designer of a very time-consuming and labor-intensive trial and error approach, which designs and evaluates a number of buildings to reach the same goal [19].

In the study of optimal life cycle cost, Wang [20] developed an LCC model, which considered cost effective design rather than only the energy performance in a single-objective function. The LCC here, which reflected a trade-off between the initial investment and the operating cost, was more advanced than other previous models. Znouda [21] developed a standard GA which included a tool (CHEOPS) that allowed load calculation for a simple building. This model produces both energy consumption and a LCC optimal model. Typically, the economic impact (LCC) and the environmental impact (energy consumption) were used as design goals in the multi-objective optimization model. Hasan et al. [22] combined the IDA ICE 3.0, energy simulation tool, and the GenOpt2.0, genetic algorithm program, to develop an LCC design process. In the multi-objective optimization model developed by Shi (2011) [23], EnergyPlus, a sophisticated and widely used BEPS software tool, was integrated into mode FRONTIER, an optimization suite which contains a series of pre-programmed optimization algorithms, to minimize the energy demand for space conditioning and insulation usage for office buildings in Southeast China. With only one insulation material, the insulation usage ishas essentially an equivalent term to the initial cost.

2.3. Tools for Fast Decision Making in the Early Design Phase

Most of the recently developed optimal LCC tools are calculated based on a very detailed database which includes the finishing paint coating and type of stone finishing [24,25]. It is quite difficult to use an optimal LCC tool that uses a detailed database at the early design stage. Therefore, data which cannot be determined and obtained at early design stage to calculate the LCC must be included in the tool [26]. This helps to minimize both time and effort needed to calculate the initial investment and operating cost to give results quickly, thereby resulting in quick decision making. Moreover, the fast automatic process of the optimized LCC analysis presents the optimal LCC or the lowest LCC that can be drawn using the given variables, providing a guideline for the designer.

3. The Framework of Decision Supporting Tools in the Early Design Phase

The basic framework of the decision supporting tools was established with an input module, an assessment module and an output module and core elements which were derived from the assessment tool comparative analysis [27].

3.1. Input Module

The items to calculate for the LCC analysis should be decided considering the analysis target. The initial investment can largely be divided into items which influence the building's operating stage, i.e., energy consumption and those not. Considering the goal of the study, the items that did not affect the energy performance had a relatively low influence. Hence, these were not included in the LCC analysis and the initial investment that was entered into the input module was set up. The initial investment cost usually comprises civil engineering, architecture, mechanical system, electric system, landscape and interior construction. Out of these, civil engineering, landscaping and interior construction were excluded because their impact on the energy performance is relatively insignificant. In the electric construction, only the lighting system needs to be considered and should be addressed in the early design phase.

3.1.1. Architecture

Extracting the variables that affect the energy consumption and cost is important. Architectural design factors at the early design phase include building shape, size, window/wall ratio, envelope and structure. Though the number of deciding factors is not large, they interact with one another, which in turn further affects the energy performance. Therefore, each factor needs to be systematically classified into two categories [28]. One is the factors which affect both the energy consumption and the architectural design, whilst the other has no relevance to the design but affects energy consumption. The factors affecting the design have to be further sorted first through literature review. After the sorting process, a sensitivity analysis should be performed to select the critical input parameters. The parameters selected through statistical analysis have relatively greater influence or sensitivity on the energy consumption than others. After sensitivity analysis, the parameters which had larger ratios were selected. A factor influencing the energy, though it is not one of the design factors, was also checked, as it had to be entered to predict the energy consumption. It is particularly important to pre-set the factors that affect energy consumption as a default database for prediction used at the early design phase even if they are not any design factor. The factors that do not include design factors but must be input are the internal loads, ventilation rate, indoor temperature set points and operating hours. If these inputs are set up as standard information, time and efforts to find and input them can be greatly reduced.

3.1.2. Mechanical and Electrical System

For mechanical systems, a Heating, Ventilating, Air-Conditioning, and Refrigeration (HVAC&R) system should be of primary interest as it is a complicated system with numerous variables and restrictions. According to past research, the HVAC&R system matrix has a combination of 16 AC sub-systems, 15 plant sub-systems and 4 transportation systems. The number of possible combinations that can be made is therefore 960 [29]. However, it was narrowed down to the realistic and universal systems which are used in Korea. Thus, a total of 56 sub-system combinations (4×14) can be produced. The present study used these 56 basic HVAC&R database sets to calculate the energy consumption of the HVAC&R system. In the electric system, only lighting system that can be decided in the early design phase needs to be considered and others should be set to a default value [30].

3.2. Assessment Module

A LCC process usually includes the following steps: definition of LCC analysis/identification of the goals, establishment of a basic assumption for the analysis, selection methodology for the LCC analysis (e.g., cost breakdown structure, identifying data sources and contingencies) and LCC modeling [31].

3.2.1. Definition of the LCC Analysis/Identification of Goals

Life cycle cost (LCC) is an approach that assess the total cost of an asset over its life cycle including initial investment, operating cost, maintenance costs. Nowadays, most builders are only concerned about the initial investment, working toward their minimization [32]. This leads to an emphasis on the initial cost, in detriment of other life cycle cost, and, for the same cases, to the supporting of solutions that require smaller investment but have higher operational costs (such as the application of less insulation resulting in higher need for heating and cooling energy) and also lower Zero Emission levels (like higher carbon emission).

Life cycle cost is an economic methodology for selecting the most cost-effective alternative over a particular time frame, taking into consideration its initial cost (construction), operational cost and maintenance cost. However, the material cost and installation cost of the initial investment as well as the energy operating cost are the most influential in reflecting the energy saving strategies into the building design. Therefore, only the costs of these two stages of the life cycle were calculated to analyze the LCC in this study (Figure 1).

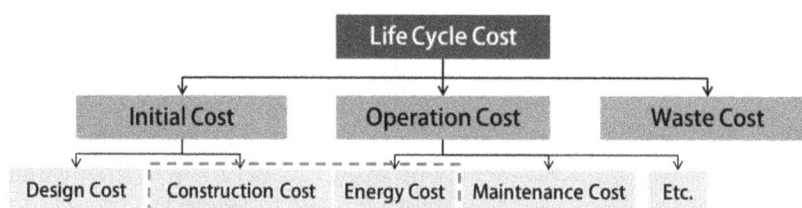

Figure 1. Scope of the Life Cycle Cost Analysis.

3.2.2. Establishment of the Basic Assumptions for Analysis

A discount rate should be established to calculate operating cost, with interest rate (10 years on average), oil, electricity and gas prices considered in the calculation [33]. South Korea is also expected to face carbon price soon which is not yet decided in the market. Therefore, this study employed a price from the CO_2 emissions trading market of the EUA (European Union Allowance) [34].

3.2.3. Selection of Methodology for the LCC analysis

The process of predicting the LCC requires calculation of the costs that occur throughout the period in question at a certain time. The global cost method [35] was used to calculate the LCC. This calculation method includes a system that affects the building's energy performance. In this study, the construction and facility costs that affect the energy performance were included in the calculation. All costs were deemed to occur when they are calculated. The global cost calculation method is as shown below:

$$\text{Life Cycle Cost } (X) = C_p\ (C_i + C_o) = C_p\ (C_m + C_E + C_{Em}) \tag{1}$$

where C_p = Conversion Factor of Present Value; C_i = Initial Cost; C_o = Operation Cost; C_m = Construction Cost; C_E = Energy Cost = $C_C \times (E_C + E_H + E_L + E_E + C_{EM})$; C_{EM} = CO_2 Emission Cost = $C_C \times E_{Emission}$, where E_C = Cooling Energy; E_H = Heating Energy; E_L = Lighting Energy; E_E = Electric Energy; $E_{Emission}$ = CO_2 Emission.

The discount rate coefficient R_d was used to refer to the replacement cost and the final value to the starting year. It is expressed as:

$$R_d = \frac{1}{(1 + R_r)^i} \tag{2}$$

where R_r is the real interest rate in the year of calculation. When the annual costs occur for many years, such as the running costs, the present value factor, which is expressed as a function of the number of years n and the interest rate R_r can be calculated using Equation (3):

$$f_{pv}(n) = ((1 + R_r) - 1)/(R_r \times (1 + R_r)^n) \tag{3}$$

This study used the present value method to calculate the value of all future costs at the current value:

$$P_F = F\frac{1}{(1 + i_r)^n} P_A = A\frac{(1 + i)^n - 1}{i_r(1 + i_r)^n} P = P_F + P_A \tag{4}$$

P_F: Present value of a lump sum; A: Annual Value; F: Future Cost; P: Present Worth Factor; i_r: interest; n: Analysis Period.

3.2.4. LCC Modeling

Table 1 presents the LCC analysis procedure for a nZEB in which the CO_2 emission cost was considered.

Table 1. LCC Analysis procedure of nZEB.

Procedure		Contents
1. Identification of goals Investigation of LCC formation items /Establishment of analysis model	→	LCC Analysis of a nearly Zero Emission Building Selection of applicable technology, Establishment of analysis model Construction stage: Initial Investment cost Operation stage: Energy Cost, Environmental cost
2. Establishment of the basic assumptions for analysis	→	Analysis period, Time point of analysis, Analytical unit, Interest rate, Inflation rate, Discount rate, Energy price escalation rate, Market-linked energy cost, Present value factor, Trading cost of CO_2 emission

Table 1. *Cont.*

Procedure		Contents		
		Initial Investment cost		
3. Selection methodology for LCC analysis	→	Energy Consumption Conversion of energy consumption and CO_2 emission occurring during the operating stage	→	Energy cost: Energy consumption, Price of energy Environmental pollution charge: CO_2 emissions, Price of CO_2 emission trading
		DCFA (Conversion into present value) → LCC estimation Application of a discount rate and prime cost interlocking, Present value factor calculation of prices and energy sources		
4. Decision making based on LCC analysis	→	LCC estimation results → Making a decision for each analyzed goal		

3.3. Output Module

A linear scaling transformation was applied to normalize the performance assessment results. The method is suitable when different attributes are mixed. The attributes consist of decision making factors which are in different units and cases where a minimization standard (e.g., expense) and a maximization standard (e.g., profit) that should be applied are mixed. Under a complicated linear scaling transformation in particular, the value is transformed between the range of the maximum and minimum value of each attribute. This means that each attribute transforms precisely to a value between 0 and 1. Hence, a complicated linear scaling transformation was applied to this study. Under the assumption that E is economic feasibility. for example, the maximum value of E is obtained when the LCC is at its lowest. Ej is the performance value of E for the alternative j. E_{max} and E_{min} represent the LCC of the baseline model and the optimized model, respectively. This suggests that E_{min} is the lowest possible life cycle cost among the selected variables, which signifies the optimal cost. If E_{max} and E_{min} are normalized using complicated linear scaling transformation, Equation (5) is obtained. When each performance result is transformed to a dimensionless score through normalization, the output value lies between 0 and 1. In this case, 0 represents the standard model, whilst 1 represents the target building with maximum performance.

- *Establishing the minimum performance*: Minimum performance means the lowest performance realized by the suggested model. Therefore, the LCC is the largest model and the value of the universal standard model that complies with the minimum legal requirements in general has the minimum performance.
- *Establishing the optimal performance*: The optimal performance implies the best performance of the suggested model conditions, meaning that the building is cost-optimal with the lowest LCC under the given condition. Therefore, an optimization design method was applied and the economic performance resulting from its best solution was set as the optimal performance. The optimization design is determined by the objective function for many decision variables:

$$Env = \frac{Env_j - Env_{min}}{Env_{max} - Env_{min}} \tag{5}$$

3.4. Framework for the Decision Tool for Use at the Early Design Stage

The basic framework is established with an input module, an assessment module and an output module and core elements (Figure 2). An assessment software program for nZEB is developed based upon this framework derived from a logical calculation flowchart.

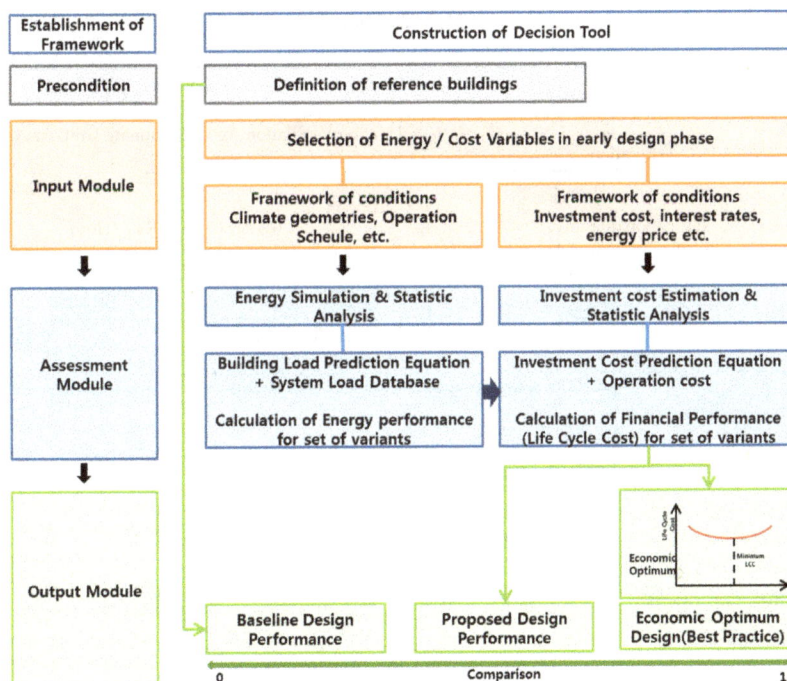

Figure 2. Framework of the Decision Tool.

4. Development of the Decision Support Tool

4.1. Input Module

4.1.1. Description of the Baseline Building

In Korea, it is mandatory to submit an "energy conservation plan" when applying for a construction permit to build an office building larger than 5000 m². In this study, the building data of 158 typical office buildings with their design drawings and energy conservation plans were thoroughly examined among these data, the contents of 30,000–50,000 m² of floor area are mentioned in this text. At this time, the Baseline is the building which is the Mean Average within the database.

The baseline buildings served as the basis for calculations applying the methodology. The baseline building was defined to reflect the typical buildings of the building stock of Seoul, Korea [36]. The characteristics of the baseline building were determined by careful examination of a typical design. Survey results showed that most of the office buildings in Korea are of basic module type consisting of a central building core. Figure in the Table 2 is a simplified model for simulation and Table 2 shows a description of the baseline model and operating features.

Table 2. Brief Description of the Baseline Model.

Category	Factors		Value
Climate Site	Climate Data		Seoul (TMY2)
	Heating/Cooling period	Heating	1/1~3/15, 10/31~12/3
		Cooling	3/15~10/31
Building	Floor area		40,000 m²
	Ceiling height		2700 mm
	Plenum height		1400 mm
	Width/Depth Ratio (%)		1:1
	Window/Wall Ratio (%)		40%
	Wall		2.48 W/m²·K
	Windows	U-Value	2.1 W/m²·K
		Visual Light Transmittance	70%
		Solar Heat Gain Coefficient	0.2

Table 2. *Cont.*

Category	Factors	Value			
System	Heating Cooling Ventilation	Heating Ventilation Air Conditioning Unit, Fan Coil Unit			
Operation Occupancy	Temperature control	Heating Cooling	24 °C 26 °C		
	Ventilation quantity Number of occupants	0.3 Air Change Rate/Hour 5 m^2/person			
	Internal heat (W)	Person	Latent Sensible	70 45	
		Equipment Lighting	10.4 W/m^2 15.1 W/m^2		
Plan of Typical Office					

4.1.2. Design Variables

In this study, the building data of 153 typical office buildings with their design drawings and energy conservation plans were thoroughly examined, and all the variables which contribute to building energy consumption were identified. Among them, 24 factors are selected as energy related design parameters based on sensitivity analysis. These variables are related to architects' early design decision and, at the same time, are able to be defined in quantifiable objective values. A number of precedent researches were also examined to verify the validity of the findings [37–39] (Table 3). Also, the practically applicable ranges of each design variable in the building were set based upon investigated data, which encompassed performance from the minimum level regulated by the building code to cutting edge commercialized technologies.

Table 3. Applicable ranges of the design variables.

Category	Energy Strategies		Ranges
Volume, Shape, Plan	gross floor area (m^2) # of stories floor height (m) width/depth ratio (%) Window/wall ratio (%)		30,000–50,000 (Mean 41,000) 20–30 (Mean 22) 3.7–4.4 (Mean 4.0) 1:1–1:2 30–60
Arrangement	Orientation		0~90 (0: South)
Others	Insulation performance (W/m^2·K)	Roof Wall Ground Floor	0.56–0.15 0.27–0.09 0.69–0.19
	Window performance	U-value Solar Heat Gain Coefficient Visual Light Transmittance	1.00–1.72 0.2–0.4 15%–70%
	Ventilation rate		0.4–0.8

4.1.3. Sensitivity Analysis

The parameters which have significant contribution at the 5% significant level (*p*-value) in ANOVA were selected as the contributing energy factors, and the non-significant terms (*p*-value greater than 0.05) were eliminated. Only nine parameters were selected as significant factors for heating energy consumption. For cooling energy consumption, 13 parameters were found to be significant. They are; Floor to Floor Height, Window Wall Ratio (Façade), Orientation, Window Wall Ratio (South), Window Wall Ratio R (North), Ventilation, U-value of the Wall, U-Value of Window (North), Solar Heat Gain Coefficient-South for heating energy. For cooling energy, Gross Floor Area, Surface Floor Ratio, Floor to Floor Height, Width Depth Ratio, Window Wall Ratio (East), Window Wall Ratio (West), Daylighting, U-Value of Wall, Window U-factor (South), Solar Heat Gain Coefficient (West), Solar Heat Gain Coefficient (South), Visual Light Transmittance (West), Visual Light Transmittance (South) will be selected (Tables 4 and 5).

Table 4. Coefficient (a) for Heating Load.

Factors	Unstandardized Coefficients B	Standard Error	ß	t	p
(Invariable)	30.6	1.99		15.392	0.000
Floor Height	1.75	0.64	0.152	2.734	0.008
Window Wall Ratio (Façade)	1.78	0.64	0.155	2.788	0.007
Orientation	1.65	0.64	0.143	2.574	0.012
Window Wall Ratio (South)	−1.92	0.64	−0.167	−3.003	0.004
Window Wall Ratio (North)	−1.47	0.64	−0.128	−2.294	0.025
Ventilation	1.51	0.64	0.131	2.361	0.021
Wall	−8.72	0.64	−0.756	−13.589	0.000
U-value (North)	−1.59	0.64	−0.138	−2.480	0.016
Solar Heat Gain Coefficient (South)	2.86	0.64	0.248	4.456	0.000

Table 5. Coefficient (a) for Cooling Load.

Factors	Unstandardized Coefficients B	Standard Error	ß	t	p
(Invariable)	46.93	1.675		28.020	0.000
Gross Floor Area	−1.62	0.45	−0.252	−3.573	0.001
SFR	−0.80	0.45	−0.125	−1.765	0.082
Floor Height	1.41	0.45	0.219	3.110	0.003
Width Depth Ratio	1.48	0.45	0.232	3.284	0.002
Window Wall Ratio (East)	1.10	0.45	0.172	2.438	0.017
Window Wall Ratio (West)	1.12	0.45	0.174	2.465	0.016
Daylighting	−1.62	0.45	−0.253	−3.580	0.001
U-value (Wall)	−1.49	0.45	−0.232	−3.295	0.002
U-value (Win-S)	0.94	0.45	0.147	2.077	0.042
Solar Heat Gain Coefficient (West)	−2.60	0.45	−0.406	−5.748	0.000
Solar Heat Gain Coefficient (South)	−2.10	0.45	−0.327	−4.631	0.000
Visual Light Transmittance (West)	0.62	0.45	0.010	0.137	0.049
Visual Light Transmittance (South)	−0.78	0.45	−0.122	−1.733	0.049

4.2. Assessment Module

4.2.1. Initial Cost

Costs related to building elements with no influence on the energy performance, such as e floor covering and wall painting, while the costs that remain the same for all measures/packages/variants, e.g., earthworks and foundations, staircases, and demolition costs can be excluded. The costs must be market-based and coherent regarding the location and time of the investment costs, running costs, energy costs, and CO_2 cost. Data sources can be an evaluation of recent construction projects, an analysis of standard offers of construction companies, or existing cost databases which have been derived from market-based data gathering.

(1) Architecture Cost Prediction Equation

It is not easy to estimate the construction cost. In particular, in the early stage of design, the cost is calculated excluding the design cost. This is because many undetermined design factors are generally present which affect the construction cost. Therefore, it becomes difficult to generate reliable cost estimates. In the current study, a construction cost prediction equation was developed for the early design stage using statistical methods and actual construction cost estimated data. The reliability of the prediction equation was increased by producing data through interpolation. Variables used in the construction cost prediction equation are the factors from the early design stage which affect energy consumption during the operating stage. The predicting equation was developed through the following four steps:

- Estimation of a first-order model based on actual data volatility
- Generation of model development data and step by step data interpolation
- Development of an initial investment cost estimation model and equation
- Correction of the model based on estimation results

- *Estimation of the first-order model based on actual data volatility*

The first-order model based on actual data volatility was estimated in the first step of the analytical process, namely validating the relationship between the construction cost and the various independent variables for the data enrichment or interpolation. Interpolation is essential for the estimation of the initial investment as data extraction for each condition is limited [40]. The construction variability based on the baseline floor area (1500, 2000 or 2500 m^2) was the most fundamental and significant part and was set as a reference point for data interpolation done later. This variability arose due to a lack of measured values when changes in the independent variables occurred along with the changes in the baseline floor area. First, an estimation equation between the construction cost and the changes in the independent variable was derived. No data were available to represent the changes in the baseline floor area and the floor height. Therefore, a regression equation for changes in the construction cost according to changes in the floor height, and the relation between the change in the floor height and the floor area was applied at 1500 m^2, 2000 m^2 and 2500 m^2. Figure 3 shows a graphical representation of the derivation of the first-order model for changes in the construction cost with changes in certain independent variables (Figure 3).

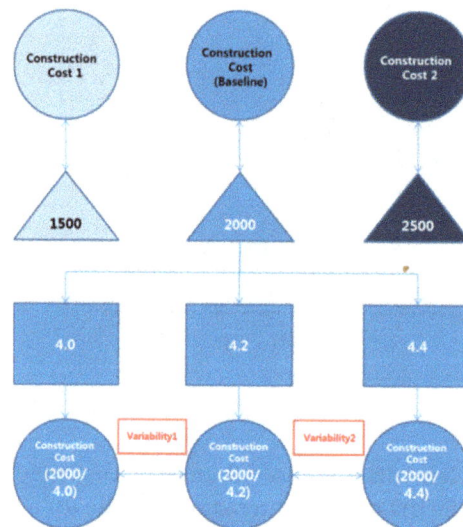

Figure 3. First-order model estimation process based on volatility.

- *Generation of model development data and step-by-step data interpolation*

This step refers to the process of filling the areas which lacked data for specific conditions through a what-if analysis, using the model created in the previous step. A what-if analysis is a process of finding an outcome when values are altered with respect to the actual values. This kind of analysis is often applied in time-series forecasting models. However, since this model does not have time-series data, a linear model was applied [41]. The result from the previous first-order model estimation process was applied to perform data interpolation based on a construction cost relations estimation equation to populate the area for which data did not exist. The results obtained from the regression estimation equation were applied and data interpolation was performed in the following order: the baseline floor area, floor height, width depth ratio, window wall ratio, U-value of wall and wall insulation. This series of steps was applied sequentially and repeatedly (Figure 4).

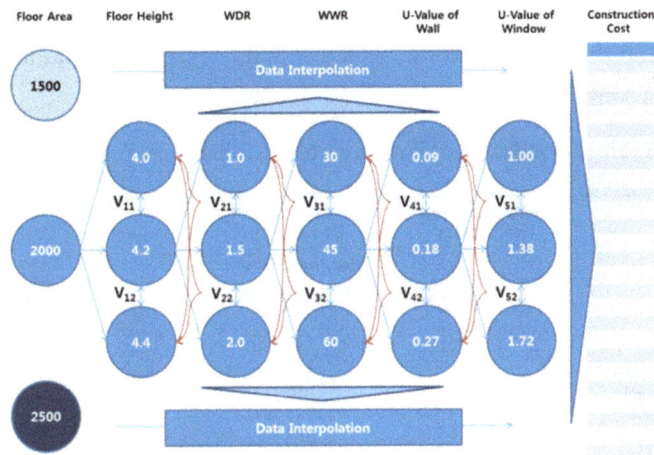

Figure 4. Model development data and step-by-step data interpolation.

- *Development of an estimation model for the initial investment cost*

Based on the integrated data extracted from the analysis process in phases 1 and 2, a linear model was established. Since time series factors were absent, the ARIMA model was not applied. Instead, a linear model, which has a high potential for expansion, was estimated [42]. The regression model was used as a basis for derivation of the Best Linear Unbiased Estimator (BLUE) from several linear models [43].

- *Correction of the model based on estimation results*

In this final step, the derived initial investment cost estimation model was examined and validated for any model adjustment. By hypothesizing a particular condition, discrepancies between the estimates derived from the model and the real values provided at the early design phase were calculated and examined. The average error rate was 3%–5%, and to improve this, the model was adjusted through random simulations of the estimated values by changing the intercept term and the beta coefficient. Thus, the model created from step 3 was adjusted to a final form to derive the initial investment cost model estimation equation (Figure 5).

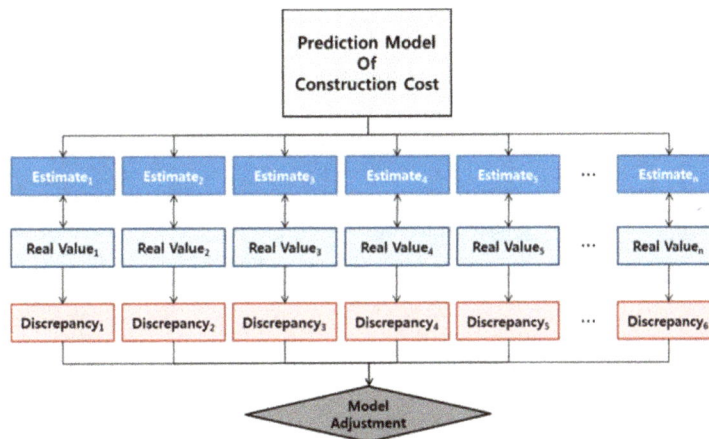

Figure 5. Final correction of the model based on estimation results.

- *Prediction Equation*

A mathematical model was developed through a multiple regression analysis (Table 6). Multiple regression analyses were conducted using the data set, and the load prediction equations were derived.

The coefficient of determination (R^2) between the regression models was 0.845 for the heating load prediction and 0.998 for initial cost prediction. As a result, the regression equations were found to have a considerable predictive power.

Table 6. Coefficient of Equations of the Initial Cost Prediction.

Factors	Unstandardized Coefficients		Standardized Coefficients	ß	t	p
	B		Standard Error			
(Invariable)	−467,611,530.097		10,486,340.385		−44.592 **	0.000
Gross Floor Area	599,921.127		950.139	0.982	631.404 **	0.000
Floor Height	142,207,711.837		2,352,457.595	0.107	60.451 **	0.000
Width Depth Ratio	25,200,179.482		943,528.129	0.053	26.708 **	0.000
Window Wall Ratio (South)	−73,285.833		11,449.391	−0.012	−6.401 **	0.000
Window Wall Ratio (North)	−85,622.281		14,210.349	−0.011	−6.025 **	0.000
Window Wall Ratio (South)	−119,988.105		13,482.439	−0.015	−8.900 **	0.000
Window Wall Ratio (North)	−113,804.377		18,499.495	−0.011	−6.152 **	0.000
U-value of Wall	−112,897,138.380		5,241,935.320	−0.028	−21.537 **	0.000
U-value of Window	−30,919,301.401		1,309,811.482	−0.036	−23.606 **	0.000

$$F = 68,492.228, R^2 = 0.998, \text{adj. } R^2 = 0.998; ** p < 0.01.$$

The equation is y = (−467,611,530.097) + 599,921.127 × Gross Floor Area + 142,207,711.837 × Floor Height + 25,200,179.482 × Width Depth Ratio + (−73,285.833) × Window Wall Ratio (East) + (−85,622.281) × Window Wall Ratio (West) + (−119,988.105) × Window Wall Ratio (South) + (−113,804.377) × Window Wall Ratio (North) + (−112,897,138.380) × U-value of Wall + (−30,919,301.401) × U-value of Window.

(2) Mechanical System Cost Prediction Equation

This study used 56 basic HVAC&R database sets, as mentioned above. For the facility system construction cost for each set, the capital cost is calculated for baseline and basic HVAC&R system, which consists of typical floor, mechanical room, and riser construction cost. The construction cost is then used to predict 56 combinations of HVAC&R system initial construction cost (Figure 6) [44].

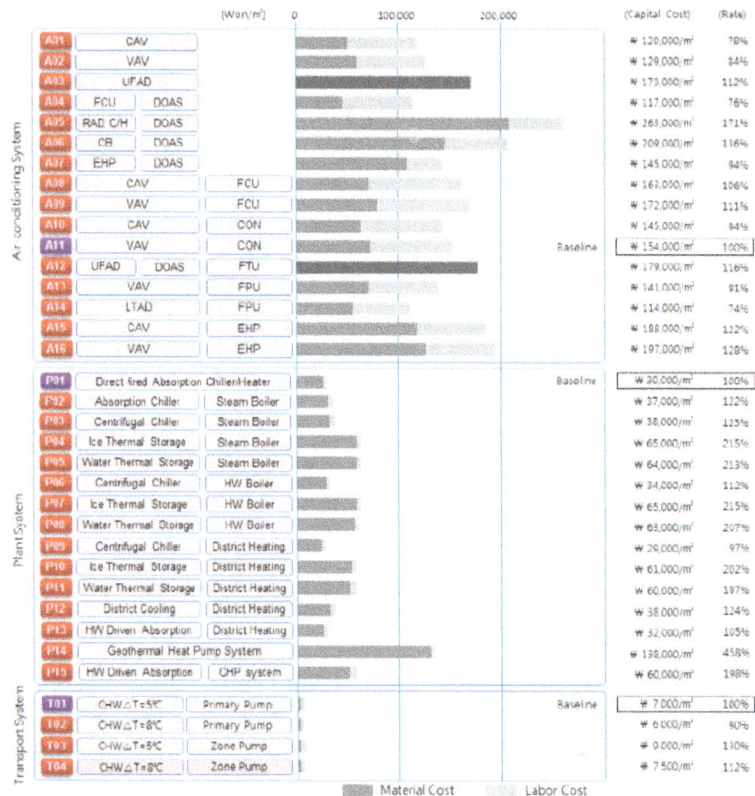

Figure 6. Analysis of construction cost of HVAC system boundary.

(3) Electrical system

Assuming that electrical construction cost is rarely an influencing factor to be decided in the early design stage, the electrical construction cost was taken as a constant, except for the lighting system and the cost of the lighting system. The constant was 178,251 won/m^2. The value for the lighting system and the cost of the lighting system was automatically selected from 48,413 won/m^2, 33,889 won/m^2 and 17,659 won/m^2, depending upon the selected lighting density [45].

4.2.2. Operating Cost

The energy cost of operating was calculated considering the expected energy consumption, the energy price inflation rate, inflation rate and the discount rate that reflected the interest rate and operational period. The building load prediction equation was developed with the variables mentioned above and the energy consumption was calculated through a connection with the disclosed equipment system database.

(1) Building Load Prediction Equation

The energy load prediction equations developed in the present study can be easily utilized to estimate heating and cooling load of office buildings in the central climatic zone of Korea during the early design stage. The number of tests needed to analyze the sensitivity was 3^{24} (=282,429,536,481), which was reduced to 81 using an orthogonal array. The tests applied an EnergyPlus simulation. The prediction equation was developed using ANOVA and multiple regression analysis (Table 7) [46]. The criterion used for the improvement of the prediction was the coefficient of determination (R^2). R^2 represents the square of the correlation between the predicted value and actual value. It is expressed as a decimal number between 0.00 and 1.00. 1.00 means perfect prediction in the model. The p-values were $R^2 = 0.753$ for heating energy consumption and $R^2 = 0.602$ for cooling energy consumption.

Table 7. Equations for Energy Consumption Prediction.

Partial R-Square	R-Square	Modification of R-Square	Standard Error
0.883	0.780	0.753	4.712

Heating Energy Consumption $(Y) = 30.692 + 1.75\,X_1 + 1.78\,X_2 - 1.65\,X_3 - 1.93\,X_4 + 1.47\,X_5 + 1.514\,X_6 - 8.17\,X_7 - 1.59\,X_8 + 2.86\,X_9$

X_1	Floor to Floor Height
X_2	Window Wall Ratio (Façade)
X_3	Orientation
X_4	Window Wall Ratio (South)
X_5	Window Wall Ratio R (North)
X_6	Ventilation
X_7	U-value of the Wall
X_8	U-Value of Window (North)
X_9	Solar Heat Gain Coefficient-South

Partial R-square	R-square	Modification of R-square	Standard Error
0.861	0.666	0.602	3.3329

Cooling Energy Consumption $(Y) = 46.94 - 1.62\,X_1 - 0.80\,X_2 + 1.41\,X_3 + 1.49\,X_4 - 1.10\,X_5 + 1.12\,X_6 - 1.62\,X_7 - 1.53\,X_8 + 0.94\,X_9 - 2.60\,X_{10} - 2.10\,X_{11} + 0.06\,X_{12} - 0.79\,X_{13}$

X_1	Gross Floor Area
X_2	Surface Floor Ratio
X_3	Floor Height
X_4	Width Depth Ratio
X_5	Window Wall Ratio (East)
X_6	Window Wall Ratio (West)
X_7	Daylighting
X_8	U-value of Wall
X_9	U-value of Window (South)
X_{10}	Solar Heat Gain Coefficient-West
X_{11}	Solar Heat Gain Coefficient-East
X_{12}	Visual Light Transmittance-West
X_{13}	Visual Light Transmittance-East

(2) System Load Prediction Equation

Energy simulation is a considerably accurate tool, but in order to use this, a professional knowledge and detailed input variables are needed. Therefore, an easily and conveniently available tool is highly desirable at the early design phase for practical purposes.

To compensate for the drawbacks of energy simulation, a database was developed based on the comparison analysis of different combinations of Heating Ventilation Air Conditioning & Refrigeration (HVAC&R) systems. This database was opened to public. As shown in Table 8, 56 possible combinations (4 Air conditioning × 14 Plant system) were chosen as the most widely used systems in Korea [44].

The energy consumed by operating the water heating and lighting systems, and the electrical distribution for the equipment use in the building was calculated by using data from an energy-consumption survey of 40 office buildings [47] (Table 9).

Table 8. Basic HVAC&R database set.

Air Conditioning System			Plant System
		P01	Directed Fire Absorption Chiller/Heater
		P02	Absorption Chiller + Steam Boiler
		P03	Centrifugal Chiller + Steam Boiler
		P04	Ice Thermal Storage + Steam Boiler
		P05	Water Thermal Storage + Steam Boiler
A03	UFAD	P06	Centrifugal Chiller + Hot Water Boiler
A06	CB + DOAS	P07	Ice Thermal Storage + Hot Water Boiler
A08	CAV + FCU	P08	Water Thermal Storage + Hot Water Boiler
A11	VAV + CON	P09	Centrifugal Chiller + District Heating
		P10	Ice Thermal Storage + District Heating
		P11	Water Thermal Storage + District Heating
		P12	District Cooling + District Heating
		P13	Hot Water Driven Absorption + District Heating
		P14	Geothermal Heat Pump System

UFAD: Underfloor air distribution, CB: Chilled Beam; DOAS: Dedicated Outdoor Air System; DOAS: Dedicated Outdoor Air System; CAV: Constant Air Volume; FCU: Fan Coil Unit; VAV: Variable Air Volume; CON: Controller.

Table 9. Lighting and Electrical Energy Consumed.

Category	Calculation Model
Electricity Energy	$y = 10.4\ x^*$
Lighting Energy	$y = 15.1\ x^*$

x^*: Area (m^2).

4.3. Output Module

Finding an optimal model with the given variables requires calculation with application of the LCC evaluation equation to the optimization methodology. To undertake this optimization methodology, objective function and constraints which are appropriate for the given question are required.

4.3.1. Optimization Methodology

A multi-objective optimization function should be applied that uses multiple local minima solutions to produce an optimal value of the building energy cost and initial investment. Solution Space incorporates the features for discrete variables [48–50]. In this context, a genetic algorithm (GA) was believed to be the most appropriate among the many optimization models for calculating the optimal value of the discrete variables [51]. The overall process of the GA has been shown below (Figure 7).

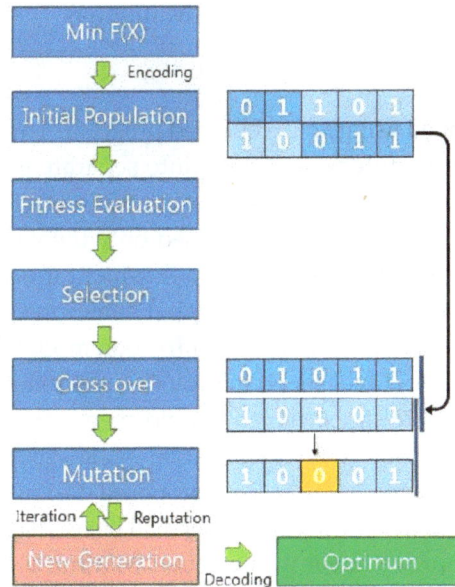

Figure 7. Process of GA algorithm.

4.3.2. Objective Function of the LCC and Constraints

As indicated earlier, an objective function was considered in the model, namely minimization of cost and energy consumption. The objective function can be expressed as shown in Equation (6):

$$\text{Objective function is } f(x) = \min (X_1) \tag{6}$$

where, X_1 = Life Cycle Cost = $C_p(C_i + C_o) = C_p(C_i + C_E + C_{Em})$; C_p = Conversion Factor of Present Value; C_i = Initial Cost; C_o = Operation Cost; C_E = Energy Cost; C_{Em} = CO_2 Emission Cost.

4.3.3. Constraints

The constraints on the design variables essentially delineate the bounds of the feasible range for each variable, which is influenced by market availability of the relevant products and the applicable codes and regulations. Constraints on the selected design variables are shown in Table 10.

Table 10. Constraints for the selected design variables.

Variables		Symbol	Constraints		
Gross floor area (m²)		X_1		$30{,}000 \leq X_1 \leq 50{,}000$	
Floor area (m²)		X_2		$1500 \leq X_2 \leq 2500$	
Floor height (ceiling 1.3 m)		X_3		$4.0 \leq X_3 \leq 4.4$	
Width depth ratio		X_4		$1 \leq X_4 \leq 2$	
Orientation		X_5	0		1
		X_6	1		0
Window wall ratio (%)	East	X_7		$30 \leq X_7 \leq 60$	
	West	X_8		$30 \leq X_8 \leq 60$	
	South	X_9		$30 \leq X_9 \leq 60$	
	North	X_{10}		$30 \leq X_{10} \leq 60$	
U-value of wall (W/m²·K)		X_{11}		$0.09 \leq X_{11} \leq 0.27$	
U-value of window (W/m²·K)		X_{12}		$1.0 \leq X_{12} \leq 1.72$	
		X_{13}		$1.0 \leq X_{13} \leq 1.72$	
		X_{14}		$1.0 \leq X_{14} \leq 1.72$	
		X_{15}		$1.0 \leq X_{15} \leq 1.72$	
Solar Heat Gain Coefficient		X_{16}		$0.3 \leq X_{16} \leq 0.6$	
		X_{17}		$0.3 \leq X_{17} \leq 0.6$	
		X_{18}		$0.3 \leq X_{18} \leq 0.6$	
Visual Light Transmittance (%)		X_{19}		$30 \leq X_{19} \leq 60$	
		X_{20}		$30 \leq X_{20} \leq 60$	
		X_{21}		$30 \leq X_{21} \leq 60$	
Lighting density (W/m²)		X_{22}	7	9.7	15

5. Development of a Fast Decision Support Tool and Validation

5.1. Fast Decision Support Tool

The LCC Model was developed using an Excel-based program. Thus, the program can be executed in a very simple manner. Each module was divided into separate Excel sheets which consisted of an Input Module, an Assessment Module and an Output module. This tool was developed for use at the early design phase. The input module comprised only the variables considered at the early design stage to reduce the time of analysis. For the assessment module, a cost/energy prediction equation was developed that does not affect the energy but allows the users to calculate the output by simply adjusting the variables. A computational optimization model calculation program (MatLab) was linked with the output module (Excel sheet) to provide an automated optimization process.

In most cases, the proposed designs, which should meet actual land conditions, designer's intentions and legal obligations, do not fit the optimization model condition. This optimization model will be used for designing and recognizing how far a proposed design lies from the optimized model. The baseline result which is shown in the output module will be used as starting point to measure how far the proposed design was improved over the baseline building. The LCC of the design proposed by the designer will probably be positioned between the LCC of the baseline building and that of the optimal building. The LCC that is the closest to the optimal building is the optimal design, that is, the lowest LCC. This can then serve as a yardstick to show what performance the proposed design can attain and also offer a guideline to the designer. As shown in Figure 8, the tool developed in the current study was output graphically when a design was compared with the optimized building and the baseline building (Figure 8).

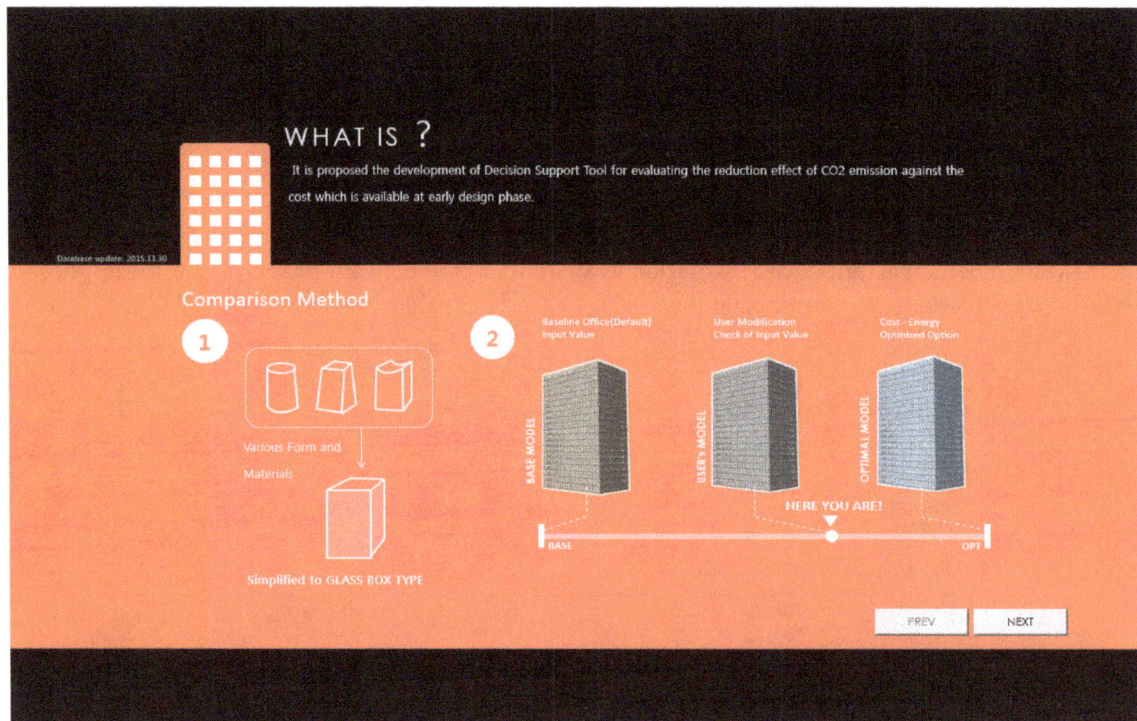

Figure 8. Fast Decision Support Tool developed using excel-based program (output module).

5.2. Case Study

The present study also verified the optimization model by running the model with a MatLab program. Case A was designed with the actual preference of the client (or designer), whereas case B was the model that allowed a significant LCC saving, which was obtained from the optimization model.

Case B, despite its high initial investment cost, provided a higher economic efficiency in terms of life cycle with a lower energy cost than that of case A. When case B increased the efficiency by selecting a chilled beam to enhance the envelope performance (a better U-value of the wall and enhanced window and air tightness), the initial investment grew by 300,000 won/m^2. The lighting facility cost, when LED lightings replaced the general ones, also rose by 300,000 won/m^2. The initial investment was finally increased by 300,000 won/m^2, and the energy consumption decreased by 300,000 won/m^2 with the enhanced envelope performance and 300,000 won/m^2 with the improved facility efficiency. This meant that with 40 years of the building's operation there will be a drop in the total operating cost by 60%. For case B, the total LCC was reduced by 35% compared to case A (Table 11). A continuous comparison between case A (proposed model) and case B (optimized model) enabled the decision makers to find an alternative which is closer to a nZEB. A comparison with the basic model is not described here.

Table 11. Evaluation Results.

Category		Case A (Proposed Model)	Case B (Optimized Model)
Architecture	Gross Area	40,000 m^2	
	Floor Area	2200 m^2	
	Floor Height	4.3 m	
	Width Depth Ratio	1:2	
	Orientation	South	
	Window Wall Ratio	40%	40%
	U-Value of Wall	0.27 W/m^2·K	0.09 W/m^2·K
	U-Value of Window	1.72 W/m^2·K	1.0 W/m^2·K
	Solar Heat Gain Coefficient	60%	30%
	Air Tightness	2.0	0.5
Mechanical	Air Distribution System	CAV + FCU	Chilled Beam + Dedicated Outdoor Air System
	Plant System	Direct Fired Absorption Chiller/Heater	
Electrical	Lighting Density	15 W/m^2	7 W/m^2
Energy Use	Mechanical	210.61 kWh/m^2·year	78.66 kWh/m^2·y
	Electric	45.27 kWh/m^2·year	13.51 kWh/m^2·y
	Process	106.05 kWh/m^2·year	106.05 kWh/m^2·y
	Total	361.97 kWh/m^2·year	294.24 kWh/m^2·y
CO_2 Emission		208.29 TCO_2/m^2·year	129.07 TCO_2/m^2·y
Cost	Investment Cost	1,780,150 won/m^2	1,809,103 won/m^2
	Energy Cost (40 years)	79,001,783 won/m^2	50,749,169 won/m^2
	Total	80,781,933 won/m^2	52,558,272 won/m^2

6. Conclusions

Multi-objective optimizations have been applied with multiple optimization targets (objective functions) through use of genetic algorithm [6], the Taguchi-ANOVA method [9] and the PSO algorithm [10] as optimization techniques. Recent studies on life cycle cost buildings published in other countries [11–15] estimated all costs, including building energy and reduction of CO_2 emissions. However, Korean studies have limited their scope to optimal construction cost, focusing on structure, materials and construction. Especially, past studies rarely used a generic algorithm (GA), which is known to be a good method to identify the cost optimization model based on discrete variable data.

Therefore, LCC evaluation and a cost optimal design tool with GA algorithm (MATLAB program) were developed in the present study which can be used in the early design phase. With this tool the client can recognize the financial benefit of nZEB rapidly in the early design phase and invest to achieve a higher performance from the nZEB. This newly developed tool can also provide the necessary guidance to a designer on the building design of a nZEB and enable them to take effective decisions without spending much time and effort.

It has been found in this study that more design variables must be considered in the early design phase to be able to easily use the tool for fast decision making. Programs for different purposes should be developed that can facilitate decision making for the design of other buildings as well as office buildings. It is essential to continue to update the database of the developed programs as the performance of the technology applied to the baseline building will improve and the unit cost will fall as the technology develops. Therefore, an up-to-date database needs to be maintained through market surveys.

Acknowledgments: This research was supported by Basic Science Research Program through the National Research Foundation of Korea (NRF) funded by the Ministry of Science, ICT & Future Planning (NRF-2013R1A1A3013119).

Conflicts of Interest: The author declares no conflict of interest.

References

1. International Energy Agency (IEA). TASK 40/Annex 52 (2008). Towards Net Zero Energy Solar Buildings, IEASHC Task 40 and ECBCS Annex 52. 2011. Available online: http://www.iea-ebc.org/fileadmin/user_upload/docs/Facts/EBC_Annex_52_Factsheet.pdf (accessed on 16 December 2016).

2. Tatari, O. Cost premium prediction of certified green buildings: A neural network approach. *Build. Environ.* **2011**, *46*, 1081–1086. [CrossRef]

3. Black, R.; Brown, G.; Diaz, K.; Gibler, K.; Grissom, T. Behavioral Research in Real Estate: A Search for the Boundaries. *J. Real Estate Pract. Educ.* **2003**, *6*, 85–112.

4. Kang, S. A Study on the Design Process of Zero Emission Buildings. Master's Thesis, Chung-Ang University, Seoul, Korea, 10 December 2007.

5. King, P.; Clare, M. *Zero Carbon Task Group Report—The Definition of Zero Carbon*; UK Green Building Council Press: London, UK, 2008.

6. MOTL. Zero Energy Building Certification Launching (2016), Korean Government Announcement. Available online: https://www.molit.go.kr/USR/NEWS/m_71/dtl.jsp?lcmspage=1&id=95078225 (accessed on 16 December 2016).

7. Torcellini, P.; Pless, S.; Deru, M. Zero energy buildings: A critical look at the definition. In Proceedings of the ACEEE Summer Study, Pacific Grove, CA, USA, 14–18 August 2006.

8. Grözinger, J.; Boermans, T.; Ashok John, J.; Wehringer, F.; Seehusen, J. *Towards Nearly Zero-Energy Buildings Definition of Common Principles Under the EPBD—Final Report*; ECOFYS Germany GmbH: Köln, Germany, 2013.

9. KICT. What Is Zero Energy Buildings? Korea Energy Agncy: Kyungki, Korea, 2016. Available online: http://nzeb.kict.re.kr/zeroEnergy/about.php (accessed on 16 December 2016).

10. Song, D. High-Performance HVAC system for Net Zero Energy Building (NZEB). *Architecture* **2014**, *58*, 26–31.

11. Kang, J.; Kim, D.; Thorsten, S.; Park, C. Assessment of energy saving potentials in terms of room's thermal characteristics using building simulation. In Proceedings of the AIK Conference, Seoul, Korea, 23–26 October 2013; pp. 371–373.

12. Shin, H.; Chang, K. Analysis of Energy Consumption and Cost based on Combination of Element Technologies for Implementing Zero-Energy House. *J. KIAEBS* **2014**, *9*, 163–170.

13. Cho, D. Policies at Home and Abroad for Dissemination of nearly Zero Energy Buildings. *Architecture* **2014**, *58*, 10–15.

14. Elkington, J. Towards the sustainable corporation: Win-win-win business strategies for sustainable development. *Calif. Manag. Rev.* **1994**, *36*, 90–100. [CrossRef]

15. Cam, C.N. From global climate change to low carbon cities the triple bottom line revisit. In Proceedings of the First International Conference on Sustainable Urbanization, Hongkong, China, 15–17 December 2010.

16. Dean, J.; VanGeet, O.; Simkus, S.; Eastment, M. *Design and Evaluation of a Net Zero Energy Low-Income Residential Housing Development in Lafayette, Colorado*; Technical Report NREL/TP-7A40-51450; National Renewable Energy Laboratory: Golden, CO, USA, 2012.

17. Reijinders, L.; van Roekel, A. Comprehensiveness and adequacy of tools for the environmental improvement of buildings. *J. Clean. Prod.* **1999**, *7*, 221–225. [CrossRef]

18. Deru, M.; Torcellini, P. Improving Sustainability of Buildings through a Performance-Based Design Approach. In Proceedings of the World Renewable Energy Congress VIII (WREC 2004), Denver, CO, USA, 29 August–3 September 2004.

19. Chidiac, S.E.; Catania, E.J.C.; Morofsky, E.; Foo, S. A screening methodology for implementing cost effective energy retrofit measurein Canadian office buildings. *Energy Build.* **2011**, *43*, 614–620. [CrossRef]

20. Wang, J.J.; Jing, Y.Y. Review on multi-criteria decision analysis aid in sustainable energy decision-making. *Renew. Sustain. Energy Rev.* **2009**, *13*, 2263–2278. [CrossRef]

21. Znouda, E.; Ghrab-Morcos, N.; Hadj-Alouane, A. Optimization of Mediterranean building design using genetic algorithms. *Energy Build.* **2007**, *39*, 148–153. [CrossRef]

22. Hamdy, M.; Hasan, A.; Siren, K. Applying a multi-objective optimization approach for Design of low-emission cost-effective dwellings. *Build. Environ.* **2011**, *46*, 109–123. [CrossRef]

23. Shi, X. Design optimization of insulation usage and space conditioning load using energy simulation and genetic algorithm. *Energy* **2011**, *36*, 1659–1667. [CrossRef]

24. European Committee for Standardization. *Sustainability of Construction Works—Sustainability Assessment of Buildings—Part 4: Framework for the Assessment of Economic Performance*; British Standards Institute: London, UK, 2012.

25. IMMOVALUE: Improving the Market Impact of Energy Certification by Introducing Energy Efficiency and Life-Cycle Costs into Property Valuation Practice, IEE/07/553. 2010. Available online: http://www.immovalue.org (accessed on 16 December 2016).

26. Hofer, G.; Herzog, B.; Grim, M. Calculating life cycle cost in the early design phase to encourage energy efficient and sustainable buildings. In Proceedings of the IEECB Focus 2010, Frankfurt, Germany, 13–14 April 2010.

27. Kang, H.; Rhee, E. A Comparative Analysis of Performance Assessment Tools for Establishing Evaluation Framework for Sustainable Buildings. *Archit. Res.* **2014**, *14*, 131–136. [CrossRef]

28. Forsberg, A.; Malmborg, V.F. Tools for environmental assessment of the built environment. *Build. Environ.* **2004**, *39*, 223–228. [CrossRef]

29. Kun, C.J. Development of an energy evaluation methodology to make multiple predictions of the HVAC&R system energy demand for office buildings. *Energy Build.* **2014**, *80*, 169–183.

30. Jung, J.R. Evaluation of Alternative for Building Service System in High-Rise Building Based on Life Cycle Cost Analysis. Ph.D. Thesis, Yonsei University, Seoul, Korea, 5 August 2002.

31. Yum, Y.S. Evaluation of Sustainable Building Technologies for Multi-Family Residential Building Based on Life Cycle Cost Analysis. Master's Thesis, Chung-Ang University, Seoul, Korea, 4 August 2008.

32. Sesana, M.M.; Salvalai, G. Overview on life cycle methodologies and economic feasibility for nZEBs. *Build. Environ.* **2013**, *67*, 211–216. [CrossRef]

33. Lho, S.W. A Study of CO_2 Emission Trading Market Focusing on EU ETS. *J. Environ. Pol. Admin* **2009**, *17*, 25–44.

34. European Emission Allowance. Available online: https://www.eex.com (accessed on 16 December 2016).

35. Lee, G.H.; Kwon, Y.C.; Rhee, E.G. A Study on the Economic Analysis Method of Building Energy Saving Measures Considering Environmental Costs. *J. Archit. Inst. Korea* **2001**, *17*, 217–224.

36. The Bank of Korea Economic Statistics System. Available online: http://ecos.bok.or.kr/jsp/use/qnamgt/QNACtl.jsp (accessed on 15 April 2016).

37. Korea Power Exchange. *Survey on Electricity Consumption Characters of Home Appliances*; Korea Power Exchange: Seoul, Korea, 2006; pp. 50–75.

38. Lam, J.C.; Hui, S.C.; Chan, A.L. Regression Analysis of High-rise Fully Air-conditioned Office Buildings. *Energy Build.* **1997**, *26*, 189–197. [CrossRef]

39. Min-Chul, K. A Study on the Character of Changes in Unit Plans Prepared Apartment Balcony Extension. Master's Thesis, Kyungpook University, Daegu, Korea, 10 December 2007.

40. Ho-Tae, S. A Study on the Development of Load Prediction Equation and Design Guidelines for the Energy Conservation of Office Buildings. Ph.D. Thesis, Department of Architecture Graduate School, Seoul National University, Seoul, Korea, 12 December 1994.

41. Hofer, G. *Integrated Planning for Building Refurbishment—Taking Life-Cycle Costs into Account (LCC-Reburb)*; European Commission: Brussels, Belgium, 2005.

42. Hasan, A.; Vuolle, M.; Siren, K. Minimisation of life cycle cost of a detached house using combined simulation and optimization. *Build. Environ.* **2008**, *43*, 2022–2034. [CrossRef]

43. Yoo, S.I. A Study on the Prediction of Required Total Cost and Avoidable Cost of a High-Rise Apartment Building. Master's Thesis, Chung-Ang University, Seoul, Korea, 26 June 2000.

44. Cho, J.K. Development of an HVAC & R Systems Energy Evaluation Methodology and Simulation Program for Office Buildings. *Korean J. Air Cond. Refrig. Eng.* **2013**, *25*, 363–370.

45.   Kim, Y.R. A Development of a Mobile Application for Building Energy Prediction Using Performance Prediction Model. Master's Thesis, Chung Ang Univeristy, Seoul, Korea, 2014.

46.   Hwan, J.J. A Study on the Development of Expert System for Selecting and Modifying Orthogonal Array in Taguchi Method. *Korea Intell. Inf. Syst. Soc.* **1997**, *1*, 350–361.

47.   Hoseon, Y. Effects of Various Factors on the Energy Consumption of Korean-Style Apartment Houses. *J. Air-Cond. Refrig.* **2002**, *40*, 972–980.

48.   Leigh, S.B.; Won, J.S.; Bae, J.I. An Energy Management Process and Prediction of Energy Use in Office Building. *J. Asian Archit. Build. Eng.* **2005**, *4*, 501–508. [CrossRef]

49.   Wright, J.; Loosemore, H.; Farmani, R. Optimization of building thermal design and control by multi-criterion genetic algorithm. *Energy Build.* **2002**, *34*, 959–972. [CrossRef]

50.   Cai, Z.; Wang, Y. A multi objective optimization-based evolutionary algorithm for constrained optimization. *IEEE Trans. Evolut. Comput.* **2006**, *10*, 658–675. [CrossRef]

51.   Thomas, B.; Jan, G.; von Bernhard, M.; Nesen, S.; Ashok, J. *Assessment of Cost Optimal Calculations in the Context of the EPBD (ENER/C3/2013-414)*; Final Report; ECOFYS Germany GmbH: Cologne, Germany, 2015.

3

A Causal and Real-Time Capable Power Management Algorithm for Off-Highway Hybrid Propulsion Systems

Johannes Schalk [1,*] and Harald Aschemann [2]

[1]   MTU Friedrichshafen GmbH, Maybachplatz 1, 88045 Friedrichshafen, Germany
[2]   Chair of Mechatronics, Rostock University, Justus-von-Liebig Weg 6, 18059 Rostock, Germany;
      harald.aschemann@uni-rostock.de
*   Correspondence: johannes.schalk@mtu-online.com

Academic Editor: William Holderbaum

Abstract: Hybrid propulsion systems allow for a reduction of fuel consumption and pollutant emissions of future off-highway applications. A challenging aspect of a hybridization is the larger number of system components that further increases both the complexity and the diversification of such systems. Hence, beside a standardization on the hardware side for off-highway systems, a high flexibility and modularity of the control schemes is required to employ them in as many different applications as possible. In this paper, a causal optimization-based power management algorithm is introduced to control the power split between engine and electric machine in a hybrid powertrain. The algorithm optimizes the power split to achieve the maximum power supply efficiency and, thereby, considers the energy cost for maintaining the battery charge. Furthermore, the power management provides an optional function to control the battery state of charge in such a way that a target value is attained. In a simulation case study, the potential and the benefits of the proposed power management for the hybrid powertrain—aiming at a reduction of the fuel consumption of a DMU (diesel multiple unit train) operated on a representative track—will be shown.

Keywords: off-highway propulsion system; hybrid electric vehicle; hybrid electric diesel multiple unit train (DMU); hybridization; power management strategy; energy management strategy

1. Introduction

1.1. Motivation

In the automotive industry, hybrid electric and full electric vehicles play a significant role in efforts to meet future legislated emission targets. Beside stricter limits for pollutant emissions of nitrogen oxides and particulates, European regulations also aim at reducing the average CO_2 emissions of passenger cars to 95 g/km [1]. For off-highway applications—such as diesel multiple unit trains (DMUs) or marine applications—so far no legislative regulation to reduce CO_2 emissions has been implemented. Nevertheless, lowering fuel consumption and thus CO_2 emissions is the major goal in developing future off-highway propulsion systems. This is comprehensible considering that the fuel share of total life cycle costs for off-highway applications can be up to over 90% and, additionally, fuel prices are very likely to increase in the mid- and long-term future [2]. Hybrid powertrain technology can help to significantly lower fuel consumption of rail and marine applications [3–6]. Compared to conventional non-hybrid powertrains, additional operational functionalities of hybrids like recuperation of kinetic energy, engine operation point shifting or electric boosting improve the system performance and reduce pollutant emissions [7]. Figure 1 shows an overview of the functionalities of hybrid powertrains and their impact on emissions, fuel consumption and system performance.

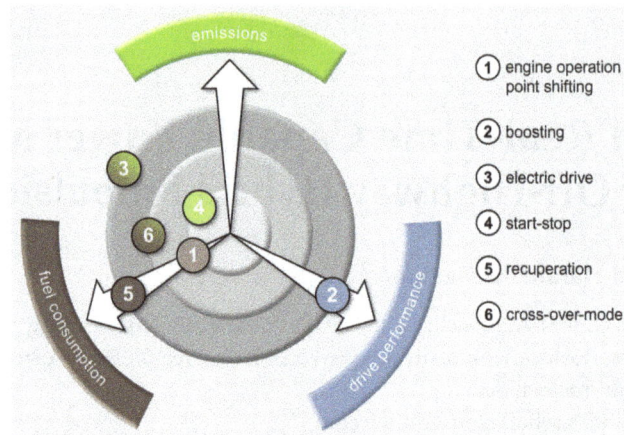

Figure 1. Overview of hybrid powertrain functionalities.

On the other hand, electrification of propulsion systems further increases the number of system components and, therefore, the complexity of already highly sophisticated off-highway architectures consisting of multiple engines, generators, gearboxes, etc. This leads to an even higher diversification of system configurations. To cope with this variety in series production, it is compulsory to standardize components of hybrid propulsion systems and to use them for various applications. Besides the hardware standardization, it is also necessary to harmonize software functions and control algorithms in order to apply them to as many different system configurations and architectures as possible. Therefore, a power management algorithm (PMA) is required to control the power split between combustion engine and electric motor in a hybrid propulsion system. The PMA has to offer the flexibility and modularity to be applicable to a vast variety of systems, including multi-engine system topologies.

1.2. Classification of Power Management Algorithms

In the past, a wide range of different academic approaches of PMAs to optimize the power split in hybrid propulsion systems for on- and non-road applications have been published in the literature. In order to classify multiple PMAs by the basis of their data, one can differentiate between causal and non-causal strategies [8,9]. In causal strategies, the decisions made by the PMAs are based upon present and past state variables. On the contrary, in a non-causal approach, knowledge of the whole drive cycle is processed. Considering data of the complete drive cycle enables an optimization of the fuel consumption for a global optimum [9,10]. For DMUs, where a detailed knowledge of the track and timetable is available, a non-causal PMA in combination with a predictive strategy is feasible. From a functional perspective, heuristic, often called rule-based or sub-optimal, and optimization-based, or also known as optimal, algorithms can be distinguished [9,11–13]. Heuristic strategies include deterministic rule-based and fuzzy logic control algorithms [14–17]. Those approaches are state of the art in most prototype and production hybrid vehicles [9]. They are real-time capable and are processed online. The optimization-based or optimal PMAs are mainly processed offline. Those strategies are normally non-causal and, therefore, process information about the whole drive cycle. The optimal PMAs are divided into numerical methods such as dynamic programming (DP) and analytical approaches, like Pontryagins minimum principle [11,18,19]. DP enables to find a global optimum but is not real-time capable and, therefore, frequently used as benchmark PMA. Other optimization-based PMAs are the so called equivalent consumption minimization strategies (ECMS) [9,12]. The approach associates the stored electric energy to a future increase or decrease of fuel consumption. Therefore, usually an equivalence factor is imposed to convert battery into fuel power, and based on that into an equivalent fuel consumption. In the meantime, many different ECMS approaches, primarily for automotive applications, were introduced. They use rule- or map-based routines as well as online adaption to determine the equivalent factor and optimize power split [20–24].

1.3. Requirements for the Off-Highway Application Power Management

In this paper a new power management algorithm (PMA) which controls the power split between combustion engine and electric motor in a hybrid propulsion system is presented. It is an online optimization-based causal strategy. Only present and past state variables are used, and no knowledge of the whole drive cycle is necessary. The PMA is developed to be applied to various applications and system configurations, including systems with multiple engines and a variety of electrical components. It is suited for use in hardware-in-the-loop (HiL) experiments as well as field testing of systems for off-highway applications. Furthermore it has to fulfill, amongst others, the following requirements:

- Modular structure
- Real-time capability
- Compatibility to series production propulsion system control units
- Suitability for different mission scenarios and velocity/load control modes (including driver-controlled traction torque demand and automatically controlled drive strategies)
- Consideration of variable auxiliary or external load requests in the PMA.

It is important to point out that some of these requirements significantly differ from demands for PMAs used in automotive applications. Especially the necessity of a high modularity and the flexibility of the algorithm for various applications, topologies and system components are challenging concerning algorithm development and code implementation.

The paper comprises a simulation case study where the PMA is exemplarily applied to control a hybrid propulsion system for a DMU application. In Section 2, boundary conditions for the simulation case study are presented, including the specification of the system hardware in Section 2.1, as well as a short description of the drive strategies in Section 2.2, which are used for the simulations. Afterwards, in Section 3, a brief overview over the simulation model approach is given along with examples for a model validation. For the model validation, measured data from a hardware-in-the-loop (HiL) test rig is used. It represents a realistic prototype of a hybrid propulsion system driving a virtual vehicle model of the DMU. In Section 4, the PMA is described comprehensively. At first, the general problem for the control of hybrid systems is defined. Section 4.2 points out the possible operation modes of parallel hybrid powertrains which are relevant for the PMA. Thereafter, Section 4.3 presents the different functions and their interaction with the PMA. The results of the simulation case study are finally discussed in Section 5.

2. Boundaries of the Simulation Case Study

2.1. Hybrid Diesel Multiple Unit Train (DMU) System Specification

The PMA introduced in this paper has been developed for a Siemens Desiro VT642 DMU dedicated for regional passenger transportation. In a simulation case study, the vehicle is equipped with two identical parallel hybrid power units (PU), which form the propulsion system. Each PU consists of a diesel internal combustion engine (ICE) and an electric motor/generator unit (MG), which is placed on the engine output shaft center, see Figure 2. The angular velocities of ICE and MG can be decoupled by a clutch (CL). The gearbox (GR) is a six-speed automatic transmission with a hydrodynamic torque-converter on its input shaft. As energy storage unit, an electrochemical lithium-ion accumulator (BAT) is used in each PU. Furthermore, Figure 2 shows the arrangement of the auxiliary loads, comprising electric auxiliaries (AUX) and mechanical power take-offs (PTO) at the ICE, like the hydraulic fan. Table 1 shows a more detailed specification of the vehicle and system components.

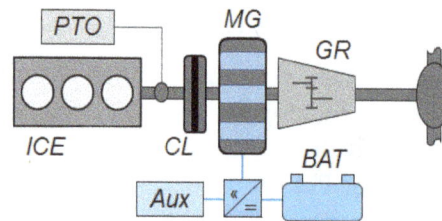

Figure 2. Scheme of a parallel hybrid power unit.

Table 1. System specification of vehicle and components.

Vehicle	type number PUs operating weight	Siemens Desiro VT642 2 83,000 kg
Engine	type number cylinders rated power rated speed	MTU 6H1800R85LP 6 390 kW 1800 rpm
Motor/Generator	type rated power rated speed	p. magnet synchron. motor 200 kW (continuous) 1600 rpm
Gearbox	type number gears	ZF EcoLife 6
Battery	type nom. voltage nom. capacity max./min. current	Lithium-ion battery 670 V 90 Ah +/-300 A

In the simulation case study, a system configuration is analyzed, where two identical PUs operate synchronously. It means, for example, that the power demand for the ICE calculated by the PMA is identical in both PUs. It is possible and, moreover, from an efficiency perspective rather beneficial to consider two PUs that are logically linked but operated individually. Such an operation strategy, however, is not state-of-the-art yet because a high voltage link between the electric circuits of both PUs would be required. Figure 3 outlines a system configuration with two PUs, which are linked on their electric circuit and share a joint battery. The second combustion engine is masked, which implies that such systems enable to restore one ICE. This is possible because—compared to a conventional non-hybrid system—the hybridization increases the system power considerably. Thus, considering track profiles and mission scenarios with a relatively low traction power demand, it is feasible to remove one of the ICEs and compensate the loss of traction power with the MGs. It becomes clear that even quite conventional propulsion system configurations of DMUs with two PUs can differ significantly by their design and operation, not to mention systems layouts with more than two PUs. This shows the necessity of a highly flexible PMA approach for off-highway hybrid systems.

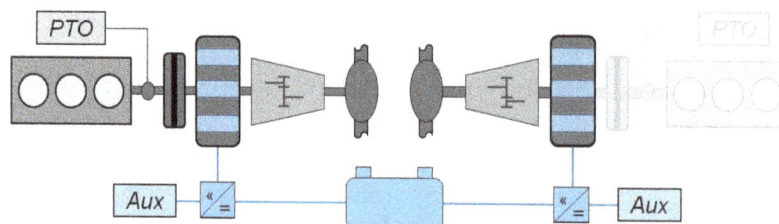

Figure 3. Scheme of a parallel hybrid propulsion system for a diesel multiple unit train (DMU) consisting of two PUs and a combined electric circuit with the potential to replace one ICE.

2.2. Drive Strategy

Optimizations of the drive strategy to operate a DMU on a given track are the object of research in many publications. Often dynamic programming (DP) algorithms are used to optimize the velocity trajectory of hybrid vehicles [19,25]. Due to the fact that DP requires a high computational power, they are not real-time capable. Recently, approaches have been proposed where the DP algorithm is outsourced and computed on a cloud computing platform. The optimized velocity profile is sent back to the vehicle afterwards [26]. Such approaches have significant potential to reduce fuel consumption.

However, for the simulation case study in this paper, a more simple drive strategy is used. Initially, a mission specification for the track, on which the DMU is operated, is defined. It specifies the driving distance, drive and stop durations as well as velocity limits of each track sequence. Based on this information, a driver simulation model, implemented in Simulink and Stateflow, computes the requested velocity profile. The objective of the driving strategy is the reduction of the DMU's energy consumption traveling on the tracks considering all the specified constraints (driving duration, velocity limits etc.). For this purpose, the algorithm aims at minimizing the cruising speed, considering the quadratic influence of velocity on the required traction force, see Equation (1). To operate the vehicle at a cruising speed as low as possible, the driving strategy asks for maximum acceleration and deceleration in the particular parts of each track sequence. Acceleration and deceleration demands are restricted by the performance capabilities of the PU, vehicle limitations or even operational boundaries, like a deceleration limit of -1 m/s^2 due to comfort considerations. In DMUs with conventional PUs, such a drive strategy, where the energy consumption is minimized, is considerably simple and leads to good results. Due to its simplicity it is fairly realistic that a human DMU operator in real world driving can easily track such a velocity profile.

For a hybrid system with energy storage unit, the possibility of a recovery of kinetic energy arises. For the hybrid drive strategy, hence, the algorithm of the driver model retards the deceleration phase. The strategy demands the vehicle to decelerate with a constant brake power, equivalent to the maximum mechanic brake power of the MG. This leads to slightly higher cruising speeds and, thus, energy consumption, but also significantly increases the amount of recuperated energy. Therefore, this drive strategy is used for hybrid operation mode.

3. System Simulation Model and Validation

The virtual vehicle and propulsion system are built up as a dynamic simulation model in the simulation environment GT-SUITE. GT-SUITE library blocks are used to model the physical components as well as to implement the basic control functions belonging to the components. Physical components and functions are combined to so-called compounds and archived in a compound library. The compounds possess standardized interfaces to enable a high degree of modularity, which allows to build simulation models of various system configurations for diverse applications. Each component provides the actual values of its physical state variables as well as their maximum and minimum values for each of them, as model outputs. More complex control algorithms, like the PMA, are realized in MATLAB/Simulink and Stateflow. The Simulink models can be compiled to dll-code and implemented in the GT-SUITE environment. The simulation models are suited for performance simulations, controls development, system parameter and configuration optimizations as well as hardware-in-the-loop testing.

3.1. Modelling System Components

This section gives a brief overview over the physical modeling theory and the equations, which the component models are based on. Furthermore, the predefined objects of the GT-SUITE library, which are used for the particular component models, are presented. Additionally, the most important model parameters are specified.

3.1.1. Vehicle Model

The dynamic vehicle simulation model uses the "GT VehicleBody", which is based on the equation of motion. The traction force F_{Tr} is equal to the sum of all drive resistance forces:

$$F_{Tr}(t) = c_r\, m\, g \cos{(\alpha)} + \frac{1}{2} c_D A\, \rho\, v(t)^2 + m\, g \sin{(\alpha)} + \left(m + \frac{\theta_{red}}{r_{dyn}^2} \right) \dot{v}(t) \qquad (1)$$

with track inclination α, air density ρ, vehicle velocity v and the reduced mass moment of inertia θ_{red}. The required input traction torque M_{Drive} to the vehicle differential is calculated by:

$$M_{Drive}(t) = \frac{F_{Tr}(t) \cdot r_{dyn}}{\eta_{Dif} \cdot i_{Dif}(t)}. \tag{2}$$

Table 2 gives an overview over the most important vehicle model parameters.

Table 2. Vehicle model parameters.

Parameter	Symbol	Unit	Value
Vehicle mass	m	kg	83,000
Track coefficient	c_r	-	0.001
Drag coefficient	c_D	-	0.8
Frontal area	A	m^2	10.8
Dynamic roll radius	r_{dyn}	m	0.38
Differential ratio	i_{Dif}	-	2.59
Differential efficiency	η_{Dif}	-	0.95

3.1.2. Engine Model

The engine simulation model used in this case study is based on the "GT EngineState" object. It computes fuel consumption using a static brake specific fuel consumption (*bsfc*) map, see Figure 4a. The fuel consumption FC_{ICE} is calculated by:

$$FC_{ICE}(t) = \int_0^t M_{ICE}(\tau) \cdot \omega_{ICE}(\tau) \cdot bsfc(\tau) \, d\tau \tag{3}$$

The angular velocity ω_{ICE} of the ICE in $\frac{rad}{sec}$ (or n_{ICE} in rpm) follows from

$$\dot{\omega}_{ICE}(t) = \frac{M_{ICE}(t) + M_{PTO}(t)}{\theta_{ICE}} \tag{4}$$

with the engine torque M_{ICE}, the mass moment of inertia θ_{ICE} of the ICE and the mechanical auxiliary torque M_{PTO}. The mechanical auxiliaries or PTOs (power-take-offs) are directly linked to the engine output shaft. In the real vehicle, M_{PTO} is a function of both angular velocity and PTO load, which depends on coolant or boost temperatures etc. In the simulation model, M_{PTO} or P_{PTO} is a function of the engine angular velocity, whereas the load is set constant, see Figure 4b.

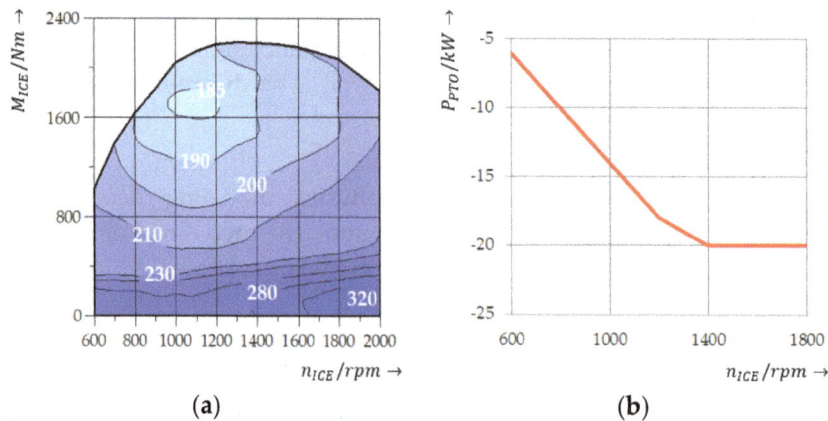

Figure 4. (a) Generic bsfc map for a diesel engine; (b) Mechanical auxiliary power P_{PTO} over engine speed n_{ICE}.

3.1.3. Gearbox Model

The gearbox is a six-speed transmission with a hydrodynamic torque converter (HTC) and transmits the input torque M_{GR} to the vehicle differential input shaft. The model consists of a continuously-variable gear set, a HTC including a model of the torque converter clutch, and the controls module. It manages the clutch actuation and gear shifts. The controls are implemented in Simulink and integrated into the GT-SUITE model as dll-file. The required gear is computed externally by one module of the power management. The gearbox output torque is calculated by:

$$M_{GR} = M_{ICE} + M_{PTO} + M_{MG}, \tag{5}$$

$$M_{GR\_out} = \eta_{GR} \cdot i_{GR} \cdot M_{GR} \tag{6}$$

(valid at closed torque converter clutch) with the MG effective torque M_{MG} and the gear ratio i_{GR}. For simplicity, the gearbox efficiency η_{GR} is set constant. The HTC enables the railway vehicle to accelerate with high traction forces. In gears with a gear ratio of $z > 1$ it is "deactivated" and locked with a torque converter clutch. To calculate the turbine torque of the HTC, M_{Turb} the impeller torque M_{Imp} is multiplied by the torque ratio i_{HTC}, which is taken from the torque ratio table in Figure 5a. The impeller torque is calculated using the capacity factor K_{HTC}.

$$M_{Imp} = \left(\frac{\omega_{Imp}}{K_{HTC}} \right)^2 \text{ with } K_{HTC} = f_1 \left(\frac{\omega_{Turb}}{\omega_{Imp}} \right), \tag{7}$$

$$M_{Turb} = i_{HTC} \cdot M_{Imp} \text{ with } i_{HTC} = f_2 \left(\frac{\omega_{Turb}}{\omega_{Imp}} \right). \tag{8}$$

3.1.4. Motor/Generator Model

To determine the electro-mechanic losses in the MG model, the efficiency η_{MG} is interpolated linearly from a map, see Figure 5b. Beside the losses of the MG itself, it also includes the electric efficiency of the electric converter. From the mechanic power P_{MG} and the efficiency, the electric power in the generator mode becomes $Pe_{Gen} = \eta_{MG} \cdot P_{MG}$, whereas the motor mode is characterized by $Pe_{Mot} = \frac{P_{MG}}{\eta_{MG}}$. Generally, electric machines are capable of running on overload for a restricted period of time. The overload duration is thermally restricted. Therefore, the MG involves a simple heat model to simulate heating and cooling sequences:

$$m' \cdot c \cdot \frac{dT}{dt} = \dot{Q}_{Loss}(t) + \dot{Q}_{Cooling}(t). \tag{9}$$

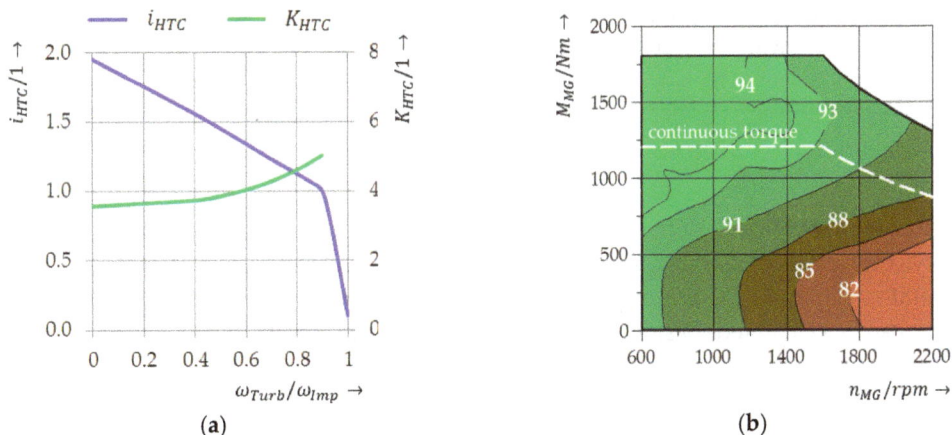

Figure 5. (a) Torque converter torque ratio i_{HTC} and capacity factor K_{HTC} over the speed ratio; (b) MG efficiency map.

The product of mass m' and specific heat c (here for iron) represents the time constant of the differential Equation (9) and depends on the MG size. The heat flows follow from:

$$\dot{Q}_{Loss}(t) = (1 - \eta_{MG}(t)) \cdot P_{MG}(t),$$ (10)

$$\dot{Q}_{Cooling}(t) = \alpha_w \cdot A' \cdot (T_{Coolant} - T(t)).$$ (11)

The parameters m', α_w, $T_{Coolant}$ and A' for this simplified approach are set to match pre-defined heating-up and cool-down times as well as to reach thermal steady state at the rated continuous power of the MG. The internal control routine of the MG model deactivates the overload mode and reduces the maximum power as soon as a temperature threshold is exceeded. MG power in overload mode is increased by a factor of 1.5 compared to the continuous operation mode, see Figure 5b.

3.1.5. Battery Model

The battery model is based on the equivalent electric circuit shown in Figure 6a. The voltage of the battery U_{BAT} follows from Kirchhoff's voltage law for the illustrated equivalent circuit:

$$U_{BAT}(t) = U_{OC}(t) - R_i(t) \cdot I_{BAT}(t)$$ (12)

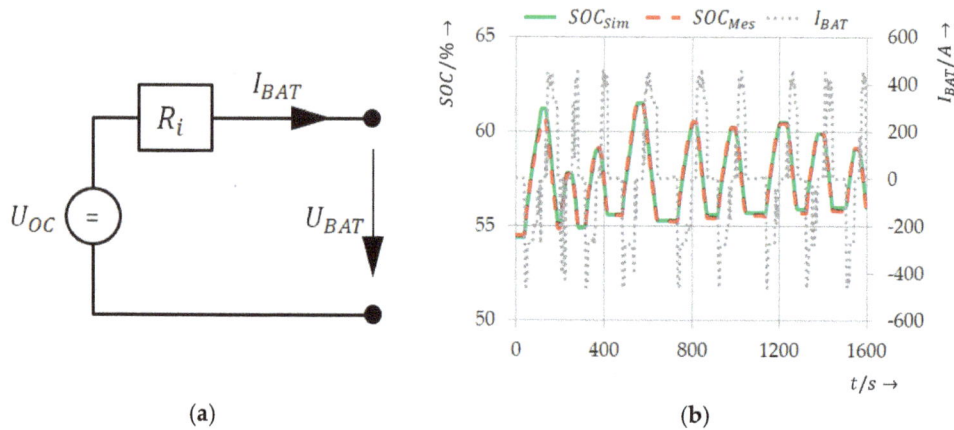

Figure 6. (a) Equivalent electric circuit for the battery; (b) Battery validation cycle: simulated SOC_{Sim} and measured SOC_{Mes} over time.

The open-voltage circuit U_{OC} represents the equilibrium potential of the battery, which corresponds to the number n_{ser} of battery cells connected in series (13). The battery stack capacity C_{BAT} follows from the number of parallel linked battery cells n_{par}, with:

$$U_{OC} = n_{ser} \cdot U_{cell},$$ (13)

$$C_{BAT} = n_{par} \cdot C_{cell}.$$ (14)

The internal resistance R_i of the battery in (12) is taken from 2-D lookup tables parametrized for each battery cell type. For charging and discharging events, different tables of $R_i(SOC)$ as a function of battery state of charge SOC are employed. From the power sum at the electric circuit, the battery current I_{BAT} can be calculated. Beside the MG, a further power sink of the electric circuit is the electric auxiliary power Pe_{Aux}:

$$I_{BAT}(t) = \frac{Pe_{MG}(t) + Pe_{Aux}(t)}{U_{OC}(t)}.$$ (15)

The integration of I_{BAT} according to (15) leads to the battery SOC. It is one of the most important input variables of the PMA of a hybrid propulsion system, because it represents the energetic status of the energy storage. It is often the reference to trigger mode switches between various system operation modes. SOC_0 is the initial state of charge value of the battery:

$$SOC(t) = SOC_0 + \int_0^t \frac{I_{BAT}(\tau)}{C_{BAT}} \, d\tau. \tag{16}$$

3.2. Model Validation

This section gives a brief overview of the model validation. The component models outlined in the previous section are used for open- and closed-loop component tests. The measured data of those tests contributes to a further improvement of the model accuracy.

3.2.1. Battery Model Validation

For model parameterization and validation purposes, measured data from open-loop battery tests was used. In the test procedure, a battery prototype on a battery test bench with a DC/DC power source was stressed with a battery current profile simulated beforehand. The current profile results from a simulation of a DMU with two hybrid PUs operating in regional passenger transportation. Battery state variables as voltage, current and SOC are measured. Afterwards, simulations of the battery were executed to optimize the internal resistance tables in the battery model for a minimum deviation of measured and simulated values for voltage and SOC. The optimization is based on Brent's algorithm [27]. Figure 6b shows a good matching between the calculated and observed SOC.

3.2.2. Powertrain Model Validation

In more extensive test campaigns of the whole propulsion system, validation data for the PU model was recorded. For this purpose, a prototype of a MTU hybrid PU was installed on a system test bench. In hardware-in-the-loop tests (HiL tests), the powertrain on the test bed is coupled with the simulation model of a DMU running on a virtual track. The test bed automation system is linked via a real-time interface to the simulation model of the vehicle and driver. In such a HiL test environment, performance tests can be executed or software functions are validated etc. Figure 7a gives a schematic overview of the experimental set-up. In Figure 7b, the simulated and measured results for the accumulated fuel consumptions over the whole drive cycle—for conventional and hybrid operation—are compared. The stated results are normalized on the fuel consumption for conventional mode. For the measurements campaign and the model validation, a previous version of the PMA, comprising a rule-based approach, was used. Obviously, the results underline and validate the achieved high accuracy of the simulation models.

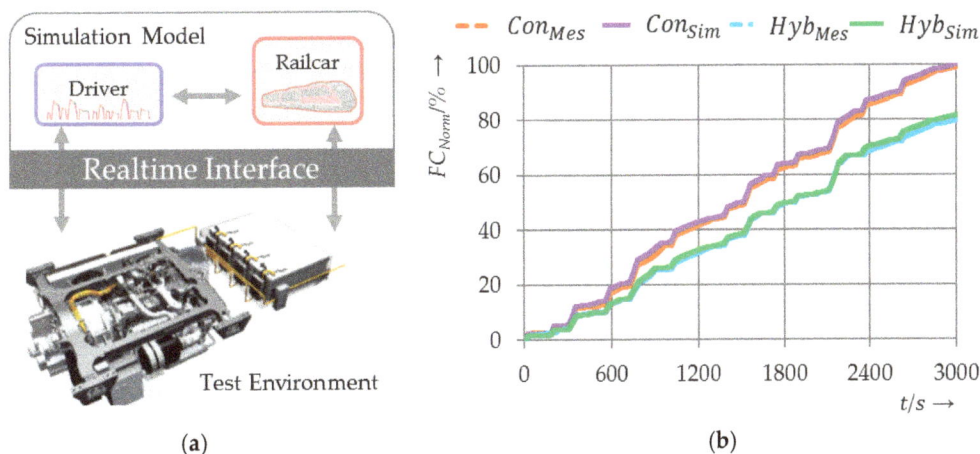

Figure 7. (a) Schematic overview over experimental HiL layout; (b) Measured and simulated normalized fuel consumption for conventional and hybrid mode.

4. Hybrid Power Management Algorithm (PMA)

4.1. Problem Definition

The PMA controls the power flow in the propulsion system and optimizes the power split between the ICE and the MG. Based on the optimal control problem [26], a suitable cost functional or performance index J is stated in (17). The cost functional J substitutes a corrected amount of fuel consumption FC_{Cor} of the system in a time interval $t \in [t_0, t_{end}]$. The objective of the PMA is to minimize J and FC_{Cor} over the given drive cycle. FC_{Cor} represents the quantity to evaluate the performance of the hybrid system according to its energy consumption and efficiency:

$$J = \phi(SOC(t_{end})) + \int_{t_0}^{t_{end}} \dot{m}_{Fuel}(u(\tau), x(\tau), \tau) \, d\tau \tag{17}$$

J is calculated by the integration of fuel mass flow $\dot{m}_{Fuel}$ in $\frac{kg}{sec}$, as a function of the control inputs $u(t)$ and state variables $x(t)$. The first term in (17) stands for a correction term to penalize the deviation of energy stored in the battery between t_0 and t_{end}:

$$\phi(SOC(t_{end})) = \lambda \cdot [SOC(t_0) - SOC(t_{end})]. \tag{18}$$

Here, λ represents a fuel equivalent factor, which converts the deviation of SOC into a value of fuel consumption in $\frac{kg}{\%}$. In contrast to some ECMS control algorithms, where an online adaption of λ is part of the PMA, λ is set constant in this case. It is determined by the amount of fuel, which is needed to balance battery charge deviation, assuming that charging is performed while the vehicle is at standstill (halt) at the end of each drive cycle:

$$\lambda = \frac{E_{BAT}^{100}}{H_u \cdot \eta_{chrg}^{op}}. \tag{19}$$

E_{BAT}^{100} stands for the maximum amount of energy, which can be stored in the battery at $SOC = 100\%$. H_u is the lower heating value of diesel fuel, and η_{chrg}^{op} denotes the charging efficiency at a predefined ICE operation point. This operating point for charging at standstill (halt) is optimized offline for maximum charging efficiency and set as a parameter.

4.2. Hybrid Operation Modes

The configuration of the parallel hybrid system, see Figure 2, enables different operation modes. Table 3 outlines the various operation modes and assigns them to drive events. The drive events result from the requested traction power P_{Trac}. This table shows the signs of ICE and MG power demands, P_{ICE} and P_{MG}, as well as the time derivative $\dot{SOC}$ of the state of charge for the different drive modes. The signs take the auxiliary powers into account: $P_{PTO} < 0$ at the ICE output shaft and $Pe_{AUX} > 0$ at the electric circuit. In Table 3 the plus sign, +, stands for a power request higher than zero. The minus sign, -, represents a negative power request, whereas, 0, refers to a constant value. The double plus and minus signs indicate very high or low power requests, which for example arise at acceleration or braking events.

During acceleration, two modes are possible. At *Pure ICE* mode, the traction power is exclusively provided by the ICE. The MG runs as generator and supplies electric auxiliaries. In the *Combined Mode*, the traction power is split between ICE and MG. During cruising, *Pure Electric* driving is possible. Furthermore, the battery can be charged by increasing the ICE load. In the case of a negative traction power demand, the deceleration event, the MG serves as generator to recuperate as much energy as possible. The clutch between ICE and MG is opened, to operate the ICE at idle speed. At halt, the vehicle velocity is zero. In *Normal Halt* mode, the ICE powers the PTO and supplies power for the AUX. If necessary, the ICE load can be increased to charges the battery. Additionally, the ICE may be switched off in *Start/Stop* mode, and the AUX power is supplied by the BAT. At deceleration and halt events, the different operation modes are triggered according to pre-defined thresholds of the SOC and other state variables. In those events, the PMA is a heuristic rule-based approach, where no optimization is executed. During acceleration and cruising, the optimization routine for P_{ICE} and P_{MG} is active.

Table 3. Drive Events and Hybrid Operation Modes.

Drive Event	Operation Mode	P_{Trac}	P_{ICE}	P_{MG}	$\dot{SOC}$
Acceleration	Pure ICE	++	++	-	0
	Combined Mode	++	+	+	-
Cruising	Pure ICE	+	+	-	0
	Pure Electric	+	0	+	-
	Combined Mode	+	+	+	-
	Charge Mode	+	++	-	+
Deceleration	Recuperation	−	+	−	+
Halt	Normal Halt	0	+	-	0
	Start/Stop	0	0	0	-
	Charge Mode	0	+	-	+

4.3. PMA Optimization Routine

Figure 8 shows a signal flow scheme of the PMA optimization routine including its most important input and output signals. Each block of the scheme represents one function of the routine, which is described in this section. In case of a DMU with more than one PU, P_{ICE} and P_{MG} represent the power demand from the particular power path. This means, for example, that if more than one engine is present, P_{ICE} substitutes the sum of engine power demands. A further function block splits P_{ICE} and distributes the particular share of power to the different engines. In case of PUs running synchronously, as assumed in this paper, power is equally split to the PUs' machines. In this section, for simplicity, the PMA refers to the ICE and MG, neglecting the fact that more than one of each component is involved.

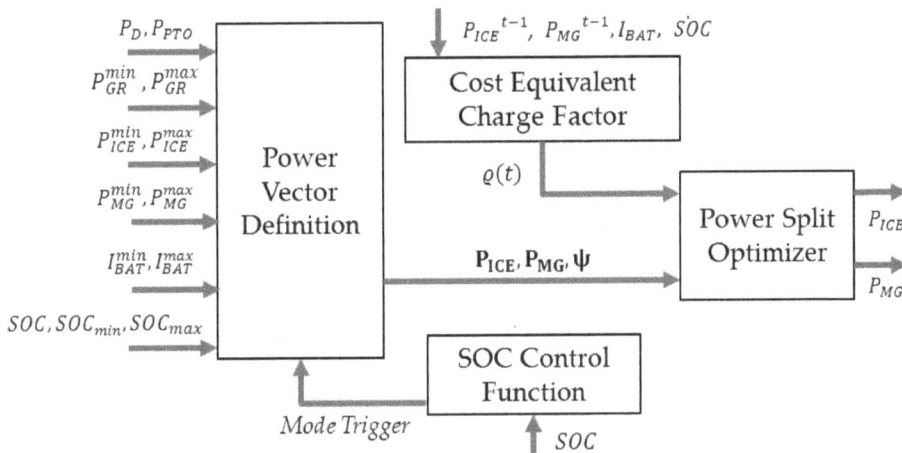

Figure 8. Signal flow scheme of the PMA optimization routine.

4.3.1. Power Vector Definition

From the traction power required and the power request of the mechanical auxiliaries P_{PTO}, this function block calculates the propelling power demand P_D. It always equals the sum of the power demands, P_{ICE} plus P_{MG}. Moreover, the gearbox input power limitations $[P_{GR}^{min}, P_{GR}^{max}]$ as well as further vehicle constraints, have to be considered. In a next functional step, the inappropriate operation modes for the current drive event are rejected. This is necessary because the battery SOC always has to be in a valid state of charge range, $SOC_{min} \leq SOC \leq SOC_{max}$. If, for instance, SOC is above its maximum value, *Charge Mode* is prohibited. Based on the feasible modes and constraints regarding the ICE, MG and BAT the maximum and minimum values for the power split ratio ψ are defined. From those extremes, the operation point vectors $\boldsymbol{\psi}$, $\boldsymbol{P_{ICE}}$ and $\boldsymbol{P_{MG}}$ for the split ratios of the length k are determined. They follow from:

$$\boldsymbol{\psi} = \begin{pmatrix} \psi_1^{max} \\ \vdots \\ \psi_k^{min} \end{pmatrix}, \tag{20}$$

$$\boldsymbol{P_{ICE}} = \boldsymbol{\psi} \cdot P_D \text{ and } \boldsymbol{P_{MG}} = (1 - \boldsymbol{\psi}) \cdot P_D, \tag{21}$$

$$\text{subject to} \begin{cases} 0 \leq P_{ICE} \leq P_{ICE}^{max} \\ P_{MG}^{min} \leq P_{MG} \leq P_{MG}^{max} \\ SOC_{min} \leq SOC \leq SOC_{max} \\ I_{BAT}^{min} \leq I_{BAT} \leq I_{BAT}^{max} \\ P_{GR}^{min} \leq (P_D + P_{PTO}) \leq P_{GR}^{max} \end{cases}. \tag{22}$$

4.3.2. Power Split Optimization

The optimization function block determines the control output signals for P_{ICE} and P_{MG}. Therefore, the algorithm calculates a power supply efficiency vector of length k, $\Omega = (\Omega_1 \ldots \Omega_k)$, where each element of the vector assigns a power supply efficiency Ω_i to the particular elements of the operation point vectors. The power supply efficiency correlates to the sum of requested powers, P_D, P_{PTO} and Pe_{AUX}, divided by the sum of power consumption, P_{Fuel} plus Pe_{BAT-} (electric discharge power):

$$\Omega^k = \frac{P_D + P_{PTO} + Pe_{AUX} + Pe_{BAT+}^k}{P_{Fuel}^k + \left(\frac{Pe_{BAT-}^k}{\varrho(t)} \right)} \tag{23}$$

with:

$$P_{Fuel}^k = \frac{\boldsymbol{P_{ICE}}(k)}{\eta_{ICE}(\boldsymbol{P_{ICE}}(k), \omega_{ICE})}. \tag{24}$$

It is differentiated between electric battery power for charging Pe_{BAT+} and discharging Pe_{BAT-}. Pe_{BAT}^k for each vector element k is calculated by:

$$Pe_{BAT}^k = -\frac{P_{MG}(k)}{\eta_{BAT}(I_{BAT}) \cdot \eta_{MG}(P_{MG}(k), \omega_{MG})}, \tag{25}$$

hence:

$$Pe_{BAT+}^k = max[0, Pe_{BAT}^k] \text{ and } Pe_{BAT-}^k = |min[0, Pe_{BAT}^k]|. \tag{26}$$

The values for the conversion efficiencies η_{ICE}, η_{MG} and η_{BAT} are taken from efficiency maps (e.g., Figure 5b) or, in case of η_{ICE}, are calculated based on lookup tables of $bsfc(\boldsymbol{P_{ICE}}(k), \omega_{ICE})$. The variable $\varrho(t)$ in (23) stands for an energy cost equivalent charge factor and is described more detailed in the next subsection. Finally, the control output signals for P_{ICE} and P_{MG} are determined by the maximum value Ω^{max} of the power supply efficiency vector Ω:

$$arg\ max\{\Omega(\boldsymbol{P_{ICE}}, \boldsymbol{P_{MG}}, P_D, P_{PTO}, Pe_{AUX})\} \rightarrow P_{ICE}, P_{MG}. \tag{27}$$

4.3.3. Energy Cost Equivalent Charge Factor

The energy cost equivalent charge factor $\varrho(t)$ accounts for the amount of fuel energy that was used in the time interval $t \in [t_0, t]$ to charge the battery. Its valid range is $\eta_{chrg}^{min} \leq \varrho(t) \leq 1$. If the battery is charged only by recuperation during braking, without any consumption of fuel, $\varrho(t)$ has the value 1. But if the energy stored in the battery is provided by the ICE, $\varrho(t)$ equals the medium efficiency of the charging procedure. Hence, for $\dot{SOC} > 0$, $\varrho(t)$ is computed by:

$$\varrho(t) = \frac{E_{BAT}^*(t)}{E_{ChrgFuel}^*(t)} \tag{28}$$

with:

$$E_{BAT}^*(t) = E_{BAT}^*(t_0) + \int_{t_0}^{t} U_{OC}(\tau) \cdot I_{BAT}(\tau) \, d\tau \tag{29}$$

and:

$$E_{ChrgFuel}^*(t) = E_{ChrgFuel}^*(t_0) + \int_{t_0}^{t} P_{ChrgFuel}(\tau) \, d\tau. \tag{30}$$

In the case of discharging, $\dot{SOC} \leq 0$, the PMA modifies (28) and holds $\varrho(t)$ constant. It follows from (28) that $\varrho(t)$ substitutes the ratio of energy E_{BAT}^*, stored in the battery, and the fuel energy needed to charge the battery $E_{ChrgFuel}^*$. Please note that $P_{ChrgFuel}$ represents the share of fuel consumed for charging. $P_{ChrgFuel}$ can be calculated from the electric charging power at the battery terminal and the product of conversion efficiencies for the ICE as well as MG, leading to:

$$P_{ChrgFuel}(t) = \frac{U_{BAT}(t) \cdot I_{BAT+}}{\eta_{ICE}(M_{ICE}(t), \omega_{ICE}(t)) \cdot \eta_{MG}(M_{MG}(t), \omega_{MG}(t))}. \tag{31}$$

The index * implies that E_{BAT}^* represents the amount of energy in between the admissible boundaries of the battery state of charge, $SOC_{min} \leq SOC \leq SOC_{max}$. This makes sense because only this partition of the battery is used during operation.

From a mathematical point of few, $\varrho(t)$ set in (23) is handled like an efficiency value, which increases the energy costs, or in other words, reduces the efficiency of the electrical path. In case of $\varrho(t)$ being close to one, it means that most of the energy in the battery was provided by "free" sources like recuperation or even external sources, for instance, external charging in a depot. As a consequence, a high portion of P_D is provided by the electric path. This results in ψ close to zero due to its higher conversion efficiency as compared to the ICE path. In applications, system configurations or even drive cycles, where not much of the consumed energy can be recuperated during the deceleration and only a relatively small BAT is used, *Charge Mode* is employed quite regularly to maintain SOC within its admissible interval. In this case, $\varrho(t)$ falls below one and, thus, increases the electric energy cost. Due to its implementation, $\varrho(t)$ adapts itself dependent on system parameters, the drive cycle and other operational boundaries.

4.3.4. SOC Control Function

There are system configurations or constraints for specific applications, where SOC does not only have to be within its bounds but have to match a specific value at the end of the drive cycle or even among single track sequences. This is the case if the additional installed electric power is intended to improve system performance or, for example, to replace one combustion engine, see Figure 3. Therefore, the PMA features a function block that determines an expected amount of energy that is recuperated during braking down from the current vehicle velocity. The computation addresses the driving resistance forces according to (1), estimates a brake trajectory in compliance with the drive strategy, see Section 2.2, and determines the recuperation duration $t_{Rec} = f(v)$. From t_{Rec} and an assumed mean electric recuperation power Pe_{Rec}^{mean}, a recuperated SOC_{Rec}^{calc} is calculated:

$$SOC_{Rec}^{calc} = F_{Rec}^i \cdot \frac{t_{rec} \cdot Pe_{Rec}^{mean}}{E_{BAT}^{100}}. \tag{32}$$

The factor F_{Rec}^i in (32) represents a correction factor and is adapted online for a better match of the predicted SOC_{Rec}^{calc} and the real SOC_{Rec}. It is calculated by a comparably simple approach according to (33) but leads to relatively good results (see Section 5):

$$F_{Rec}^i = F_{Rec}^{i-1} + dF_{Rec} \text{ with } dF_{Rec} = \frac{SOC_{Rec}}{SOC_{Rec}^{calc}} - F_{Rec}^{i-1}, \tag{33}$$

$$\text{subject to } -0.15 \leq dF_{Rec} \leq 0.15. \tag{34}$$

The computation of F_{Rec}^i is processed at discrete states i, after each recuperation event, where the velocity is reduced from cruising speed to a halt. Based on SOC_{Rec}^{calc} and a target value for SOC_{Trg}, a finite-state machine sets a trigger to initiate the *Charge Mode*. The PMA strives for equality of SOC and SOC_{Trg} at the start of the next cycle sequence. Hence, the *Charge Mode* is activated until $SOC \geq (SOC_{Trg} - SOC_{Rec}^{calc})$ is achieved.

The SOC control function is a straightforward approach to control the state of charge of the battery BAT in order to attain a predefined target value. By a proper choice of the target value, it allows for implementing different operating strategies for hybrid systems. If SOC_{Trg} is close to the upper SOC limit, a performance oriented strategy is applied. Here, it is ensured that at any time a high battery SOC is available for electrically boosted acceleration events. For a low SOC_{Trg}, the PMA allows a wider range of possible operation modes and, moreover, to optimize a power split only from an efficiency point of view. Furthermore, this charge triggering function is intended to be the interface for further optional predictive algorithms, which process the knowledge of the whole drive cycle.

5. Case Study Results

In the simulation case study, the DMU specified in Section 2 with two identical PUs is employed. The vehicle parameters are shown in Table 2. The electric auxiliary power is set to $Pe_{Aux} = 30$ kW for the vehicle. In the reference case, where the DMU runs on *Pure ICE* mode, the auxiliaries are supplied by a generator driven by the ICE. For evaluation purposes, the corrected fuel consumption FC_{Cor}, calculated with (15)–(17), is used. The DMU drives on a track which was generated from an evaluation of various DMUs' field data. This track is representative for the real world operation of such a vehicle.

The fuel consumption of the reference diesel-driven vehicle adds up to 32.8 L. In hybrid operation, the DMU consumes 25.8 L, which denotes a reduction of FC_{Cor} by 21.3%. Figure 9 shows the vehicle velocity profile over time for the diesel and hybrid system as well as the altitude profile. The velocity profiles differ due to the modified drive strategies and their particular brake trajectories (Section 2.2).

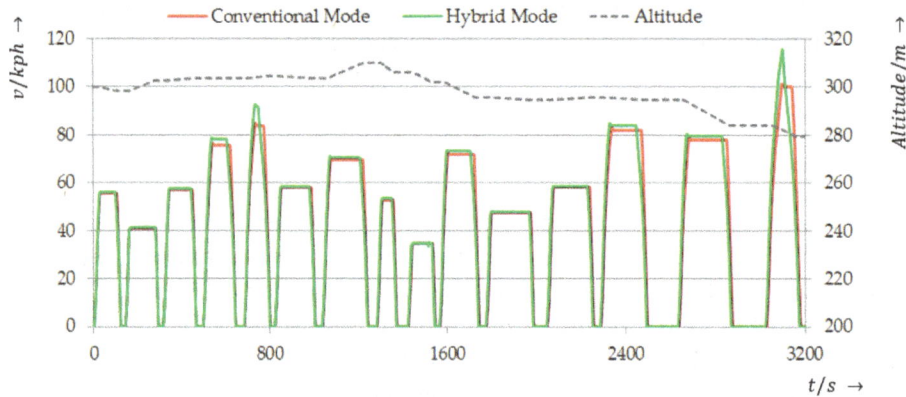

Figure 9. Altitude and velocity profiles for conventional and hybrid mode.

In Figure 10a, the battery SOC is shown. Due to the activated SOC control function, the SOCs at the end of the drive cycle and also at the end of each cycle sequence are fairly close to the target value SOC_{Trg} of 55%. The maximum deviation of the SOC and its target value at halt events occurs after the seventh drive sequence and adds up to 0.59%. Considering the comparatively simple approach from Section 4.3.4, Figure 10b shows the energy cost-equivalent charge factor $\varrho(t)$ over cycle time. It becomes obvious that $\varrho(t)$ falls during cruising and raises throughout recuperation. This is the typical behavior and points out that *Charge Mode* is active quite often. The fact that $\varrho(t)$ is not below one during the whole cycle shows that a relatively high amount of electric charging power for the battery is gained by recuperation. As a result, *Pure Electric* operation is executed rather often during cruising, because electric energy is quite "cheap". Figure 11a shows the percentages of the individual operation modes during acceleration and cruising. The *Combined Mode* extensively occurs during vehicle acceleration. The reason is that a combined use of ICE and MG increases the vehicle performance and, therefore, results in lower cruising speeds. About 56% of the time, the *Pure Electric* mode is activated, whereas the ICE is running on idle speed decoupled from the MG by the clutch. This leads to the question, whether it is feasible to completely switch off the ICE during driving. The employment of *Start/Stop* mode at halt further reduces the fuel consumption by 2.5%, which leads to a FC_{Cor} of 25.0 L. If *Start/Stop* mode is expanded to cruising events as well, the FC_{Cor} is further lowered by 1.2%, see Figure 11b.

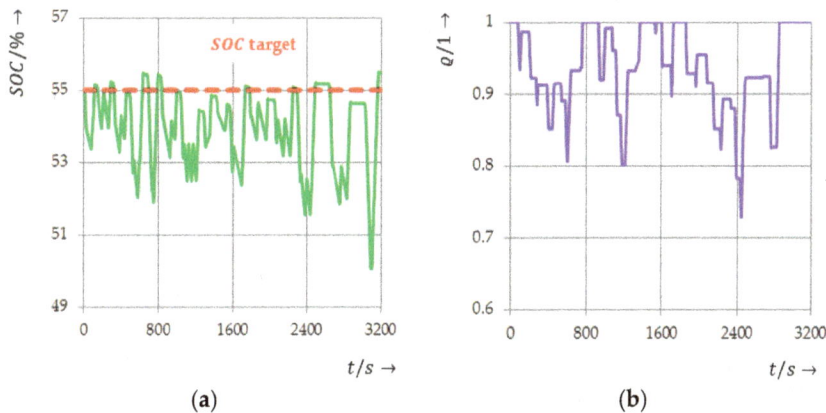

Figure 10. (a) *SOC* and its target value over cycle time; (b) Energy cost equivalent charge factor over cycle time.

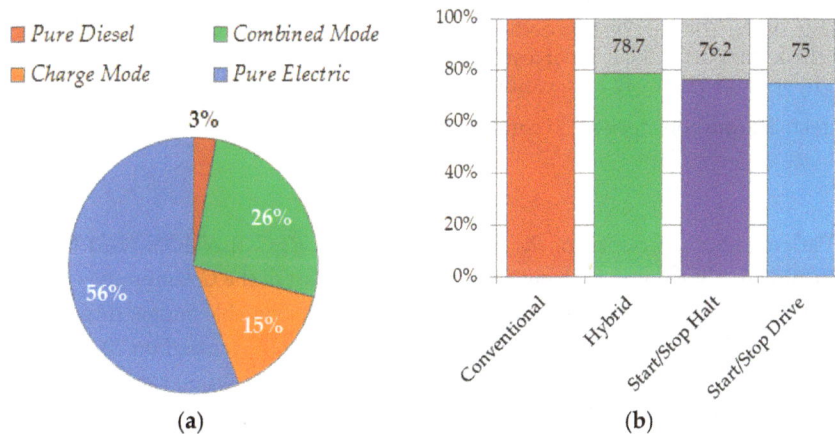

Figure 11. (a) Percental proportion of operation modes during acceleration and cruising (hybrid without *Start/Stop*); (b) Overview over percental reductions of corrected fuel consumption.

6. Conclusions and Outlook

In this contribution, an optimization-based causal PMA for hybrid propulsion systems is presented. It optimizes the power split between ICE and MG according to vectors of possible operation points. For each vector element, which represents one feasible operational power distribution, the corresponding element of a power supply efficiency vector is calculated. The power split that is related to the highest efficiency determines the power demand for ICE and MG as control output. The computation of the power supply efficiency uses an energy-cost-equivalent charge factor, which accounts for the energy consumption to charge the battery. It differentiates between regenerative charging during braking and "energy expensive" charging via the ICE. Moreover, as an optional feature, the power split can be controlled in such a way that the battery *SOC* matches a pre-defined target value at the end of the drive cycle. In a case study, the presented PMA is employed for the control a hybrid propulsion system of a DMU consisting of two identical PUs. Thereby, the DMU was operated on a track which was generated from logging field data of comparable vehicles. In comparison to the conventional diesel-driven system, a fuel consumption reduction of 21.3% was achieved. Employing *Start/Stop* at halt and, additionally, during *Pure Electric* driving, the fuel consumption was further reduced by 3.7%. It was stated that the PMA leads to high proportion of *Pure Electric* driving during cruising events. This has a significant influence on the engine load profile because especially areas of low engine loads, where engine efficiency is weak, disappear. This is even more true as soon as the engine is switched off during *Pure Electric* driving. Beside the shown fuel consumption benefits, the *Pure Electric* mode leads to shorter engine operation time and extends the time-before overhaul (TBO), which further reduces operational costs for the system operator. Another aspect which has to be considered is the influence of hybridization on exhaust emissions. DMUs' modern diesel engines are compliant to EPA Tier 4i emission standards. Those are commonly equipped with SCR (selective catalytic reduction) exhaust after-treatment systems. Investigations in the HiL test environment show that the changed engine operation in hybrid systems, compared to conventional ones, considerably influence exhaust after-treatment performance, hence real driving emissions [7]. The implementation of *Start/Stop* during *Pure Electric* driving are one possibility to influence exhaust after-treatment temperatures and, for example, prevent the SCR-catalyst from cooling down too quickly during idling.

In future investigations, the introduced PMA will be applied to system configurations with multiple PUs, where the ICEs and MGs are operated individually. Furthermore, the PMA's suitability for other applications will be proven. To demonstrate the "real-world" capability of the PMA, it will be applied to a PU prototype of a DMU propulsion system in hardware-in-the-loop testing campaigns on a system test bench. In those experiments, the impact and the benefits of the proposed PMA and its different operation modes on fuel consumption and emissions will be investigated for different DMU propulsion system configurations.

Author Contributions: Johannes Schalk developed the algorithm and the simulation model. Furthermore, he executed the case study. The HiL-Test procedure had been conducted at a test facility at MTU Friedrichshafen, where Johannes Schalk was part of the team of test engineers. Harald Aschemann supervised the work, contributed his know-how in hybrid Off-Highway applications and co-wrote the paper.

Conflicts of Interest: The authors declare no conflict of interest.

References

1. European Parliament and Council of the European Union. REGULATION (EG) No 443/2009 Setting Emission Performance Standards for New Passenger Cars as Part of the Community's Integrated Approach to Reduce CO_2 Emissions from Light-Duty Vehicles, April 2009. Available online: http://eur-lex.europa.eu/legal-content/EN/TXT/PDF/?uri=CELEX:32009R0443&from=EN (accessed on 20 August 2016).
2. Baffes, J.; Kose, M.A.; Ohnsorge, F.; Stocker, M. *The Great Plunge in Oil Prices: Causes, Consequences and Policy Responses*; Policy Research Note; World Bank Group: Washington, DC, USA, 2015.
3. Dittus, H.; Hülsebusch, D.; Ungethüm, J. Reducing DMU fuel consumption by means of hybrid energy storage. *Eur. Transp. Res. Rev.* **2011**, *3*, 149–159. [CrossRef]
4. Oszfolk, B.; Radke, M.; Ibele, Y. Hybridantrieb stellt Marktreife unter Beweis. *ETR Eisenbahntechnische Rundsch.* **2015**, *9*, 44–49.
5. Yuan, L.C.W.; Tjahjowidodo, T.; Lee, G.S.G.; Chan, R. Equivalant consumption minimization strategy for hybrid all-electric tugboats to optimize fuel savings. In Proceedings of the American Control Conference (ACC), Boston, MA, USA, 6–8 July 2016.
6. Jayaram, V.; Khan, M.Y.; Welch, W.A.; Johnson, K.; Miller, J.W.; Cocker, D.R. A generalized approach for varifying the emission benefits of off-road hybrid mobile sources. *Emiss. Control Sci. Technol.* **2016**, *2*, 89–98. [CrossRef]
7. Hass, C.; Oszfolk, B.; Schalk, J. Emissions of a hybrid propulsion system for regional trains as an example for innovative non-road propulsion systems. In Proceedings of the 8th Emission Control, Dresden, Germany, 2–3 June 2016.

8.   Hofman, T.; Steinbuch, M.; Van Druten, R.; Serrarens, A. Rule-based energy management strategies for hybrid vehicles. *Int. J. Electr. Hybrid Veh.* **2007**, *1*, 71–94. [CrossRef]

9.   Guzella, L.; Sciarretta, A. *Vehicle Propulsion Systems: Introduction to Modeling and Optimization*; Springer: Berlin/Heidelberg, Germany, 2013.

10. Goerke, D. *Untersuchungen zur Kraftstoffoptimalen Betriebsweise von Parallelhybridfahrzeugen und Darauf Basierende Auslegung Regelbasierter Betriebsstrategien*; Springer: Wiesbaden, Germany, 2016.

11. Guzella, L.; Sciarretta, A. Control of hybrid electric vehicles. *IEEE Control Syst.* **2007**, *27*, 60–70.

12. Helbing, D.I.M.; Uebel, D.I.S.; Tempelhahn, D.I.C.; Bäker, I.B. Bewertender Überblick von Methoden zur Antriebsstrangsteuerung in Hybrid- und Elektrofahrzeugen. *ATZelektronik* **2015**, *10*, 66–71. [CrossRef]

13. Karbaschian, M.A.; Söffker, D. Review and comparison of power management approaches for hybrid vehicles with focus on hydraulic drives. *Energies* **2014**, *7*, 3512–3536. [CrossRef]

14. Lampe, A. Regelbasierte Betriebsstrategien zur Vorauslegung von Hybridantriebssträngen. *ATZ Automobiltech. Z.* **2016**, *116*, 76–82. [CrossRef]

15. Hanho, S.; Hyunsoo, K. Development of near optimal rule-based control for plug-in hybrid electric vehicles taking into account drivetrain component losses. *Energies* **2016**, *9*, 420. [CrossRef]

16. Paschero, M.; Storti, G.L.; Rizzi, A.; Mascioli, F.M.F. Implementation of a fuzzy control system for a parallel hybrid vehicle powertrain on compactrio. *Int. J. Comput. Theory Eng.* **2013**, *5*, 273. [CrossRef]

17. Khoucha, F.; Benbouzid, M.; Kheloui, A. An optimal fuzzy logic power shift strategy for parallel hybrid electric vehicles. In Proceedings of the IEEE Vehicle Power and Propulsion Conference, Lille, France, 1–3 September 2010.

18. Yuan, Z.; Teng, L.; Fengchun, S.; Peng, H. Comparative study of dynamic programming and pontryagin's minimum principle on energy management of a parallel hybrid electric vehicle. *Energies* **2013**, *6*, 2305–2318. [CrossRef]

19. Leska, M.; Aschemann, H. Fuel-optimal combined driving strategy and energy management for a parallel hybrid electric railway vehicle. In Proceedings of the 20th International Conference on Methods and Models in Automation and Robotics, Miedzyzdroje, Poland, 24–27 August 2015.

20. Sivertsson, M.; Sundström, C.; Eriksson, L. *Adaptive Control of a Hybrid Powertrain with Map-Based ECMS*; IFAC World Congress: Prague, Czech Republic, 2011.

21. Mustardo, C.; Rizzoni, G.; Guezennec, Y.; Staccia, B. A-ECMS: An adaptive algorithm for hybrid electric vehicle energy management. *Eur. J. Control* **2005**, *11*, 509–524. [CrossRef]

22. Nüesch, T.; Cerofolini, A.; Mancini, G.; Cavina, N.; Onder, C.; Guzzella, L. Equivalent consumption minimization strategy for control of real driving NOx emissions of a diesel hybrid electric vehicle. *Energies* **2014**, *7*, 3148–3178. [CrossRef]

23. Onori, S.; Serrao, L. On adaptive-ECMS strategies for hybrid electric vehicles. In Proceedings of the International Scientific Conference on Hybrid and Electric Vehicles, Rueil-Malmaison, France, 6–7 December 2011.

24. Winkler, M.; Geulen, S.; Josevski, M.; Tegethoff, M.; Abel, D.; Vöcking, B. Online parameter tuning methods for adaptive ECMS control strategies in hybrid electric vehicles. In Proceedings of the FISTA World Automotive Congress, Maastricht, The Netherlands, 2–6 June 2014.

25. Katsargyri, G.-E.; Kolmanovsky, I.V.; Michelini, J.; Kuang, M.L.; Phillips, A.M.; Rinehart, M.; Dahleh, M.A. Optimally controlling electric hybrid vehicles using path forecasting. In Proceedings of the 2009 American Control Conference, St. Louis, MO, USA, 10–12 June 2009.

26. Geering, H.P. *Optimal Control with Engineering Applications*; Springer: Berlin/Heidelberg, Germany, 2007.

27. Gamma Technologies. *GT-SUITE User Manual. GT-SUITE Optimisation Manual*, version 2016; Gamma Technologies: Westmont, IL, USA, 2015.

4

Theoretical Analysis of Shrouded Horizontal Axis Wind Turbines

Tariq Abdulsalam Khamlaj * and Markus Peer Rumpfkeil

300 College Park Kettering Labs, University of Dayton, Dayton, OH 45469-0238, USA;
mrumpfkeil1@udayton.edu
* Correspondence: khamlajt1@udayton.edu

Academic Editor: Frede Blaabjerg

Abstract: Numerous analytical studies for power augmentation systems can be found in the literature with the goal to improve the performance of wind turbines by increasing the energy density of the air at the rotor. All methods to date are only concerned with the effects of a diffuser as the power augmentation, and this work extends the semi-empirical shrouded wind turbine model introduced first by Foreman to incorporate a converging-diverging nozzle into the system. The analysis is based on assumptions and approximations of the conservation laws to calculate optimal power coefficients and power extraction, as well as augmentation ratios. It is revealed that the power enhancement is proportional to the mass stream rise produced by the nozzle diffuser-augmented wind turbine (NDAWT). Such mass flow rise can only be accomplished through two essential principles: the increase in the area ratios and/or by reducing the negative back pressure at the exit. The thrust coefficient for optimal power production of a conventional bare wind turbine is known to be 8/9, whereas the theoretical analysis of the NDAWT predicts an ideal thrust coefficient either lower or higher than 8/9 depending on the back pressure coefficient at which the shrouded turbine operates. Computed performance expectations demonstrate a good agreement with numerical and experimental results, and it is demonstrated that much larger power coefficients than for traditional wind turbines are achievable. Lastly, the developed model is very well suited for the preliminary design of a shrouded wind turbine where typically many trade-off studies need to be conducted inexpensively.

Keywords: nozzle diffuser augmented; wind turbine; wind lens; momentum theory; Betz limit

1. Introduction

There are numerous unresolved problems if one wants to increase the power production of a conventional wind turbine by simply increasing the diameter of the rotor, e.g., transportation, installation and maintenance, to name a few. Due to this fact, integration of wind generators into a national or regional power grid can be inhibited by the unacceptable reliability of very large units or the economic liability of many smaller units of comparable total power output. These technical factors interact with the economic constraints associated with matching supply and demand schedules in variable wind, the low power density of wind, the high development risk of new system concepts and the capital intensive nature of wind power systems.

Many of these capital and performance challenges of conventional wind turbine systems could be potentially reduced or eliminated by enclosing the wind turbine in a suitably-shaped duct. A duct structure typically provides a diffuser section behind the rotor that produces a power augmentation of considerable magnitude (typically 1.5- to two-fold) for a given size rotor, as well as the dampening of gusts, reducing performance sensitivity to wind directionality and raising the level of axial velocity significantly [1].

The actuator disk theory for open flow wind turbines has been established for about 90 years, while shrouded horizontal axis wind turbines or diffuser-augmented wind turbines (DAWT) have been in development for more than five decades with no large commercial success to date. Unfortunately, one influential early investigator, Betz [2], concluded that diffusers were not economical for contemporaneous applications. Although this result was based on correct theory, it made the overly restrictive assumption that the diffuser exit pressure is equal to the ambient or free stream atmospheric pressure.

An experimental approach for ducted wind turbines was undertaken by Lilly and Rainbird [3] in the 1950s, and in the 1970s, a significant number of experiments were carried out by Foreman [1,4] from the Grumman Aerospace department. However, conclusions from these experiments varied significantly. The Lilly and Rainbird study concluded that no performance improvements were conceivable when the performance is normalized by the exit area of the diffuser, whereas Foreman [1] confirmed a power increase by a factor of four compared to the same rotor operating as a conventional or bare wind turbine. These research works performed not only experiments, but also developed theoretical models to verify their results. However, these models lacked a complete explanation of the major flow phenomena occurring in DAWTs. Shrouded horizontal axis wind turbines were a subject undergoing intense study at the Wind Energy Innovative System Conference in 1979. One of the major conclusions from this conference was that the economical applications of such turbines seemed not feasible at the time because of the high cost of the shroud, although the power augmentation was interesting. The expeditious development of bare horizontal axis wind turbines at around the same time led to the disappearance of DAWTs from research agendas.

Recently, however, an increase in the number of publications on the topic and attempts to commercialize the idea indicate a renewed interest in shrouded turbines. Researchers have come to the agreement that there is significant potential for improvements in this concept, and understanding the details of the flow physics is one of the keys. An investigation by Hansen [5] using both low-fidelity momentum theory and computational fluid dynamics (CFD) demonstrates that the power augmentation of a shrouded turbine is proportional to the increase in mass flow rate through the turbine blades. Furthermore, throughout the past decade, the research group of Ohya [6] has performed extensive experimental and computational work on this topic, which has led to the development of a high performance "flanged diffuser", as they call it.

The current work complements the analysis first introduced by Foreman [1] and backed by Lawn [7], but derives the results in a rather more general way. However, due to the complexity of the force on a diffuser, a closed theory cannot be established, and the analysis needs to be augmented with empirical data. The remainder of the article is organized as follows. Section 2 gives an overview of other analytical methods in the literature. Section 3 derives the newly-extended theory, and Section 4 verifies and validates the theoretical results with experimental and CFD data. Section 5 provides a comparison between augmentation with a diffuser only and with a converging-diverging nozzle. Section 6 shows performance predictions of NDAWT systems by changing area ratios of the nozzle and diffuser, diffuser and nozzle efficiencies, velocity ratios and back pressure coefficients. Finally, Section 7 concludes this article.

2. Review of Previous Analytical Models

Researchers have developed several basic 1D theoretical models to predict the power production of wind turbines, including the well-known one-dimensional momentum analysis of the flow over a horizontal axis wind turbine leading to the famous Betz limit. Predictions for the power output of a shrouded wind turbine have been published by, amongst others, Foreman (1978) [1], Lawn (2003) [7], van Bussel (2007) [8], Jamieson (2008) [9], as well as Werle and Presz (2008) [10]. However, the results of these models vary; for instance, the predictions of the thrust coefficient at the maximum power coefficient differ based on the underlying theory, as explained in more detail below. In addition, some lack a complete description of the major flow phenomena, rendering most of the models either valid for only short diffusers or entirely invalid [11]. Next, the various theories and their shortcomings are discussed briefly.

Van Bussel (2007) [8] developed a theoretical model based on Betz's theory. The continuity equation, as well as the momentum equation were used in the analysis. However, it was assumed that just downstream of the physical diffuser, the velocity is the average of the velocity far upstream and far downstream, i.e., the relation that is valid for the velocity at the rotor plane for the case of a conventional wind turbine. This assumption is only valid for the case of a short diffuser, i.e., if the diffuser exit plane is not too far off the rotor plane for which the assumption that was made regarding the average velocity is still somewhat justified, since it is certainly not true in long diffusers. Thus, the one-dimensional flow theory of van Bussel [8], though consistent with the case of the conventional wind turbine without diffuser, is not valid for longer diffusers [11].

Jamieson (2008) [9] derived a one-dimensional theory similar to the one of van Bussel. However, the diffuser is taken into account by assuming an induction factor at zero thrust. This decoupling is not ideal, since the diffuser performance is influenced by the thrust exerted by the rotating blades. He concludes that a DAWT operates optimally at the same conditions as a bare turbine with a thrust coefficient of 8/9. According to Konijn [11], the theory leads to an incorrect power coefficient of 32/27 (instead of the correct 16/27) when regarding the diffuser with an area ratio between the downstream exit and rotor plane of two.

Werle and Presz (2008) [10] use a different models for the diffuser. They assume that the effect of the diffuser can be modeled by a force on the fluid pointing against the direction of the flow. This force is modeled to be proportional to the rotor resistance. The rest of their derivation for the power coefficient is very similar to the one leading to the Betz limit, except for the presence of the force exerted by the diffuser on the fluid. The one-dimensional flow theory by Werle and Presz is questionable since the force is not necessarily proportional to the rotor resistance. For instance, the theory breaks down when the value of the loading coefficient is equal to two [11].

In summary, all of the methods mentioned above predict a maximum thrust coefficient equal to 8/9 for a DAWT; however, this is incorrect due to false assumptions made in the respective derivations. Foreman (1978) [1] and Lawn (2003) [7], on the other hand, show that optimal operating conditions occur at thrust coefficient values either higher or lower than 8/9 depending on the loading coefficient, back pressure coefficient and diffuser efficiency.

3. 1-D Theory of Nozzle Diffuser Augmented Wind Turbines

3.1. *Assumptions and Geometry*

The theory by Betz–Joukowski, as well as the theory of shrouded diffuser wind turbines is based on the premise that the turbine can be represented as an actuator disk, as given in Equation (1). The flow over the disc is considered ideal, meaning frictionless, as well as having no rotational velocity component in the wake. The flow is also assumed to be steady and incompressible.

Figure 1 shows that the actuator disc appears as a drag device decelerating the wind speed in a continuous manner from V_∞ far upstream of the rotor, to V_1 at the rotor and, finally, V_3 far downstream in the wake. The retardation of flow gives rise to a divergence in the stream tube passing through the periphery of the disc. There is an associated pressure rise on the upstream side of the rotor to a value p_1. Across the actuator disc, there is a discontinuous pressure drop from p_1 to p_2. Downstream of the rotor, there is a gradual pressure recovery until the level returns to atmospheric, p_∞, in the far wake.

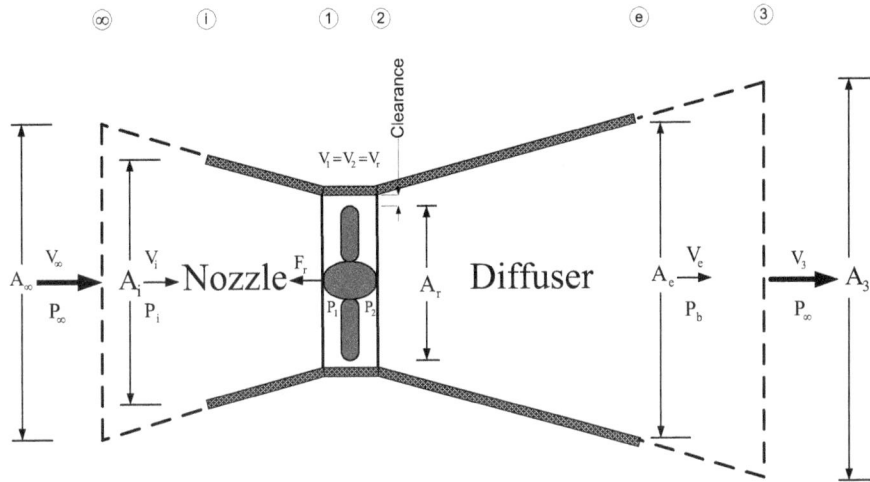

Figure 1. Stream tube passing through a shrouded wind turbine with nozzle and diffuser.

3.2. 1-D Theory for Bare Wind Turbines

Investigations of the one-dimensional theory for a conventional or bare wind turbine by Bergey [12] indicated that Lanchester (1915), Betz (1920) and Joukowski (1920) might all have independently established the maximum efficiency of an energy extraction device in an open flow. A recent study by Okulov and van Kuik [13] suggests that the ascription to Lanchester is likely inappropriate. Therefore, the Betz limit may conceivably be called the Betz–Joukowski limit; however, for shortness and familiarity, the reference as simply the Betz limit is retained [14] here.

Across the rotor, the thrust, F_r, that is the force in the streamwise direction resulting from the pressure drop across the turbine, can be written as follows [15]:

$$F_r = (p_1 - p_2)A_r \tag{1}$$

where A_r is the rotor area and p_1 and p_2 are the pressure before and after the rotor plane, respectively.

The ideal power output of the actuator or the rate of energy loss across the rotor is obtained by multiplying Equation (1) with V_r, where $V_r = V_1 = V_2$. Hence, the ideal power from the actuator disk can be written as follows:

$$P_{ideal} = (p_1 - p_2)Q, \qquad Q = V_1 A_1 \tag{2}$$

showing that the power generated by the turbine is proportional to the product of the pressure loss across it and the volumetric flow rate. However, in practice, only a fraction of the power $P < P_{ideal}$ can be harvested; hence, the turbine efficiency can be defined as follows:

$$\eta_T = \frac{P}{(p_1 - p_2)Q} \tag{3}$$

The loading coefficient for the flow caused by some factors, e.g., the degree of surface smoothness, angles of the exit and inlet section, rounded inlet section, separation layer (boundary layer), etc., is defined as the ratio of the drop in static pressure across the turbine and the dynamic pressure at the throat [1]:

$$\psi = \frac{(p_1 - p_2)}{\frac{1}{2}\rho V_1^2} \tag{4}$$

The standard definition of the thrust coefficient for a turbine is one referenced to the free stream velocity V_∞ [15].

$$C_t = \frac{p_1 - p_2}{\frac{1}{2}\rho V_\infty^2} \tag{5}$$

The velocity ratio between the rotor plane and free stream can be defined as:

$$V_R = \frac{V_1}{V_\infty}, \qquad V_1 = V_2 \tag{6}$$

Therefore, substituting Equation (4) into (5) yields:

$$C_t = \psi V_R^2 \tag{7}$$

Rewriting Equation (3):

$$P = \eta_T (p_1 - p_2) V_1 A_1 \tag{8}$$

and multiplying and dividing the right-hand side by $\frac{1}{2}\rho V_1^2$ gives:

$$P = \frac{1}{2} \eta_T \psi \rho V_1^3 A_r \tag{9}$$

where η_T is the turbine efficiency, and V_1 will be diminished as ψ increases. Similarly, the standard definition of the power coefficient is given with reference to the rotor area:

$$C_P = \frac{P}{\frac{1}{2}\rho V_\infty^3 A_r} \tag{10}$$

Substituting Equations (9) into (10) results in the following handy relation:

$$\frac{C_P}{\eta_T} = C_t V_R \tag{11}$$

The changes in pressure and velocity upstream and downstream of the actuator disc can be described by Bernoulli's equation under the assumptions mentioned earlier. Therefore, the equation for the stream tube beginning far upstream and ending at Station 1 just in front of the rotor is given by:

$$C_{p_{1\infty}} = \frac{p_1 - p_\infty}{\frac{1}{2}\rho V_\infty^2} = \left[1 - V_R^2\right] \tag{12}$$

Similarly, the Bernoulli equation holds for the flow downstream of the turbine:

$$C_{p_{23}} = \frac{p_\infty - p_2}{\frac{1}{2}\rho V_2^2} = \left[1 - \frac{V_3^2}{V_2^2}\right] \tag{13}$$

Considering the stream tube shown in Figure 1, the axial momentum balance can be written in the following form [16]:

$$V_3 \rho V_3 A_3 - V_\infty \rho V_\infty A_\infty = -F_r \tag{14}$$

Substituting the definition that was introduced for the thrust Equation (1) into Equation (14) gives:

$$V_3 \rho V_3 A_3 - V_\infty \rho V_\infty A_\infty = -(p_1 - p_2) A_r \tag{15}$$

Hence, by simplifying Equation (15), the final result will be in the following form:

$$\psi = 2 \left[\frac{V_\infty}{V_1} - \frac{V_3}{V_2}\right] \tag{16}$$

Summing the pressure differences along the streamwise direction and equating to zero leads to:

$$(p_1 - p_\infty) + (p_\infty - p_2) - (p_1 - p_2) = 0 \tag{17}$$

Substituting Equations (4), (12) and (13) into (17) yields:

$$V_R = \frac{V_1}{V_\infty} = \frac{4}{\psi + 4} \tag{18}$$

Now, since the thrust and power coefficients are functions of the velocity ratio V_R, it is easy to find these coefficients. To do so, substitute Equations (18) into (7), resulting in:

$$C_{t_C} = \frac{16\psi}{(\psi + 4)^2} \tag{19}$$

Similarly, substituting Equation (18) into (11) yields:

$$\frac{C_{P_C}}{\eta_T} = \frac{64\psi}{(\psi + 4)^3} \tag{20}$$

The maximum theoretical power coefficient of the conventional wind turbine can be obtained by differentiating $\frac{C_{P_C}}{\eta_T}$ with respect to ψ in Equation (20) and solving for the root(s). The result yields:

$$\psi_{max} = 2 \tag{21}$$

$$V_{R_{max}} = \frac{2}{3} \tag{22}$$

$$C_{t_{C_{max}}} = \frac{8}{9} \tag{23}$$

$$\frac{C_{P_{C_{max}}}}{\eta_T} = \frac{16}{27} = 59.3\% \tag{24}$$

Using Bernoulli's equation for the flow upstream of the rotor, the pressure difference between Section 1 and section ∞ relative to the dynamic pressure in the upstream flow for a conventional wind turbine operating at the Betz limit can be calculated by using Equation (12) with the maximum velocity ratio $V_{R_{max}} = \frac{2}{3}$ yielding:

$$C_{p_f} = \frac{p_1 - p_\infty}{\frac{1}{2}\rho V_\infty^2} = \left[1 - V_R^2\right] = \frac{5}{9} \tag{25}$$

Similarly, for the back pressure of the bare wind turbine, one obtains:

$$C_{p_b} = \frac{p_2 - p_\infty}{\frac{1}{2}\rho V_\infty^2} = C_{p_f} - C_t = -\frac{1}{3} \tag{26}$$

3.3. 1-D Theory for Shrouded Turbines with Nozzle and Diffuser

The results of one-dimensional momentum theory applied to a diffuser-augmented wind turbine or DAWT were first presented by Foreman [1] and re-derived by Lawn [7]. Here, their approach is expanded to include the effect of a nozzle incorporated into the shroud called a nozzle diffuser-augmented wind turbine or NDAWT.

The area ratio of the nozzle and diffuser can be defined from the continuity equation:

$$\frac{V_e}{V_2} = \frac{A_2}{A_e} = \lambda_D, \qquad \frac{V_i}{V_1} = \frac{A_1}{A_i} = \lambda_N \tag{27}$$

Referring to Figure 1, the pressure coefficient at the nozzle inlet can be expressed by applying Bernoulli's equation:

$$C_{p_f} = \frac{p_i - p_\infty}{\frac{1}{2}\rho V_\infty^2} = \left[1 - \lambda_N^2 \frac{V_1^2}{V_\infty^2}\right] \tag{28}$$

Similarly, the coefficient at the exit of the diffuser or back pressure is given by:

$$C_{p_b} = \frac{p_b - p_\infty}{\frac{1}{2}\rho V_\infty^2} = \frac{V_3^2}{V_\infty^2} - \lambda_D^2 \frac{V_2^2}{V_\infty^2} \tag{29}$$

The nozzle efficiency can be introduced by using Bernoulli's equation between the inlet of the nozzle and rotor:

$$\eta_N = \frac{p_i - p_1}{\frac{1}{2}\rho V_1^2 - \frac{1}{2}\rho V_i^2} \tag{30}$$

Similarly, the diffuser efficiency can be defined as [1,7,17]:

$$\eta_D = \frac{p_b - p_2}{\frac{1}{2}\rho V_2^2 - \frac{1}{2}\rho V_e^2} \tag{31}$$

Summing the pressure difference along the streamwise direction and equating to zero results in:

$$(p_i - p_1) - (p_i - p_\infty) + (p_1 - p_2) - (p_b - p_2) + (p_b - p_\infty) = 0 \tag{32}$$

Substituting Equations (4), (28), (29), (30) and (31) into (32) yields:

$$V_{R_{ND}} = \frac{V_1}{V_\infty} = \sqrt{\frac{1 - C_{p_b}}{\psi + \eta_N + \lambda_N^2 (1 - \eta_N) - \eta_D (1 - \lambda_D^2)}} \tag{33}$$

Substituting Equations (33) into (7) yields the thrust coefficient for the NDAWT:

$$C_{t_{ND}} = \psi V_R^2 = \psi \left[\frac{1 - C_{p_b}}{\psi + \eta_N + \lambda_N^2 (1 - \eta_N) - \eta_D (1 - \lambda_D^2)} \right] \tag{34}$$

Recall Equation (29); hence, the velocity far downstream is given as follows:

$$\frac{V_3}{V_\infty} = \sqrt{C_{p_b} + \lambda_D^2 \frac{V_1^2}{V_\infty^2}} \tag{35}$$

Similarly, substituting Equations (33) into (11) gives the power coefficient for the NDAWT:

$$\frac{C_{P_{ND}}}{\eta_T} = C_t V_R = \psi \left[\frac{1 - C_{p_b}}{\psi + \eta_N + \lambda_N^2 (1 - \eta_N) - \eta_D (1 - \lambda_D^2)} \right]^{\frac{3}{2}} \tag{36}$$

The main insight of Equation (36) is that the power available to a perfect ducted turbine can be increased by using a smaller rotor/nozzle and rotor/exit area ratio, λ_N and λ_D, high nozzle and diffuser efficiencies, η_N and η_D, and a large negative back pressure coefficient C_{p_b}.

The maximum theoretical power output of the NDAWT can be found by differentiating $\frac{C_{P_{ND}}}{\eta_T}$ with respect to ψ in Equation (36) and finding the root(s). The result leads to the following optimal loading coefficient:

$$\psi_{ND_{max}} = 2 \left(\eta_N + \lambda_N^2 (1 - \eta_N) - \eta_D (1 - \lambda_D^2) \right) \tag{37}$$

Substituting Equations (37) into (33), (34) and (36) results in:

$$V_{R_{ND_{max}}} = \sqrt{\frac{1 - C_{p_b}}{3\eta_N + 3\lambda_N^2 (1 - \eta_N) - 3\eta_D (1 - \lambda_D^2)}} \tag{38}$$

$$C_{t_{ND_{max}}} = 2 \left[\frac{1 - C_{p_b}}{3} \right] \tag{39}$$

$$\frac{C_{P_{NDmax}}}{\eta_T} = 2 \left[\frac{\left(1 - C_{p_b}\right)^3}{27 \left(\eta_N + \lambda_N^2 \left(1 - \eta_N\right) - \eta_D \left(1 - \lambda_D^2\right)\right)} \right]^{\frac{1}{2}} \tag{40}$$

Thus, the greatest power augmentation compared to conventional wind power generators is obtained for [1]:

- The largest possible negative value of the exit plane pressure coefficient (i.e., diffuser exit pressure is reduced well below atmospheric pressure).
- The largest possible diffuser and nozzle efficiencies.
- A unique relation of turbine disk loading to diffuser pressure recovery in which high recovery favors low power loading by inducing greater volume flow through the disk.

Because coefficients, such as C_{p_b}, η_D and η_N must be empirically determined, the one-dimensional theoretical performance indicated by Equation (36) is, in practice, a semi-empirical theory for which measured quantities of existing NDAWTs need to be introduced for proper quantitative evaluation.

It is prudent to verify that in the limit, one recovers the traditional Betz results presented in Section 3.2 above. One can remove the effects of the nozzle and diffuser by setting $\lambda_D = \lambda_N = 1$, as well as $\eta_D = \eta_N = 1$ and using $C_{p_b} = -\frac{1}{3}$ (semi-empirical theory) in Equations (37)–(40), yielding the familiar results. Table 1 gives a succinct summary of the theoretical equations governing the performance of bare and augmented wind turbines.

Table 1. Summary results of bare wind turbine and the current method. NDAWT, nozzle diffuser-augmented wind turbine.

Parameter	Bare Wind Turbine	NDAWT
Upstream wind speed	V_∞	V_∞
Wind speed at rotor plane	$V_\infty \cdot \frac{4}{\psi+4}$	$V_\infty \cdot \sqrt{\frac{1-C_{p_b}}{\psi+\eta_N+\lambda_N^2(1-\eta_N)-\eta_D(1-\lambda_D^2)}}$
Far wake wind speed	$V_\infty \cdot \frac{4-\psi}{\psi+4}$	$V_\infty \cdot \sqrt{C_{p_b} + \lambda_D^2 \frac{V_1^2}{V_\infty^2}}$
Power coefficient C_P	$\frac{64\psi}{(\psi+4)^3}$	$\psi \left[\frac{1-C_{p_b}}{\psi+\eta_N+\lambda_N^2(1-\eta_N)-\eta_D(1-\lambda_D^2)} \right]^{\frac{3}{2}}$
Thrust coefficient C_t	$\frac{16\psi}{(\psi+4)^2}$	$\psi \left[\frac{1-C_{p_b}}{\psi+\eta_N+\lambda_N^2(1-\eta_N)-\eta_D(1-\lambda_D^2)} \right]$

4. Validation

4.1. Validation with Field Experimental Results

The research institution at Kyushu University [6] tested a shrouded wind turbine with a brim shown in Figure 2. The parameters used for the field experiment of that wind turbine are provided in Table 2.

Table 2. Parameters of a shrouded wind turbine with a brim [6].

Parameter	Value
Turbine diameter D	0.7 m
Diffuser diameter D_e	1.072 m
Nozzle diameter D_N	0.78 m
Diffuser length L_t	1.029 m
The brim height h	0.35 m
Density ρ	1.225 kg/m^3
The back pressure of the diffuser C_{p_b}	−0.6
The nozzle efficiency η_N	85%
The diffuser efficiency η_D	85%

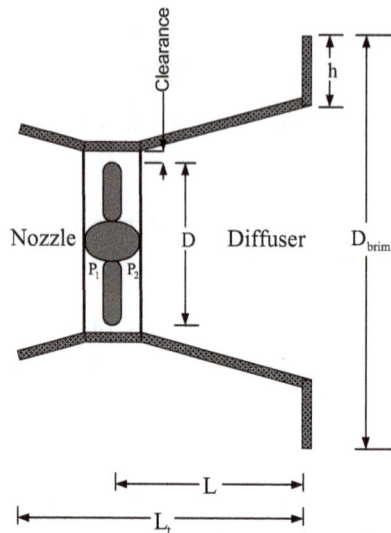

Figure 2. Wind turbine equipped with a brimmed diffuser [6].

Equations (20) and (36) are used to obtain the theoretical maximum power coefficients displayed as a function of the loading coefficient in Figure 3. In this case, the maximum power coefficient for the shrouded wind turbine with diameter $D = 0.7$ m and a brim is found to be $C_{P_{ND}} = 1.519$, which is almost three-times larger than the Betz limit for a bare wind turbine, $C_{P_C} = 0.593$. The corresponding value for the thrust coefficient of the shrouded turbine at the same optimal loading coefficient is $C_{t_{ND}} \simeq 1.1$.

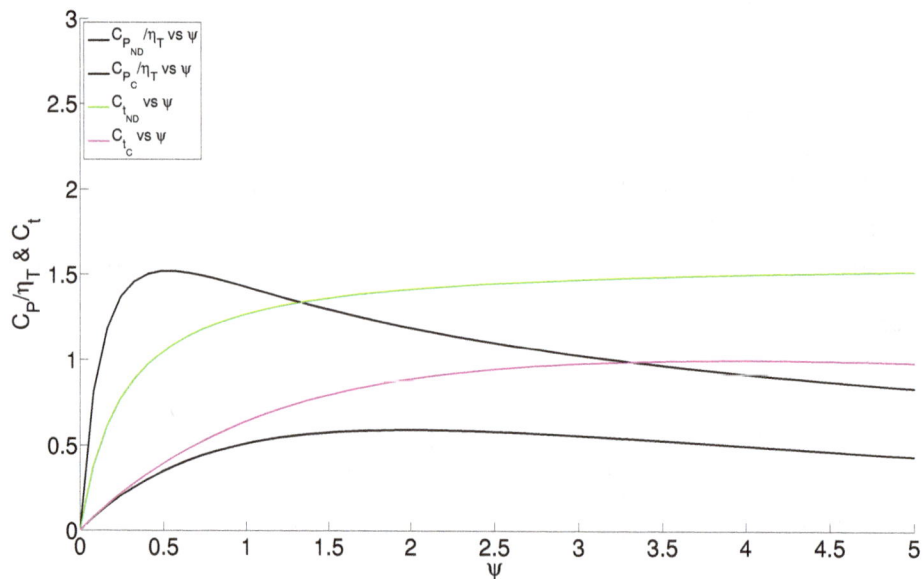

Figure 3. Power and thrust coefficients of the NDAWT using the data in Table 2 and in comparison with a bare wind turbine.

From Figure 4, one can observe that while the 1D theory of the bare wind turbine developed by Betz [15] shows a considerable deviation from the obtained experimental results as the velocity increases, the 1D theory for the NDAWT exhibits less deviation as the velocity increases. This can be explained due to the fact that the 1D theory for Betz is a completely inviscid theory, whereas for the NDAWT, the 1D theory contains semi-empirical relations, namely the pressure coefficient of the diffuser C_{p_b} that was obtained from the corresponding experimental results, as well as the nozzle and diffuser efficiencies η_N and η_D.

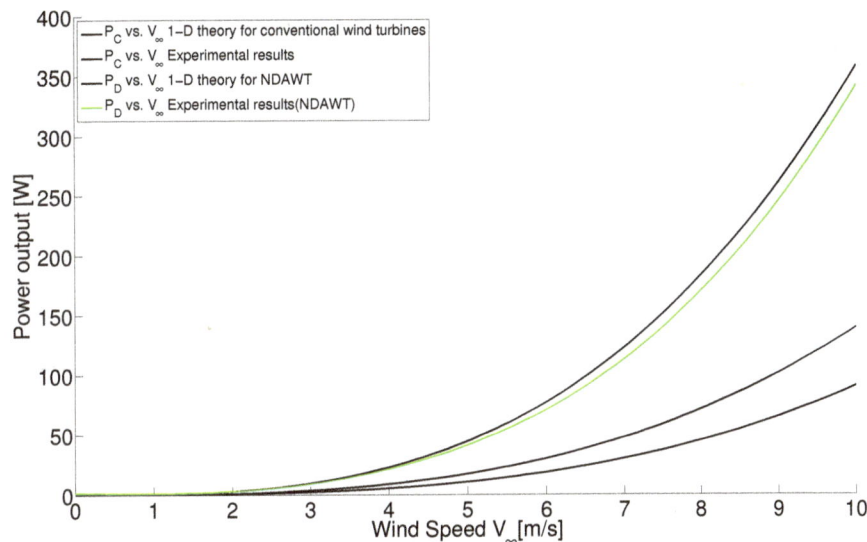

Figure 4. Comparison between the theoretical and experimental results [6] for the NDAWT and bare wind turbine.

4.2. Validation with CFD Results

Another good avenue to validate the developed semi-empirical formulations presented in Equations (33)–(36) is to make comparisons to higher fidelity CFD results. Comprehensive and complex viscous CFD computations were conducted by Hansen et al. [5,18] employing an actuator disk model to simulate the pressure drop across a wind turbine. They computed results for both a conventional and shrouded wind turbine, whose geometry is shown in Figure 5.

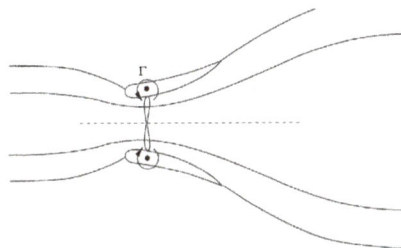

Figure 5. The duct shape used in Hansen's study [18].

It was found in Hansen's [5] study for a diffuser with an aggressive area ratio $\lambda_D = 0.54$ that the diffuser efficiency η_D is equal to 83%. Furthermore, the overall diffuser pressure recovery coefficient or back pressure C_{p_b} is estimated to be $C_{p_b} = -0.38$ for zero thrust. Using these values in Equations (20) and (36) for an unshrouded and shrouded wind turbine, respectively, the results shown in Figure 6 were calculated.

The comparison shows that the simple inviscid one-dimensional analysis for a conventional wind turbine is in good agreement with the more complex and expensive CFD results over the full range of the blade thrust. Likewise, for the shrouded wind turbine, the one dimensional inviscid analysis agrees well with the CFD results, but not surprisingly with a slightly higher maximum-power level, due to the considerable viscous losses encountered for such an aggressive diffusion area ratio ($\lambda_D = 0.54$). Furthermore, it can be noticed from Figure 6 that the NDAWT method is in good agreement with the viscous CFD results by Hansen [5]. However, at a thrust coefficient value of approximately 0.9, the new theory shows a noticeable deviation compared to the CFD results. This deviation likely occurs due to high turbulent effects at those operating conditions that cannot be captured using a simple 1D inviscid theory.

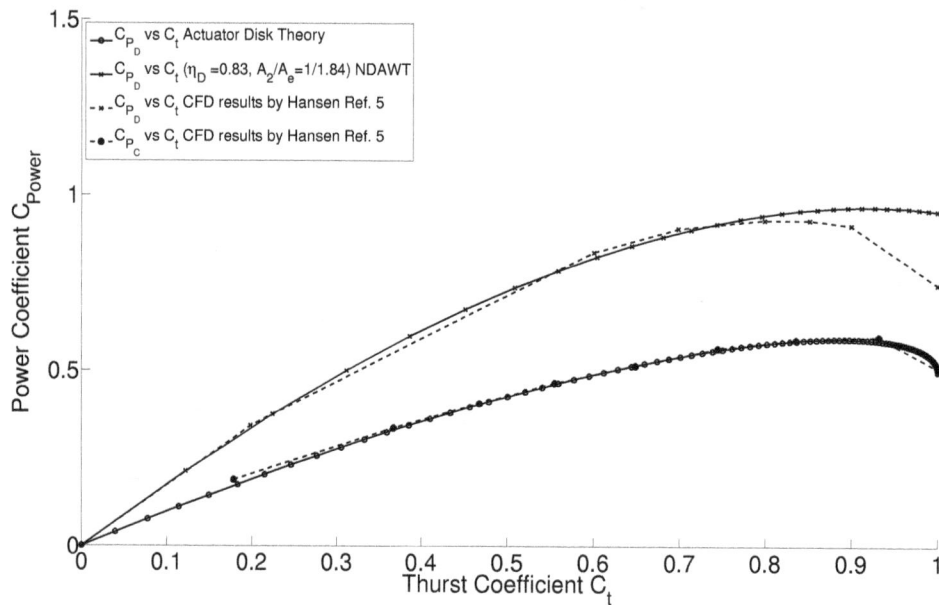

Figure 6. Theoretical calculations compared to the CFD results by Hansen et al. [5].

5. Diffuser Only vs. Converging-Diverging Nozzle

The main contribution of this article is to account for the physical effect of the addition of a nozzle to a diffuser shape, i.e., the consideration of a converging-diverging (C-D) nozzle for the shroud. In order to motivate the importance of this addition, Figures 7–9 have been created using the newly-developed generalized 1D momentum theory for the C-D nozzle, whereas the old momentum theory by Foreman et al. [1] (which is included as a limiting case in the new theory) is used for the diffuser only results. Figure 7 illustrates the velocity distribution, and one can notice the same velocity distribution for the diffuser only and C-D nozzle case if the efficiencies of both the nozzle and diffuser are equal to 100%. However, for efficiencies less than 100% and the same back pressure coefficient, one can observe that the C-D nozzle performs always better than the diffuser only and that adding a nozzle could lead to a 12% increase in the power production, as illustrated in Figures 8 and 9. In addition, based on viscous CFD simulations conducted by the authors, the nozzle has a secondary effect of decreasing the back pressure by up to 15%, as well leading to an even greater increase in power generation. Thus, the authors believe that one should use C-D nozzles as the design starting point for shrouds, and thus, the theory by Foreman et al. has been extended.

(a) Diffuser Only $\eta_D = 100\%$.

(b) C-D nozzle $\eta_N = \eta_D = 100\%$.

(c) Diffuser Only $\eta_D = 80\%$.

(d) C-D nozzle $\eta_N = \eta_D = 80\%$.

(e) Diffuser Only $\eta_D = 60\%$.

(f) C-D nozzle $\eta_N = \eta_D = 60\%$.

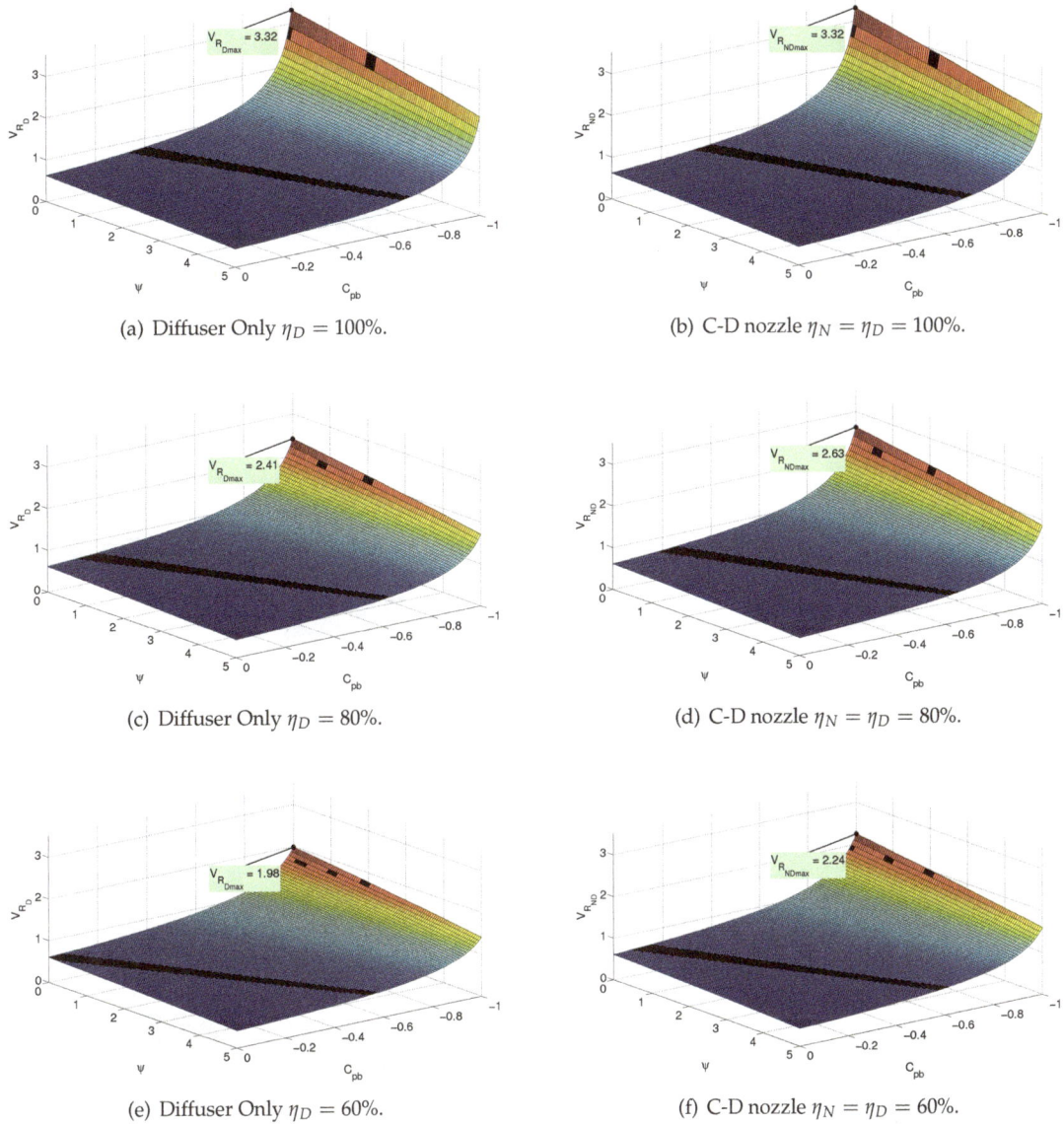

Figure 7. The effect of the pressure and loading coefficients on the velocity distribution of diffuser only and and converging-diverging (C-D) nozzle.

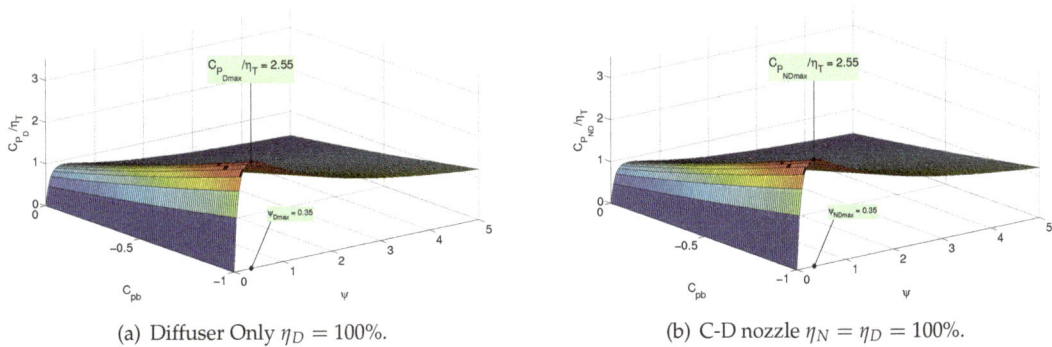

(a) Diffuser Only $\eta_D = 100\%$.

(b) C-D nozzle $\eta_N = \eta_D = 100\%$.

Figure 8. *Cont.*

(c) Diffuser Only $\eta_D = 80\%$.

(d) C-D nozzle $\eta_N = \eta_D = 80\%$.

(e) Diffuser Only $\eta_D = 60\%$.

(f) C-D nozzle $\eta_N = \eta_D = 60\%$.

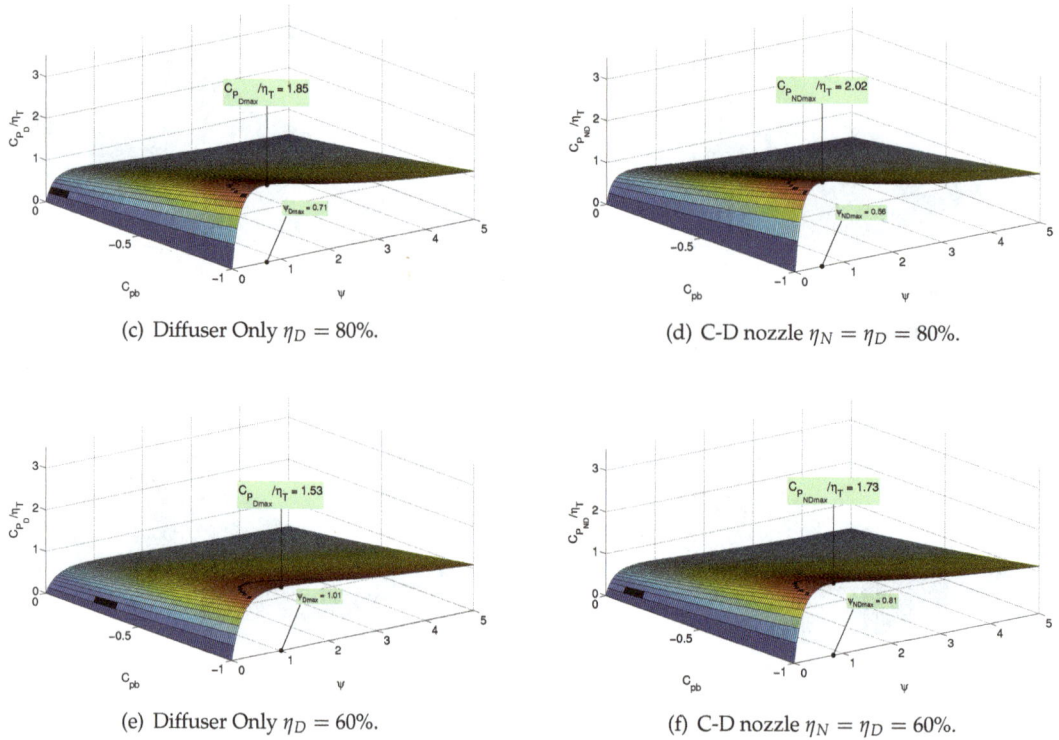

Figure 8. The effect of the pressure and loading coefficients on the power coefficient of diffuser only and C-D nozzle.

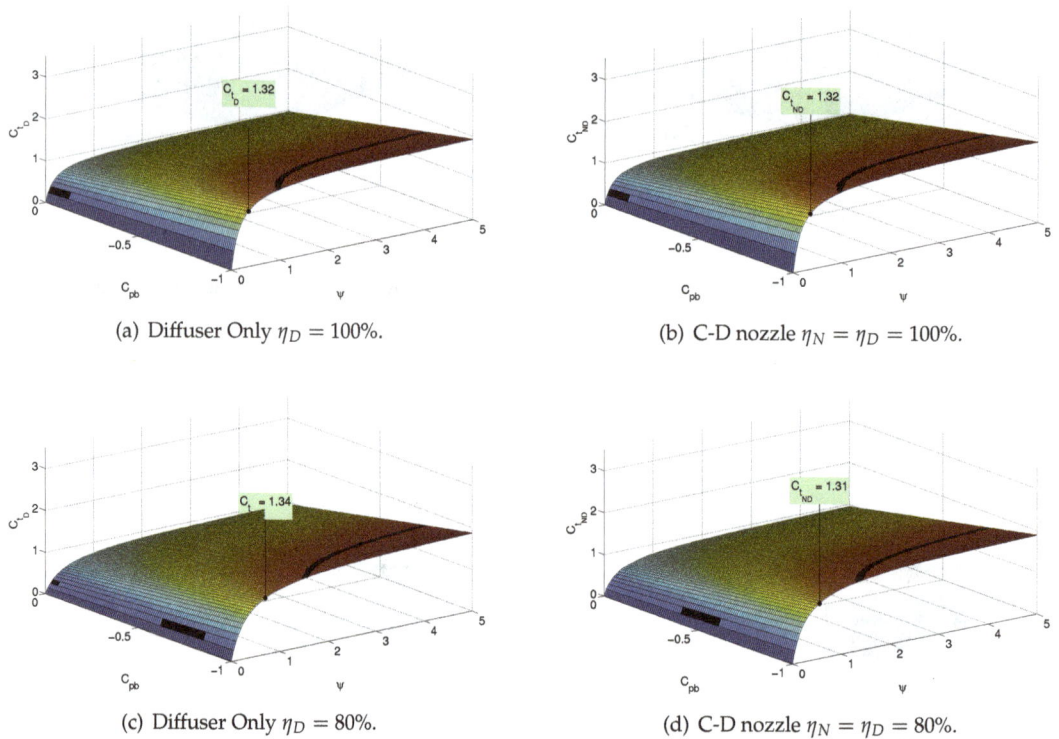

(a) Diffuser Only $\eta_D = 100\%$.

(b) C-D nozzle $\eta_N = \eta_D = 100\%$.

(c) Diffuser Only $\eta_D = 80\%$.

(d) C-D nozzle $\eta_N = \eta_D = 80\%$.

Figure 9. *Cont.*

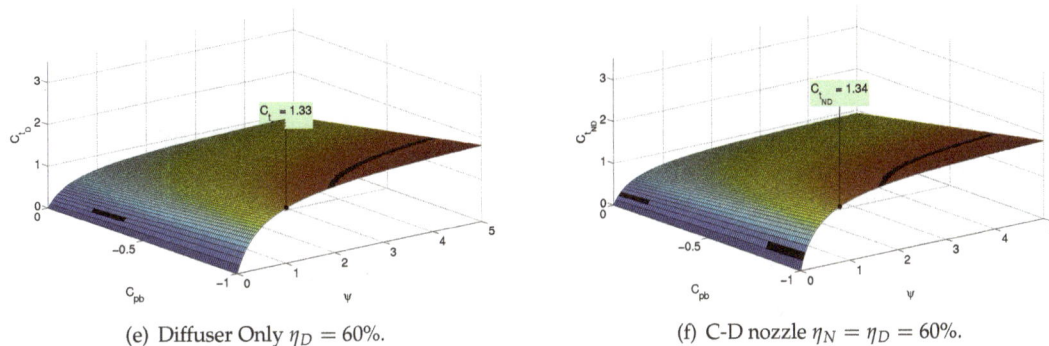

(e) Diffuser Only $\eta_D = 60\%$.

(f) C-D nozzle $\eta_N = \eta_D = 60\%$.

Figure 9. The effect of the pressure and loading coefficients on the thrust coefficient of diffuser only and C-D nozzle.

6. Results and Discussions

Equation (33) allows the relative velocity ratio, V_1/V_∞, through a shrouded wind turbine with nozzle to be computed as a function of the loading coefficient ψ, the back pressure C_{p_b}, as well as diffuser and nozzle efficiencies η_D, η_N, as illustrated in Figure 10. As can be inferred from Figure 10a, for perfect efficiencies of both the diffuser and nozzle, values above two for the velocity ratio can be obtained. It can also be noticed that higher values are achieved if the the wind turbine is operating at lower loading coefficients and lower back pressures. However, due to flow separation in the diffuser for larger exit areas, values for the back pressure coefficient below -0.7 can likely not be achieved. Similar behavior can also be observed at lower efficiencies, albeit with lower velocity ratios.

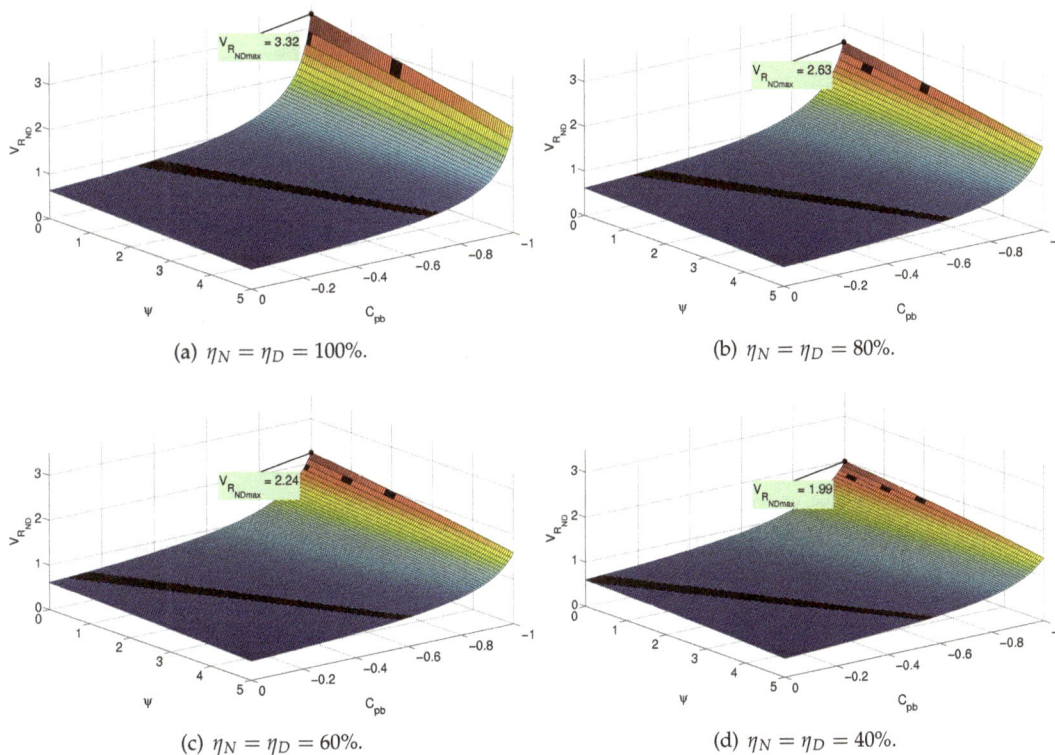

(a) $\eta_N = \eta_D = 100\%$.

(b) $\eta_N = \eta_D = 80\%$.

(c) $\eta_N = \eta_D = 60\%$.

(d) $\eta_N = \eta_D = 40\%$.

Figure 10. Plot of relative velocity ratios using $D_i = 0.76$ m, $D_1 = 0.7$ m and $D_e = 1.072$ m for different efficiencies using Equation (33).

In Figure 11a, cross-sectional view of the plots in Figure 10 are shown using $C_{p_b} = -1/3$, which corresponds to the back pressure coefficient for the Betz limit. One can observe how the nozzle and diffuser can increase the velocity ratio above the maximum of one for a bare wind turbine given by Equation (18).

As expected, when the area of the diffuser outlet and nozzle inlet area equal the rotor area Cases (2) and (3), the velocity ratio reduces approximately to one. On the other hand, when both areas of the nozzle inlet and the diffuser outlet are infinite in the limit of no load on the rotor Case (4), the velocity ratio could be theoretically infinite since the convergent-divergent nozzle or shrouded turbine is collecting from and expanding to an infinite area. In practice viscous effects such as flow separation make this behaviour impossible. The figure also shows the effect of the area ratio on the velocity ratio through Cases (6)–(7) for realistic efficiencies. It can be noticed that for values of the loading coefficients less than 0.75, it is conceivable to obtain higher velocity ratios [7].

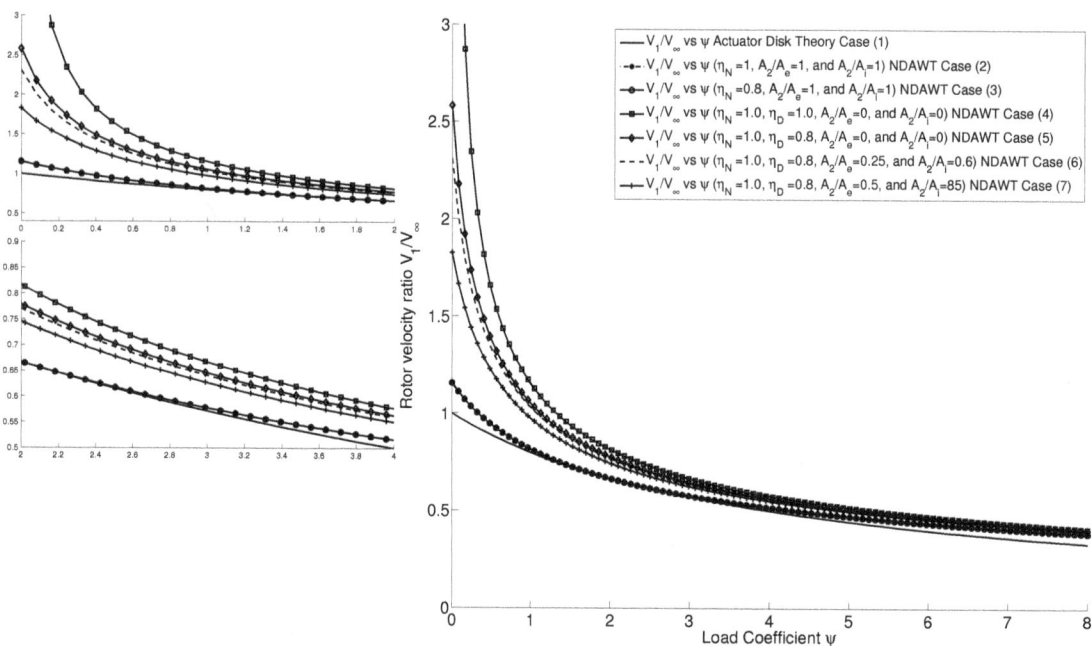

Figure 11. Cross-sectional view of the relative velocity ratio at $C_{p_b} = -1/3$.

Figure 12 provides the effect of the diffuser and nozzle efficiencies, the back pressure, as well as the loading coefficient on the power coefficient based on Equation (36). A significant observation is that the maximum power coefficient occurs at lower loading coefficients than that for the Betz limit of two. For realistic back pressure values, the power coefficient can be much higher than the one for a bare wind turbine of 16/27.

Figure 13 shows cross-sectional views of the plots in Figure 12 using again $C_{p_b} = -1/3$. It is easy to see that for a maximum power extraction for a given turbine blade area, a much more lightly loaded design should be chosen for the ducted case [7] compared to a conventional wind turbine and that the power extraction can be much higher.

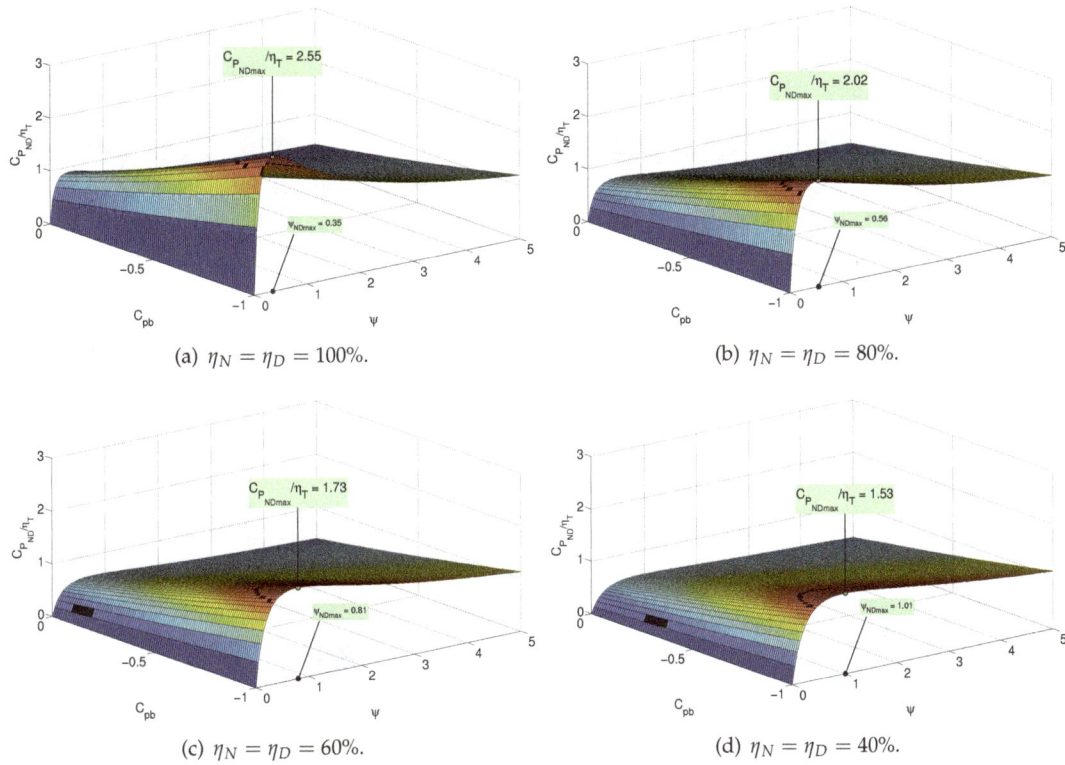

(a) $\eta_N = \eta_D = 100\%$.

(b) $\eta_N = \eta_D = 80\%$.

(c) $\eta_N = \eta_D = 60\%$.

(d) $\eta_N = \eta_D = 40\%$.

Figure 12. Plot of power coefficient using $D_i = 0.76$ m, $D_1 = 0.7$ m, and $D_e = 1.072$ m for different efficiencies using Equation (36).

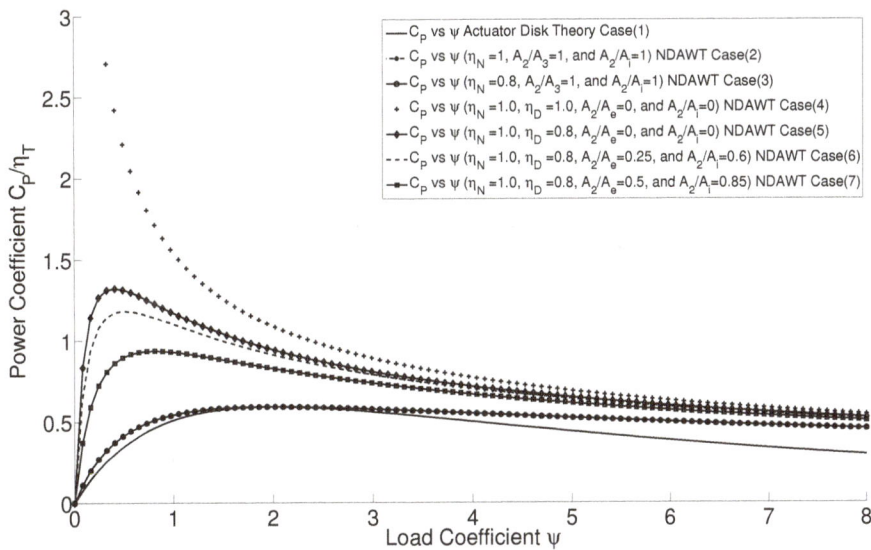

Figure 13. Cross-sectional view of the power coefficient at $C_{p_b} = -1/3$.

In Figure 14, the thrust coefficient is plotted for different values of the loading coefficient, the back pressure coefficient, as well as the nozzle and diffuser efficiencies η_D and η_N using Equation (34).

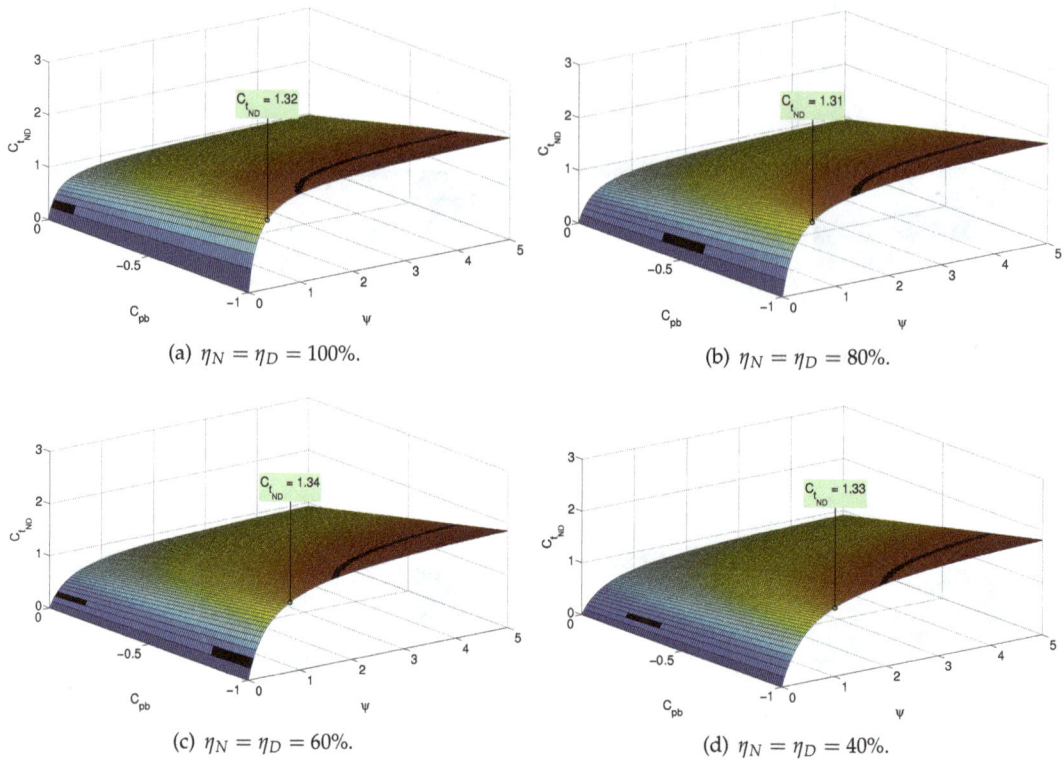

(a) $\eta_N = \eta_D = 100\%$.   (b) $\eta_N = \eta_D = 80\%$.

(c) $\eta_N = \eta_D = 60\%$.   (d) $\eta_N = \eta_D = 40\%$.

Figure 14. Plot of thrust coefficient using $D_i = 0.76$ m, $D_1 = 0.7$ m and $D_e = 1.072$ m for different efficiencies using Equation (34).

For unrealistic low values of the back pressure of $C_{p_b} = -1$, Figure 14a shows that for a loading coefficient of $\psi = 3$, the thrust coefficient could reach two instead of 8/9, as in the Betz limit. Furthermore, for realistic values of the back pressure of $C_{p_b} = -0.5$ and a loading coefficient of $\psi = 0.5$, it can be noticed that the thrust coefficient can be lower than 8/9. This makes sense since one obtains higher velocities at the throat, which lowers the pressure. Therefore, the thrust would be lower, since it is a function of the pressure. Finally, it can be observed through Figure 14a–d that as the efficiencies for both the diffuser and nozzle decrease, the thrust coefficient tends to decrease, as well.

Figure 15 displays cross-sectional views of the plots in Figure 14 using again $C_{p_b} = -1/3$. One should keep in mind that the thrust coefficient is the product of the velocity ratio times the loading coefficient. One can see that in Case (1), the maximum thrust happens at $\psi = 2$ at a value of 8/9. One can also notice that for infinite values of the nozzle (inlet) and diffuser (outlet) areas Case (4), a maximum thrust coefficient of 1.35 is obtained. Using realistic values of the area ratio in Cases (6) and (7), we see that values higher than 8/9 can be experienced by the wind turbine for higher loading coefficients. However, it is also possible to achieve a thrust coefficient lower than 8/9 for smaller loading coefficients. The reason to work with a lower thrust coefficient is to obtain higher velocities at the rotor plane and lower pressures at the exit plane [7].

Lastly, for preliminary design purposes, Figure 16 shows how the maximum power coefficient obtained at the optimal loading varies as a function of back pressure and diffuser area ratio. The designer should try to decrease the back pressure and diffuser area ratio as much as possible until viscous effects, such as flow separation, take over and render these results invalid.

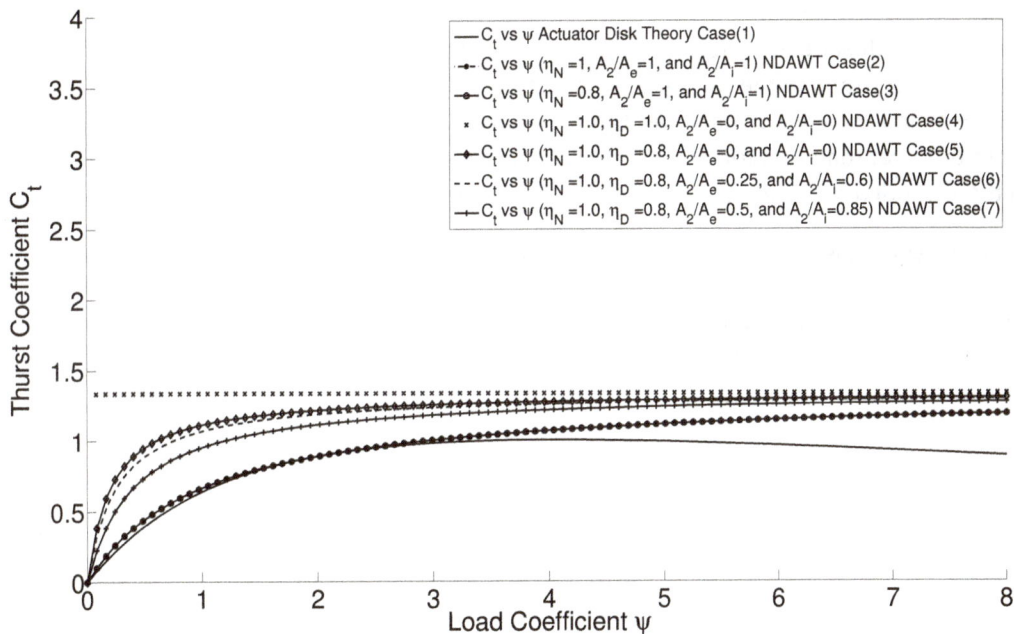

Figure 15. Cross-sectional view of the thrust coefficient at $C_{p_b} = -1/3$.

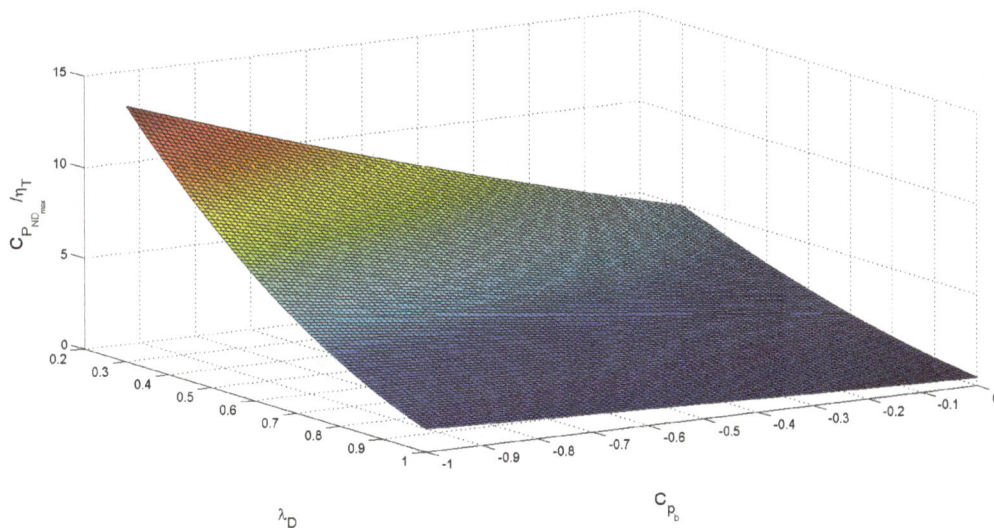

Figure 16. The maximum power coefficient as a function of back pressure and diffuser area ratio.

7. Conclusions

Numerous theoretical models were examined to predict the flow through shrouded turbines. It was found that some of the theories are only valid for short diffusers or predict incorrect values for certain area ratios due to incorrect assumptions. A more generalized theory based on work by Foreman [1] was developed in this article to be able to incorporate C-D nozzles into the shroud design. The developed generalized one-dimensional momentum theory showed reasonable agreement with experimental field data and CFD results from the literature.

It is evident from the developed equations that the nozzle and diffuser efficiencies, η_N and η_D, as well as the back pressure coefficient, C_{pb}, have the most significant impact on a nozzle diffuser-augmented wind turbine (NDAWT). Since these coefficients must be empirically determined, the developed generalized one-dimensional theory is, in practice, a semi-empirical theory. However, the theory provides a clear path for the preliminary design of NDAWT geometries, and the design focus should be on maximizing the nozzle and diffuser efficiencies, as well as lowering the back pressure at the diffuser exit with the help of, for example, flanges or brims.

Finally, according to the newly-developed generalized theory, it is shown that larger power coefficients than for conventional wind turbines can be achieved and that the maximum thrust could be higher or lower than that for a conventional wind turbine depending on the loading coefficient at which the shrouded turbine would operate.

Acknowledgments: This research was supported by the Libyan government of higher education. We are thankful to our colleagues Eric Lang and Kevin Hallinan who provided expertise that greatly assisted the research.

Author Contributions: Tariq Abdulsalam Khamlaj conducted the so-called NDAWT theory and wrote the article. Markus Peer Rumpfkeil proposed the idea and supervised the work.

Conflicts of Interest: The authors declare no conflict of interest.

References

1. Foreman, K.; Gilbert, B.; Oman, R. Diffuser augmentation of wind turbines. *Sol. Energy* **1978**, *20*, 305–311.
2. Betz, A. Energieumsetzungen in venturidüsen. *Naturwissenschaften* **1929**, *17*, 160–164.
3. Lilley, G.; Rainbird, W. *A Preliminary Report on the Design and Performance of Ducted Windmills*; College of Aeronautics College of Aeronautics Cranfield: Bedfordshire, UK, 1956.
4. Oman, R.; Foreman, K.; Gilbert, B. A progress report on the diffuser augmented wind turbine. In Proceedings of the 3rd Biennal Conference and Workshop on Wind Energy Conversion Systems, Washington, DC, USA, 9–11 June 1975; pp. 819–826.
5. Hansen, M.O.L.; Sørensen, N.N.; Flay, R. Effect of placing a diffuser around a wind turbine. *Wind Energy* **2000**, *3*, 207–213.
6. Ohya, Y.; Karasudani, T. A shrouded wind turbine generating high output power with wind-lens technology. *Energies* **2010**, *3*, 634–649.
7. Lawn, C. Optimization of the power output from ducted turbines. *Proc. Inst. Mech. Eng. Part A J. Power Energy* **2003**, *217*, 107–117.
8. Van Bussel, G.J. *The Science of Making More Torque from Wind: Diffuser Experiments and Theory Revisited*; IOP Publishing: Bristol, UK, 2007; Volume 75, p. 012010.
9. Jamieson, P. Generalized limits for energy extraction in a linear constant velocity flow field. *Wind Energy* **2008**, *11*, 445–457.
10. Werle, M.J.; Presz, W.M. Ducted wind/water turbines and propellers revisited. *J. Propuls. Power* **2008**, *24*, 1146–1150.
11. Konijn, F.B.J.; Hoeijmakers, H.W.M. *One Dimensional Flow Theory for Diffuser Augmented Wind Turbines*; University of Twente: Enschede, The Netherlands, 2010; Volume 217, p. 9.
12. Bergey, K. The Lanchester-Betz limit (energy conversion efficiency factor for windmills). *J. Energy* **1979**, *3*, 382–384.
13. Okulov, V.L.; van Kuik, G.A. The Betz–Joukowsky limit: On the contribution to rotor aerodynamics by the british, german and russian scientific schools. *Wind Energy* **2012**, *15*, 335–344.
14. Jamieson, P. *Innovation in Wind Turbine Design*; John Wiley & Sons: Hoboken, NJ, USA, 2011.
15. Manwell, J.F.; McGowan, J.G.; Rogers, A.L. *Wind Energy Explained: Theory, Design and Application*; John Wiley & Sons: Hoboken, NJ, USA, 2010.
16. Fox, R.W.; McDonald, A.; Pitchard, P. *Introduction to Fluid Mechanics*; John Wiley: Hoboken, NJ, USA, 2004.
17. Phillips, D.G. An Investigation on Diffuser Augmented Wind Turbine Design. Ph.D. Thesis, ResearchSpace@ Auckland, University of Auckland, Auckland, New Zealand, 2003.
18. Hansen, M.O. *Aerodynamics of Wind Turbines*; Routledge: London, UK, 2015.

5

A Dynamic Control Strategy for Hybrid Electric Vehicles Based on Parameter Optimization for Multiple Driving Cycles and Driving Pattern Recognition

Zhenzhen Lei [1,2], Dong Cheng [1], Yonggang Liu [1,2,*], Datong Qin [1], Yi Zhang [3] and Qingbo Xie [1]

[1] State Key Laboratory of Mechanical Transmissions & School of Automotive Engineering, Chongqing University, Chongqing 400044, China; zhenlei@umich.edu (Z.L.); chengdong_1991@163.com (D.C.); dtqin@cqu.edu.cn (D.Q.); xie_qingbo@126.com (Q.X.)

[2] Key Laboratory of Advanced Manufacture Technology for Automobile Parts, Ministry of Education, Chongqing University of Technology, Chongqing 400054, China

[3] Department of Mechanical Engineering, University of Michigan-Dearborn, Dearborn, MI 48128, USA; anding@umich.edu

* Correspondence: andylyg@umich.edu

Academic Editor: Hailong Li

Abstract: The driving pattern has an important influence on the parameter optimization of the energy management strategy (EMS) for hybrid electric vehicles (HEVs). A new algorithm using simulated annealing particle swarm optimization (SA-PSO) is proposed for parameter optimization of both the power system and control strategy of HEVs based on multiple driving cycles in order to realize the minimum fuel consumption without impairing the dynamic performance. Furthermore, taking the unknown of the actual driving cycle into consideration, an optimization method of the dynamic EMS based on driving pattern recognition is proposed in this paper. The simulation verifications for the optimized EMS based on multiple driving cycles and driving pattern recognition are carried out using Matlab/Simulink platform. The results show that compared with the original EMS, the former strategy reduces the fuel consumption by 4.36% and the latter one reduces the fuel consumption by 11.68%. A road test on the prototype vehicle is conducted and the effectiveness of the proposed EMS is validated by the test data.

Keywords: hybrid electric vehicles (HEVs); energy management strategy (EMS); particle swarm optimization (PSO); multiple driving cycles; driving pattern recognition

1. Introduction

To meet user demands for vehicle power performance, the parameters of hybrid electric vehicles (HEVs) are optimized to maintain the battery state of charge (SOC) and reduce the vehicle fuel consumption. This is not only related to the design parameters of the power system, but also the control parameters of the energy management strategy (EMS). To improve HEV performance in terms of fuel economy and ensure excellent driving performance, the simultaneous optimization for the main parameters of powertrain components and control system is necessary [1]. Recently, numerous works have been proposed to find the best solution. The genetic algorithm is used for the optimization of HEV control parameters which effectively improves the fuel economy [2–5]. The energy management algorithms based on adaptive multi-operating modes proposed in [6] solve the problem that different driving cycles should be provided with different control algorithms. Besides, the matching method of the powertrain based on driving cycles is presented for fuel cell

HEVs [7]. The above optimization algorithms are used to optimize the parameters of the power system or energy control strategy of HEVs. Several algorithms have been employed to optimize the parameters of both the power system and control strategy, such as the particle swarm optimization (PSO) algorithm [8–10] and multi-objective genetic algorithm [11]. A genetic algorithm with simulated annealing is proposed in [12] to balance between economy and dynamic performance. The DIRECT algorithm global optimization method has been used for calibrating the parameters of the vehicle EMS from the perspective of fuel economy [13]. Compared with the mentioned optimization algorithms, simulated annealing particle swarm optimization (SA-PSO) has the advantages of achieving a global optimal solution [14]. It is difficult but necessary to develop a set of global optimal solutions for the simultaneous optimization of power system and control parameters.

It's well known that the effectiveness of EMS for HEVs is greatly influenced by the driving patterns. However, the optimized parameters of HEVs based on a certain driving pattern may not maintain the battery SOC balance in other patterns, not to mention the best fuel consumption [15]. Therefore, energy management strategies based on driving pattern recognition have recently been put forward in the literature [16,17]. To optimize the vehicle performance on a random driving pattern, a multi-mode driving control algorithm using driving pattern recognition is developed for HEVs [18,19]. An intelligent energy management for parallel HEV based on driving cycle identification is proposed using a fuzzy logic controller or fuzzy neural network [20–22]. The machine-learning methods intelligently and automatically discriminate between the driving conditions [23,24]. To solve the multi-objective optimization problem for the longevity and energy efficiency of the energy storage system, a new optimization framework for determining an instantaneously optimized power management strategy has been proposed by Zhang et al. in [25], which shows excellent real-time power optimization performance against unknown diving cycles and operating conditions.

In these studies, the parameter optimization of the energy storage system, which is also very important for the effectiveness of EMS, is not taken into consideration. As mentioned above, the advantage of SA-PSO compared with other optimization algorithms is that it can obtain global optimization results, so it is meaningful to utilize the SA-PSO to realize the parameter optimization based on multiple driving cycles. Meanwhile the EMS based on driving pattern recognition should take advantage of the optimized parameters. However, few works have comprehensively analyzed how to combine the optimized parameters with the EMS based on driving pattern recognition. Besides, the EMS based on driving pattern recognition should emphasize more the influence of the variation range of battery SOC while focusing on the vehicle fuel economy. In general, the simultaneous optimization for parameters of power system and control strategy on this premise of maintaining balance of the battery SOC is worth studying and meaningful to improve the fuel economy.

In this paper, a new methodology for parameter optimization using a SA-PSO algorithm is proposed to pursue the best fuel consumption without impairing the dynamic performance. The parameters of the power system and control strategy for HEV are both optimized based on multiple driving cycles. In addition, an algorithm of the dynamic EMS based on driving pattern recognition is proposed in this paper. Twenty-three typical driving cycles from ADVISOR (2002, National Renewable Energy Laboratory, Golden, CO, USA) have been selected and classified according to the clustering analysis method through the *Euclidean distance*. Furthermore, the *Euclid approach degree* is used to realize the driving pattern recognition. The control parameters have been optimized at each class of driving patterns based on the optimization of multiple driving cycles. The proposed energy management strategies based on parameter optimization under multiple driving cycles and driving pattern recognition are both simulated on the Matlab/Simulink (R2010a, MathWorks, Natick, MA, USA) platform under the comprehensive driving cycles. Furthermore, road tests of the prototype vehicle with the proposed control strategy are conducted. The results of both the simulation and road tests validate the effectiveness of the proposed control strategies.

2. Hybrid Electric Vehicle (HEV) Rule-Based Energy Management Control Strategy

The hybrid power system considered in this paper is a typical parallel Integrated Starter and Generator (ISG) hybrid system, as shown in Figure 1. The engine and ISG motor are connected through a master clutch, and either of them can drive the vehicle alone. The ISG motor can also be used as a generator to charge the battery.

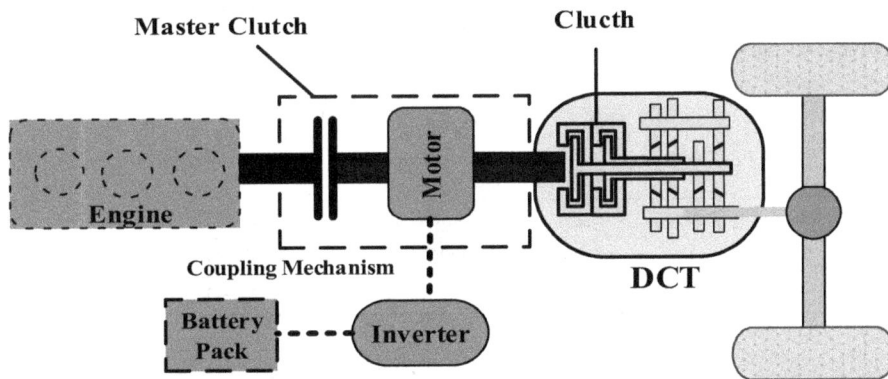

Figure 1. Configuration of the integrated starter and generator (ISG) type hybrid electric vehicle (HEV).

As shown in Figure 2, the basic control strategy in this paper is a rule-based logic threshold EMS which relies on several modes or states of operation and its decision to change modes is dependent on the power requirement of acceleration or deceleration, the SOC of the energy storage unit, and the vehicle speed [26,27]. In order to ensure that the engine operates more in high efficiency regions, in this paper, the coefficients of the engine torque in high efficiency regions (F_{up} and F_{low}) are designed to obtain the maximum and minimum engine torques based on the existing results presented in [28]. As shown in Figure 2a, when the battery SOC is higher than the low limit SOC_{low} and if the required speed is less than a certain value V_1, the vehicle will operate at pure electric mode. When the battery SOC is lower than SOC_{low} in Figure 2b, an additional torque T_{chg} is required from the engine to charge the battery. Therefore, the revised rule-based EMS is proposed as shown in Table 1. The parameters of the control strategy are shown in Table 2.

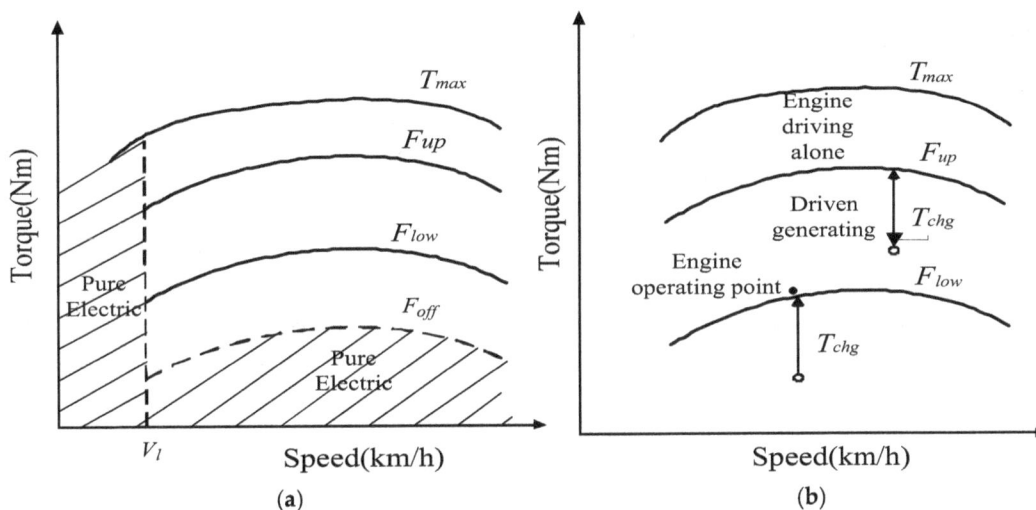

Figure 2. Logic diagram of control strategy. (**a**) $SOC > SOC_{low}$; (**b**) $SOC \leq SOC_{low}$.

Table 1. Revised rule-based energy management strategy (EMS).

Operating Mode	Constraint Condition	Torque Distribution
Electric Driving Mode	$0 < T_v \leq T_{off}$ $SOC_{up} > SOC > SOC_{low}; V > V_l$ $SOC_{up} > SOC > SOC_{low}; V \leq V_l$ $SOC > SOC_{up}$	$T_m = T_v; T_e = 0$
Driving & Charging Mode	$T_{off} < T_v \leq T_{low}$ $SOC_{low} \leq SOC \leq SOC_{up}; V > V_l$ $0 \leq T_v \leq T_{up}$	$T_e = T_{low}$ $T_m = T_v - T_{low}$ $T_e = T_{up}$ $T_m = \max(T_v - T_{up}, T_{chg\ max})$
Engine Driving Mode	$T_{low} \leq T_v \leq T_{up}$ $SOC_{up} > SOC > SOC_{low}; V > V_l$ $T_{up} \leq T_v; SOC \leq SOC_{low}$	$T_e = T_v; T_m = 0$
Motor Driving Mode	$T_{up} < T_v; SOC > SOC_{low}; V > V_l$	$T_e = T_{up}; T_m = T_v - T_{up}$
Regenerative Braking Mode	$T_v \leq 0; SOC < SOC_{up}$	$T_v = T_m + T_{mechanic}$
	$T_v \leq 0$ and $SOC > SOC_{up}$	$T_v = T_{mechanic}$

Table 2. Parameters of control strategy.

Name	Unit	Description
SOC_{up}	-	Maximum expectation of battery SOC
SOC_{low}	-	Minimum expectation of battery SOC
V	km/h	Current speed
V_l	km/h	Speed floor. When $SOC > SOC_{low}$ and $V < V_l$, pure electric mode starts
T_{max}	Nm	Maximum steady-state torque of engine
F_{off}	-	Engine off torque coefficient, $T_{off} = T_{max} \times F_{off}$
F_{low}	-	Minimum torque coefficient of engine in high efficiency regions, $T_{low} = T_{max} \times F_{low}$
F_{up}	-	Maximum torque coefficient of engine in high efficiency regions, $T_{up} = T_{max} \times F_{up}$
T_{chg}	Nm	Active charging torque of ISG motor. $T_{chg\ max}$ is the maximum charging torque of motor
T_v	Nm	Vehicle demand torque
T_m	Nm	Output torque of the ISG
T_e	Nm	Output torque of the engine
$T_{mechanic}$	Nm	Mechanic braking torque

3. Power System and Control Strategy Parameter Optimization Based on Multiple Driving Cycles

3.1. Basic Idea for Parameter Optimization

To pursue the best fuel consumption under actual driving cycle conditions, the parameter optimization of the power system and control strategy of HEV based on multiple driving cycles has been proposed. Six types of typical cycles are employed, considering the influence of urban congestion, suburban and highway conditions. The constraints of dynamic performance for the vehicle are shown in Table 3. The six types of driving cycles are shown in Table 4. The parameters of the vehicle's power system are shown in Table 5. The optimization method for the main parameters of power system and control strategy based on multiple driving cycles is generalized as follows:

(1) The assumption that the revised rule-based EMS is used for HEV.
(2) The initial parameters of power system and control strategy are selected and their values are chosen.
(3) Six types of driving cycles are selected and combined into a comprehensive driving cycle.
(4) The simultaneous optimization for the main parameters of power system and control strategy is carried out using SA-PSO algorithm with vehicle performance constraints.
(5) The optimal power system and control parameters are applied to the HEV EMS.

Table 3. Constraints of dynamic performance for the HEV.

Max. Speed		Max. Slope of Climb	Acceleration Time from 0 to 100 km/h
km/h		%	s
160 (Engine Driving Mode)	≥ 50 (Electric Driving Mode)	≥ 30 (Engine Driving Mode)	≤ 12 (Hybrid Driving Mode)

Table 4. Six types of typical driving cycles.

Mode	FTP	LA92	SC03	UDDS	HWFET	US06_HWY
Type	urban congestion		suburban		highway	

Table 5. Power source parameters of the HEV.

Description	Engine (P_{IC})	ISG Motor (P_{ISG})
Max Power (kW)	72	30
Max Torque (Nm)	137	115

The diagram of optimization method based on multiple driving cycles is shown in Figure 3.

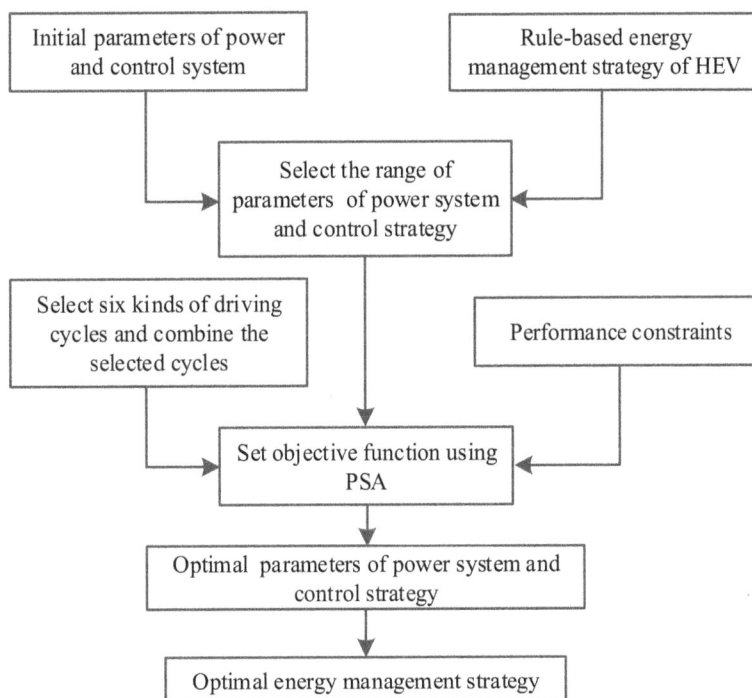

Figure 3. Diagram of optimization method under multiple driving cycles.

3.2. Parameter Definition of Power System and Control Strategy

The parameters of the power system and control strategy in terms of engine power (P_{IC}) and ISG power (P_{ISG}) are optimized in this paper to make sure that the engine and motor work in high efficiency regions on the premise of satisfying the requirements of vehicle dynamic performance. The variation of each design parameter of the power system (P_{IC} and P_{ISG}) is considered $\pm 70\%$ about the initial values, according to the results presented in [15]. The control parameters (F_{low}, F_{up}, F_{off}, SOC_{low}, SOC_{up} and V_l) are designed to ensure that the engine can work in high efficiency regions without interference with each other, as shown in Table 6. The initial values of selected parameters are obtained from the prototype vehicle.

Table 6. Variation of each parameter.

Optimal Variable	Initial Value	Variation Range
P_{IC} (kW)	72.0	21.6–122.4
P_{ISG} (kW)	30.0	9–51
F_{low}	0.6	0.43–0.73
F_{up}	0.9	0.75–0.93
F_{off}	0.235	0.2–0.4
SOC_{low}	0.25	0.2–0.4
SOC_{up}	0.8	0.75–0.9
V_l	32	10–50

3.3. State of Charge-Fuel Consumption Correction Method

In order to eliminate the influence of SOC on the vehicle fuel consumption evaluation, the battery SOC correction method should be used to correct fuel economy in the case initial and final battery SOC are not the same during a driving cycle. The SOC-fuel consumption correction method used in this paper is as follows:

$$\Delta fuel = \frac{\Delta SOC \cdot Q_{cap} \cdot \overline{U_{bat}} \cdot \overline{\eta_{eng\_chg}}}{1000 \cdot \rho} \tag{1}$$

where $\Delta fuel$ is the equivalent fuel consumption (L), ΔSOC is the variation of battery SOC between the starting and ending points, Q_{cap} is the total battery capacity (Ah), $\overline{U_{bat}}$ is the average battery bus voltage during drive cycles (V), $\overline{\eta_{eng\_chg}}$ is the average the engine power efficiency (g/kWh), and ρ is the gasoline density (g/L).

3.4. Optimization Objective Function

Taking the characteristics of different driving cycles into consideration, the target of parameter optimization of the power system and energy management control strategy is to achieve a set of optimal parameters to reduce fuel consumption as much as possible without impairing the dynamic performance. The fuel consumption is the optimization objective with the dynamic performance as the constraint. In order to prevent the excessive variation of battery SOC (ΔSOC), and specifically avoid exceeding the lower limit of SOC range, the weight coefficient of $\Delta fuel$ under different driving cycles is set to enable the motor to drive alone. The fitness function is as follows:

$$\text{Min } f(x) = \int Fuel_{use(t)} dt + \sum_{i=1}^{6} w_i \cdot |\Delta fuel_i| \tag{2}$$

$$s.t. \; u_j(x) \geq 0 \quad j = 1, 2, 3, ..., m$$
$$x_i^l \leq x_i \leq x_i^k \quad i = 1, 2, 3, ..., n$$

where $u_j(x)$ are the constraint conditions of vehicle dynamic performance (e.g., maximum speed and accelerating ability) as shown in Table 3, n is the number of optimization variables, which equals 8 in this study, x_i^l and x_i^k are the upper and lower bounds on the optimization variables respectively.

Considering the difference of the speed range and mileage of each driving cycle, the weight coefficients w_i of $\Delta fuel$ under driving cycles HWFET, FTP, LA92, US06_HWY, UDDS, SC03 through enumerative technique based on experience and simulation are chosen as 1.0, 1.0, 1.5, 1.3, 1.3 and 1.0, respectively.

3.5. Parameter Optimization for HEV Based on Simulated Annealing Particle Swarm Optimization Algorithm

The SA-PSO algorithm, firstly introduced by Metropolis et al. [29], is an optimization algorithm which combines the PSO with the Simulated Annealing method. This method has high efficiency in searching the global minimum value and the characteristics that it is easily realizable and has the advantages of both SA and PSO algorithms [30]. The particle swarm will gravitate towards the optimum solution after continuous iterations. All particles' positions and velocities are updated according to the following formulas:

$$v_i^{t+1} = w(t)v_i^t + c_1 r_1 (p_i^t - x_i^t) + c_2 r_2 (p_{gi}^t - x_i^t) \tag{3}$$

$$x_i^{t+1} = x_i^t + v_i^{t+1} \tag{4}$$

where p_i^t is the individual best optima for particle i after t iterations, p_{gi}^t is the group optima after t iterations, $w(t)$ is the inertia weight, c_1 and c_2 are two positive constants, $r_1 \in [0, 1]$ and $r_2 \in [0, 1]$ are two random parameters independent of each other, v_i^t is the velocity of particle i in iterative t, and x_i^t is the position of particle i in iterative t.

Based on the above analysis, the complete SA-PSO algorithm flowchart is shown in Figure 4.

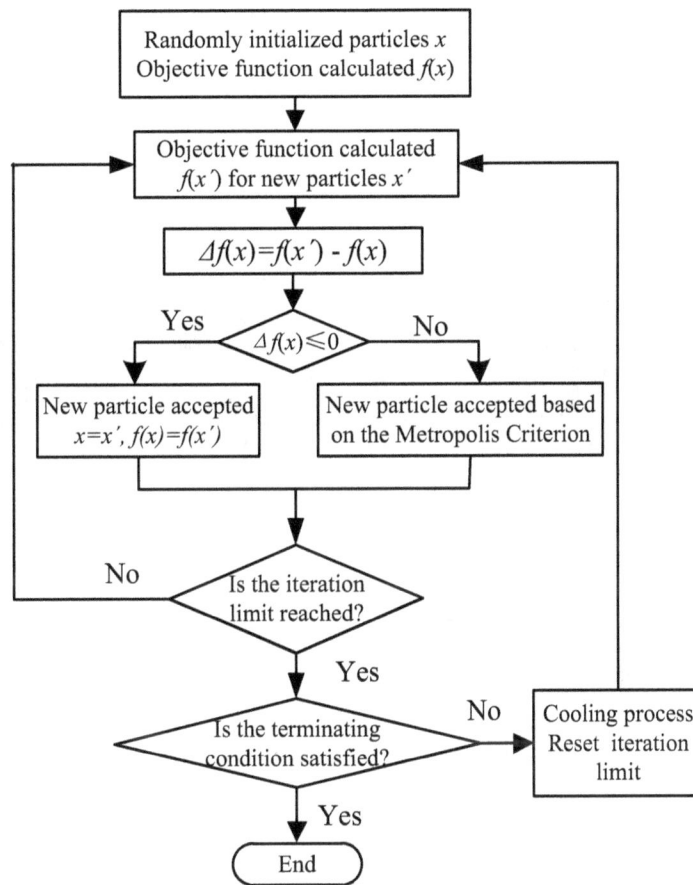

Figure 4. Optimization model based on the simulated annealing particle swarm optimization algorithm.

The detailed procedure of SA-PSO algorithm for parameter optimization is explained as follows:

Step 1: Initialize a group of random particles. The inertia should be chosen to provide a balance between the global and local exploration. The initialization consists of the following major parameters:

- Generation number: 25 Constants; c_1 and c_2: 2.05, 2.05; Initial temperature T: 9000 °C; Final temperature T_0: 0.05 °C; Anneal speed K: 0.9.

Step 2: Calculate and update the fitness function $f(x)$ of all particles. Determine p_i^t and p_{gi}^t of the current generation. Update new velocities and positions of each particle according to Equations (3) and (4).

Step 3: Calculate the difference between the optimal and non-optimal function value $\Delta f(x)$. Accept the optimal solution if $\Delta f(x)$ is greater than 0, otherwise generate a random number r within $(0, 1)$. When r is lower than $\min[1, \exp(-\Delta f(x)/t)]$, accept the optimal solution and go to Step 2, or go to the next step.

Step 4: Introduce the simulated annealing mechanism. Stop the program and output the optimal solution if the convergence criteria is satisfied, otherwise carry out the annealing process and the command "$T = 0.9 \times T$".

3.6. Simulation of Optimal Parameters Based on Multiple Driving Cycles

The simulation studies for the vehicle fuel economy are carried out using the Matlab/Simulink platform. The selected six types of driving cycles (HWFET, FTP, LA92, US06_HWY, UDDS, SC03) are successively combined into a comprehensive cycle according to driving cycles. The time-speed relationship of the comprehensive driving cycle is shown in Figure 5. The eight parameters of the power system and control strategy are optimized by the SA-PSO algorithm based on the comprehensive driving cycle, and the optimization results of the parameters are shown in Table 7.

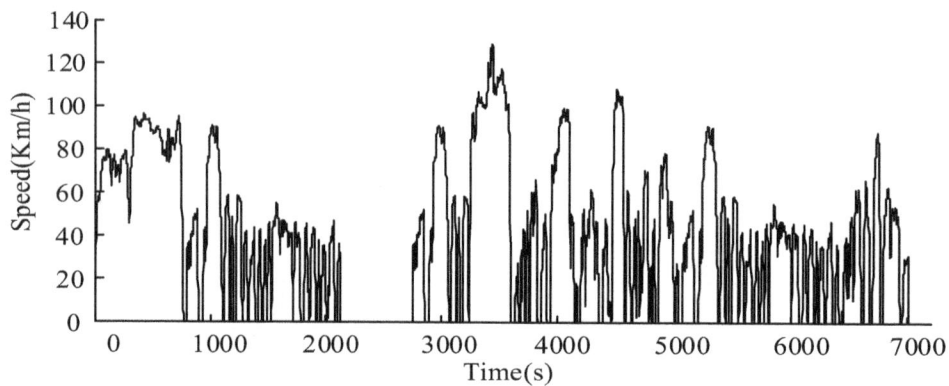

Figure 5. Time-speed relationship of the comprehensive driving cycle.

Table 7. Comparison of optimization results.

Optimal Variable	Initial Value	Optimal Value
P_{IC} (kW)	72.0	67.0
P_{ISG} (kW)	30.0	26.0
F_{low}	0.6	0.48
F_{up}	0.9	0.90
F_{off}	0.235	0.23
SOC_{low}	0.25	0.30
SOC_{up}	0.8	0.78
V_l	32	35.06

The optimized parameters satisfy the requirements of vehicle dynamic performance. The variation of the battery SOC and engine operation points of HEVs are simulated under the comprehensive cycle conditions, as shown Figures 6 and 7. The variation of battery SOC stays within 0.05 which meets the requirements for HEV in terms of the battery SOC consistency. Meanwhile, the battery SOC always fluctuates around the initial SOC value, which enables the battery to work in its high charging/discharging efficiency region. Furthermore, the engine can work in its high efficiency region and the vehicle can be driven in the electric driving mode with low speed and torque, which effectively improves the overall efficiency of whole system.

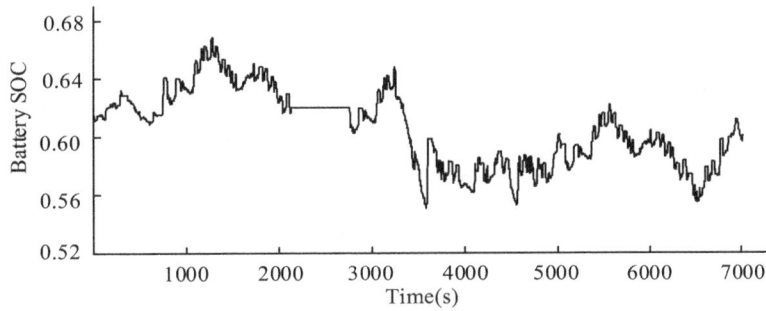

Figure 6. Variation of battery *SOC* under the comprehensive cycle conditions.

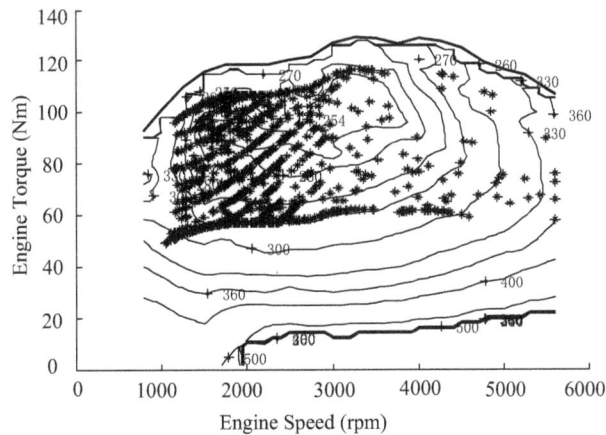

Figure 7. Engine operating points under the comprehensive cycle conditions.

4. HEV Dynamic Control Strategy Based on Driving Pattern Recognition

As mentioned above, the control parameters optimization based on multiple driving cycles is analyzed under known driving cycle conditions. However, in practice, the vehicle actual driving cycle is a random and uncertain process. In order to achieve better fuel economy, the EMS of HEVs based on driving pattern recognition is proposed after the parameter optimization under multiple driving cycles, which can optimize the control parameters in vehicle real-time control.

The diagram of EMS for HEVs based on driving pattern recognition is shown in Figure 8. Firstly, the characteristic parameters of different typical driving cycles are picked up, which are used for the clustering analysis. The control parameters of each class of the driving cycle are optimized offline based on multiple driving cycles as mentioned in Section 3. The driving pattern recognition has been realized using the *Euclid approach degree*. At last, the dynamic energy management control strategy for HEVs based on driving pattern recognition is achieved for vehicle real-time control.

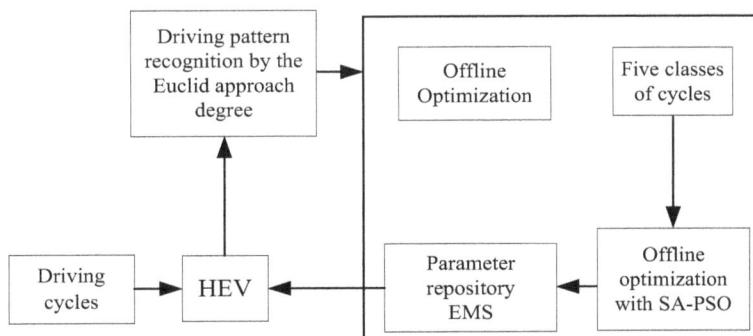

Figure 8. Diagram of EMS based on driving pattern recognition.

4.1. Selection and Classification of Characteristic Parameters for Typical Patterns

In view of the variety and complexity of vehicle driving patterns, it is significant to take all types of driving patterns into account. However, this is impractical due to the massive workload and limitation of calculation ability. Therefore, twenty-three typical driving cycles from ADVISOR are used as the research object. These driving cycles shown in Table 8 are Mode 1: JPN1015; Mode 2: ARTERIAL; Mode 3: CBD14; Mode 4: CBDTRUCK; Mode 5:COMMUTER; Mode 6: ECE_EUDC; Mode 7: HL07; Mode 8: LA92; Mode 9: MANHATTAN; Mode 10: NYCC; Mode 11: NYCCOMP; Mode 12: NYCTRUCK; Mode 13: NurembergR36; Mode 14: REP05; Mode 15: SC03; Mode 16: UDDS; Mode 17: UDDSHDV; Mode 18: US06_HWY; Mode 19: WVUCITY; Mode 20: WVUSUB; Mode 21: ARB02; Mode 22: ECE; Mode 23: IM240.

Table 8. Related characteristic parameters of twenty-three typical cycles.

Mode	v_{max}	v_{avg}	a_{max}	d_{max}	a_{avg}	d_{avg}	r_i
1	69.97	22.68	0.79	0.83	0.57	0.65	0.32
2	64.37	39.70	1.07	2.01	0.60	1.79	0.22
3	32.19	20.42	0.98	2.06	0.81	1.79	0.214
4	32.19	14.85	0.36	0.62	0.29	0.56	0.187
5	88.51	70.28	1.03	2.01	0.28	1.89	0.122
6	119.99	32.11	1.05	1.39	0.54	0.79	0.277
7	128.75	85.75	3.58	2.55	1.29	0.80	0.097
8	108.15	39.61	3.08	3.93	0.67	0.75	0.163
9	40.72	10.98	2.06	2.50	0.54	0.67	0.362
10	44.58	11.41	2.68	2.64	0.62	0.61	0.351
11	57.94	14.10	4.11	3.88	0.48	0.54	0.331
12	54.72	12.15	1.96	1.87	0.55	0.65	0.52
13	53.70	14.34	1.88	2.11	0.58	0.55	0.31
14	129.23	82.88	3.79	3.19	0.44	0.50	0.034
15	88.19	34.50	2.28	2.73	0.50	0.60	0.195
16	91.25	31.51	1.48	1.48	0.51	0.58	0.189
17	93.34	30.32	1.96	2.07	0.48	0.58	0.333
18	129.23	97.91	3.08	3.08	0.34	0.41	0.033
19	57.65	13.58	1.14	3.24	0.30	0.39	0.303
20	72.10	25.86	1.30	2.16	0.33	0.42	0.252
21	129.20	70.03	3.53	3.62	0.66	0.70	0.075
22	49.99	18.26	1.06	0.83	0.64	0.74	0.33
23	91.23	47.07	1.47	1.56	0.44	0.68	0.05

There have been some works in the literature about the selection of characteristic parameters for typical cycles [20–24]. Based on the relative importance of each parameter in driving pattern recognition, seven parameters are chosen as the characteristic parameters of driving pattern recognition in this paper. They are the maximum vehicle speed v_{max}, average vehicle speed v_{avg}, maximum acceleration a_{max}, maximum deceleration d_{max}, average acceleration a_{avg}, average deceleration d_{avg} and engine idle time ratio r_i, respectively. The *clustering analysis* is used for classification of the typical cycles. The distance between each two driving patterns of the twenty-three typical ones is calculated by the characteristic parameters with *Euclidean distance*, as expressed by Equation (5):

$$\begin{cases} \|y_i - y_j\| = \sqrt{\left(y_{i1} - y_{j1}\right)^2 + \left(y_{i2} - y_{j2}\right)^2 + ... + \left(y_{im} - y_{jm}\right)^2} \\ \qquad = \sqrt{\sum_{m=1}^{10} \left(y_{im} - y_{jm}\right)^2} \\ i \neq j \cap i, j \in Z^+ \cap i, j \in [1, 23] \end{cases} \qquad (5)$$

Before the calculation, the feature matrix of driving cycles and under-recognition cycles should be dealt with by min–max *Normalization* due to the inconsistency between the physical dimension and quantity of feature vectors of driving cycles, as described in Equation (6):

$$y_i' = \frac{y_i - y_{min}}{y_{max} - y_{min}} \tag{6}$$

where y_i is the original variable, y_{min} is the minimum value of unscaled variable and y_{max} is the maximum value of an unscaled variable. The feature vectors are scaled to the closed interval [0, 1].

The clustering-feature tree shown in Figure 9 is obtained through Statistical Product and Service Solutions (SPSS) software (20.0, IBM SPSS, New York, NY, USA). As the clustering scale of samples decreases and the sample space is more subtly divided, the driving cycles of each category become higher. In this paper, in order to ensure the similarity of each driving cycle and the accuracy of the classes, the twenty-three types of typical driving patterns are divided into five classes when the scale of clustering distance is 0.057 (the first class includes 6, 9–13, 19; the second class includes 1, 8, 17, 20, 22; the third class includes 4, 21, 23; the forth class includes 3, 7, 14; the five class includes 5 and 18).

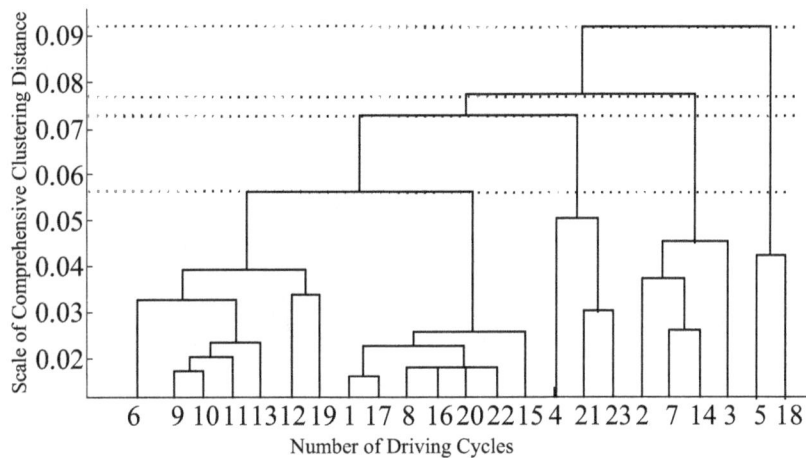

Figure 9. Clustering-feature tree of twenty-three typical patterns.

4.2. Recognition and Parameter Optimization of Driving Patterns

Although the actual vehicle driving patterns are random and uncertain, one of the twenty-three typical patterns can be selected to represent the actual driving pattern with the maximum similarity as the recognition result, and this is the basic idea of the dynamic control strategy for HEVs.

The driving pattern recognition is achieved using the *Euclid approach degree*. The representative feature vector $\mathbf{A}_n$ ($n = 1, 2, \ldots, 23$) stands for the selected twenty-three reference driving patterns, and each vector contains seven characteristic parameters of the reference driving patterns shown in Table 8. The vector $\mathbf{B}$ also contains seven characteristic parameters of the driving patterns. The distance between the feature vector of actual driving pattern and reference feature vectors is calculated by the *Euclidean distance* $\sigma(\mathbf{A}_n, \mathbf{B})$:

$$\sigma(\mathbf{A}_n, \mathbf{B}) = 1 - \frac{1}{\sqrt{m}} \left(\sum_{k=1}^{m} (\mathbf{A}_n(k) - \mathbf{B}(k))^2 \right)^{\frac{1}{2}} \tag{7}$$

where m is the number of the characteristic parameters ($m = 7$). In order to eliminate the deviation caused by different parameter units, parameters are standardized using the method of *Maximum magnitude* of 1.

The driving pattern showing the maximum similarity is recognized as the reference driving cycle as expressed:

$$\sigma(\mathbf{B}, \mathbf{A}_i) = \max\{\sigma(\mathbf{B}, \mathbf{A}_1), \sigma(\mathbf{B}, \mathbf{A}_2), ..., \sigma(\mathbf{B}, \mathbf{A}_n)\} \tag{8}$$

As shown in Equation (8), the result of driving pattern recognition means that the historical actual driving pattern $\mathbf{B}$ belongs to the driving pattern $\mathbf{A}_i$. In order to verify the effectiveness of driving pattern recognition, a comprehensive test driving cycle is established to represent the actual driving pattern. The comprehensive test driving cycle consists of five different types of typical cycles including NEDC, LA92, HWFET, UDDS and US06, as shown in Figure 10. An algorithm for real-time driving pattern recognition is proposed based on the assumption that the driving pattern will not change suddenly within a short period of time. This real-time driving pattern recognition algorithm can predict future driving cycles through the past sampling data analysis within a short time window. The time window for the information extraction of characteristic parameters is 120 s based on the research as presented in [31,32]. The recognition of driving patterns for each time window is realized using the *Euclid approach degree*. The result of driving pattern recognition under the comprehensive test driving cycles is shown in Figure 11.

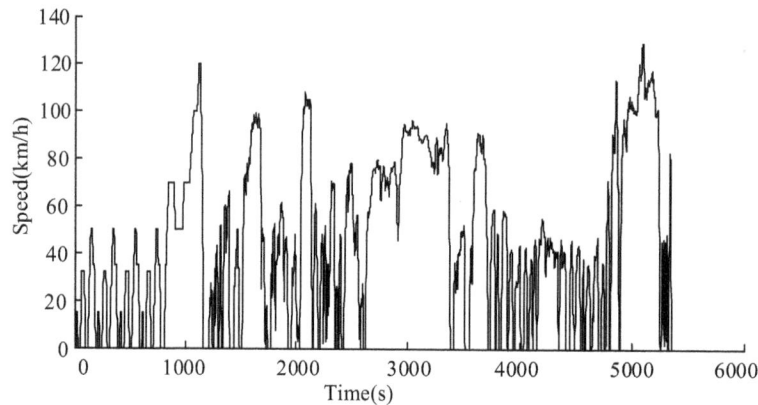

Figure 10. Time-speed relationship of the comprehensive test cycle conditions.

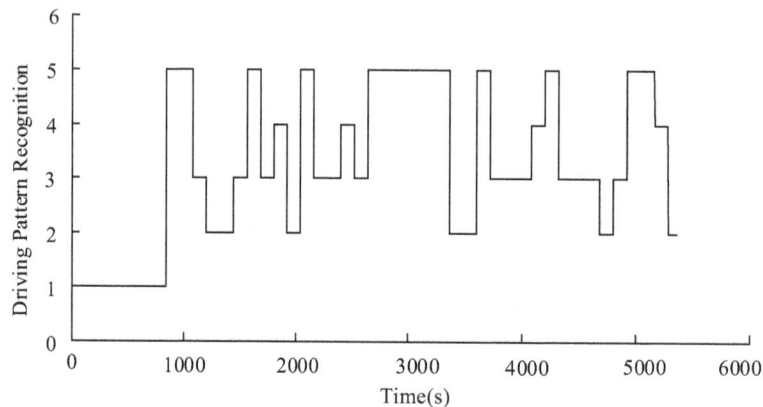

Figure 11. Result of driving pattern recognition under the comprehensive test cycle conditions.

4.3. Optimization of Control Parameters Based on Driving Pattern Recognition

In this section, the control parameters (F_{low}, F_{up}, F_{off} and V_1) of each class have been optimized based on multiple driving cycles, which has been introduced in Section 3 in detail. The optimization results of control parameters of each class are shown in Table 9.

Table 9. Optimization results of control parameters.

Classes	F_{low}	F_{high}	F_{off}	V_l
First	0.50	0.80	0.25	23.66
Second	0.48	0.90	0.23	35.06
Third	0.63	0.83	0.32	11.77
Forth	0.59	0.917	0.33	23.64
Fifth	0.68	0.80	0.40	15.98

In order to verify the effectiveness of the control parameter optimization, the driving cycles of the first class is taken as an example. The seven typical driving cycles of the first class are set as a comprehensive driving cycle (ECE_EUDC + MANHATTAN + NYCC + NYCCOMP + NYCTRUCK + NurembergR36 + WVUCITY) as shown in Figure 12. The variation of battery SOC (ΔSOC) in the first class of driving cycles is shown in Figure 13 and Table 10, where the control parameters are effective in controlling the variation of battery SOC.

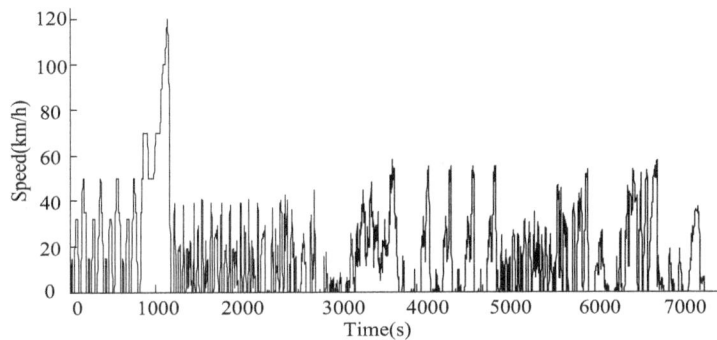

Figure 12. Time-speed relationship in the first class of driving cycles.

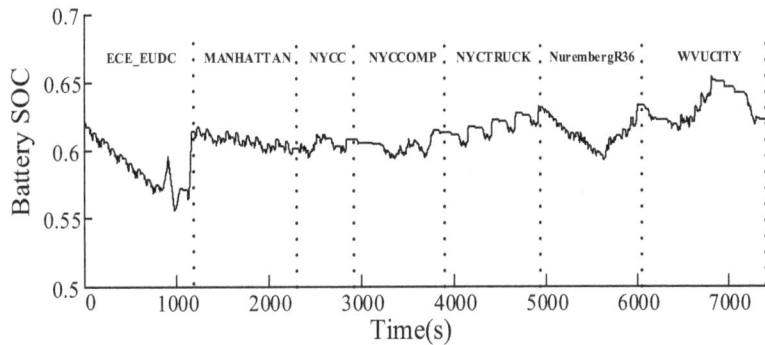

Figure 13. Variation of SOC in the first class of driving cycles.

Table 10. Variation of state of charge.

Mode	ECE_EUDC	MAN-HATTAN	NYCC	NYC-COMP
ΔSOC	−0.007	−0.017	0.01	0.009

Mode	NYC-TRUCK	NuremberR36	WVU-CITY	Comprehensive
ΔSOC	0.02	0.006	−0.015	0.005

5. Simulation

The proposed dynamic control strategy for HEVs based on parameter optimization at multiple driving cycles and driving pattern recognition has been simulated using the Matlab/Simulink platform under the comprehensive driving cycle (NEDC + LA92 + HWFET + UDDS + US06).

As shown in Figure 14, the variation of battery *SOC* stays within 0.01 under comprehensive driving cycle conditions with the proposed EMS based on driving pattern recognition. Meanwhile, the battery *SOC* always fluctuates around the initial *SOC* during the whole process, which enables the battery maintain to work in the high efficiency region. Compared with the EMS without driving pattern recognition, the battery *SOC* variation is more reasonable.

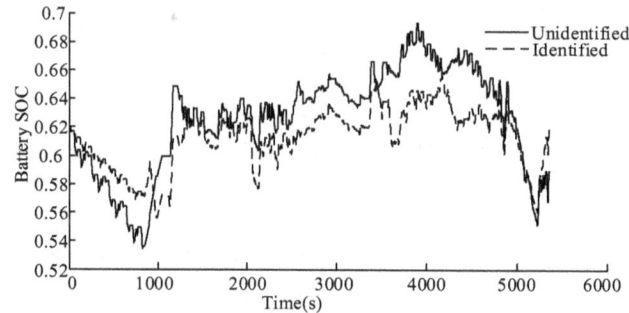

Figure 14. Variation of *SOC* under comprehensive driving cycle conditions.

Besides, the engine output power with the proposed EMS is generally larger than that with the EMS without driving pattern recognition, as shown in Figure 15. Therefore, the load of the engine is improved, which means that the engine will operate in higher efficient regions. The motor output torque at the comprehensive driving cycle is shown in Figure 16. The proposed EMS based on driving pattern recognition can adjust the control parameters to drive the vehicle in pure electric driving mode with low speed and torque, which prevents the engine from working in the low efficiency region and reduces fuel consumption. The reduction of the engine fuel consumption is shown in Figure 17.

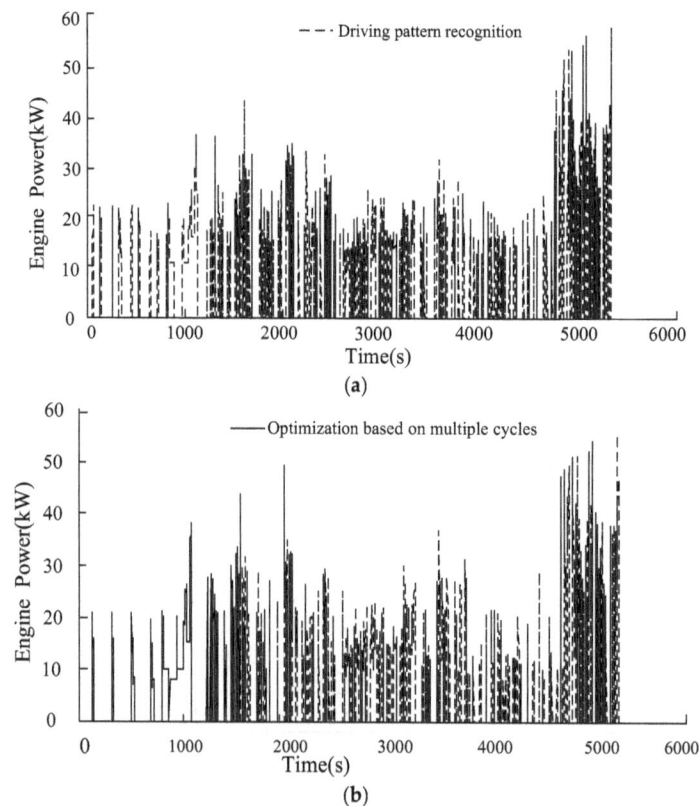

Figure 15. Engine power under the comprehensive driving cycle. (**a**) Engine power based on driving pattern recognition; (**b**) engine power based on multiple driving cycle optimization.

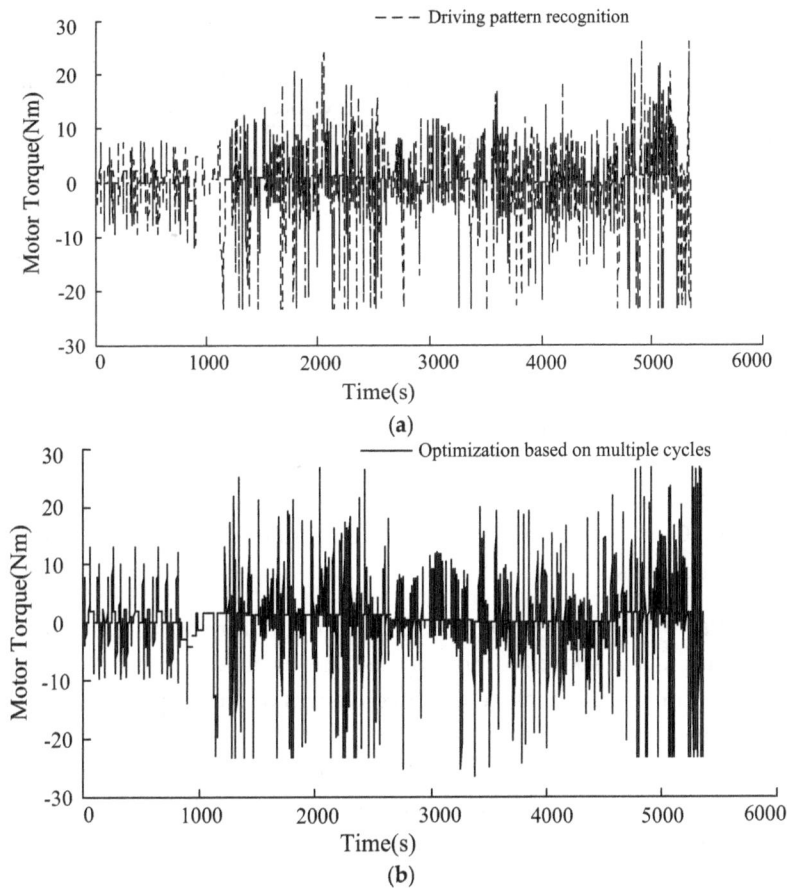

Figure 16. Motor torque under comprehensive driving cycle. (**a**) Motor power based on driving pattern recognition; (**b**) motor power based on multiple driving cycle optimization.

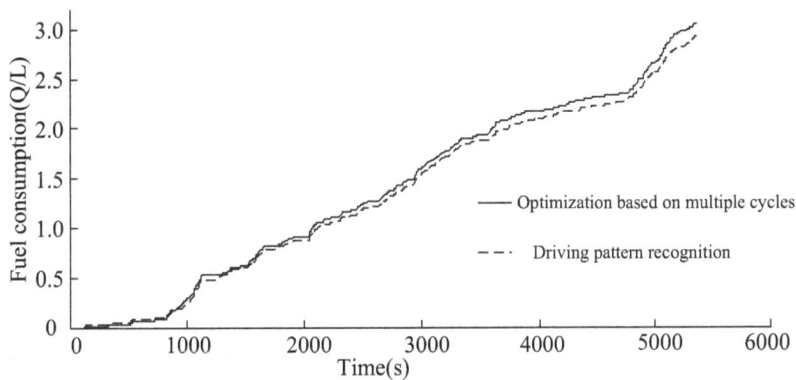

Figure 17. Fuel consumption under comprehensive driving cycle conditions.

The comparison results of fuel economy among control strategies of rule-based, multiple driving cycles optimization and driving pattern recognition are shown in Table 11 where Q_{100} is the fuel consumption of 100 km. For fuel economy comparison, SOC correction is very necessary. Therefore the SOC correction method in the SAE standards [33] is applied to compensate for the SOC difference. Compared with the rule-based control strategy, the fuel consumptions of energy management strategies based on multiple driving cycle optimization and driving pattern recognition are improved by 4.36% and 11.68%, respectively. Meanwhile the variation of battery SOC becomes smaller, which effectively improves the economic performance of the HEV vehicle.

Table 11. Fuel economy comparison results.

Factor	Rule-Based Control Strategy	Multiple Driving Cycles Optimization	Driving Pattern Recognition
Fuel Consumption (L)	3.23	3.10	2.96
Corrected Fuel Consumption (L)	3.43	3.28	3.03
Q_{100} (L/100 km)	4.75	4.55	4.36
Corrected Q_{100} (L/100 km)	5.05	4.83	4.46
Fuel Saving	-	4.24%	8.24%
Fuel Saving (SOC corrected)	-	4.36%	11.68%
ΔSOC	−0.28	−0.250	−0.098

6. Road Test on the Prototype Vehicle

The proposed dynamic control strategy for HEVs based on driving pattern recognition has been experimentally validated on a prototype HEV. The specifications of the prototype vehicle are shown in Table 12. The vehicle control software is developed on the Development to Production (D2P, DEV+PROD, Germany E.ON, Essen, Germany) and Matlab/Simulink platforms. The experiment is performed under the following conditions:

(1) Since the prototype HEV can only be tested on campus, for the sake of safety, the road test is only carried out at the low speed. Although the campus condition is only classified as an urban driving cycle, it is still valid to analyze the effectiveness of the proposed optimization method of HEV control strategy.
(2) The required torque during the whole test is too small compared with the maximum capacity of the HEV power system. To ensure that the vehicle operates in each mode without loss of generality, the parameters F_{off} = 0.20, F_{low} = 0.44, F_{up} = 0.64, V_l = 15 are designed as the optimal control strategy parameters according to the actual test conditions.

Table 12. Specifications of the prototype vehicle.

Main Parameter	Value
Curb weight (kg)	1350
Rated payload (kg)	1875
Effective radius (m)	0.295
Frontal area (m²)	2.28
Maximum engine torque (Nm)	137
Nominal motor power (kW)	20
Rated voltage (V)	288

The results of the road test have been presented in Figure 18 where the vehicle speed ranges from 0 to 45 km/h. The operation modes include the electric driving mode, driving & charging mode, engine driving mode and hybrid driving mode. The prototype HEV operates at electric driving mode during the starting process, and the small required torque prevents the engine from working in the low efficiency region. The engine driving mode is mostly activated during cruising (30–35 km/h).

The effectiveness of the control strategy proposed in this paper is well verified, as seen in Figure 18. The engine can operate in the designed operating region, which effectively improves the system efficiency. Meanwhile, the battery SOC fluctuates smoothly and the magnitude of SOC variation is only 0.005, which well meets the requirements to keep the battery SOC as constant as possible. The engine is able to work in tandem with the motor, so as to improve the vehicle economy.

In order to show the effectiveness of the proposed algorithm better, the comparison results of road tests among different control strategies have shown in Table 13. However, during the different road rests, the vehicle can't be ensured to operate under the same working conditions among the several road tests with different control strategies. Therefore, these comparison results are roughly taken as a reference.

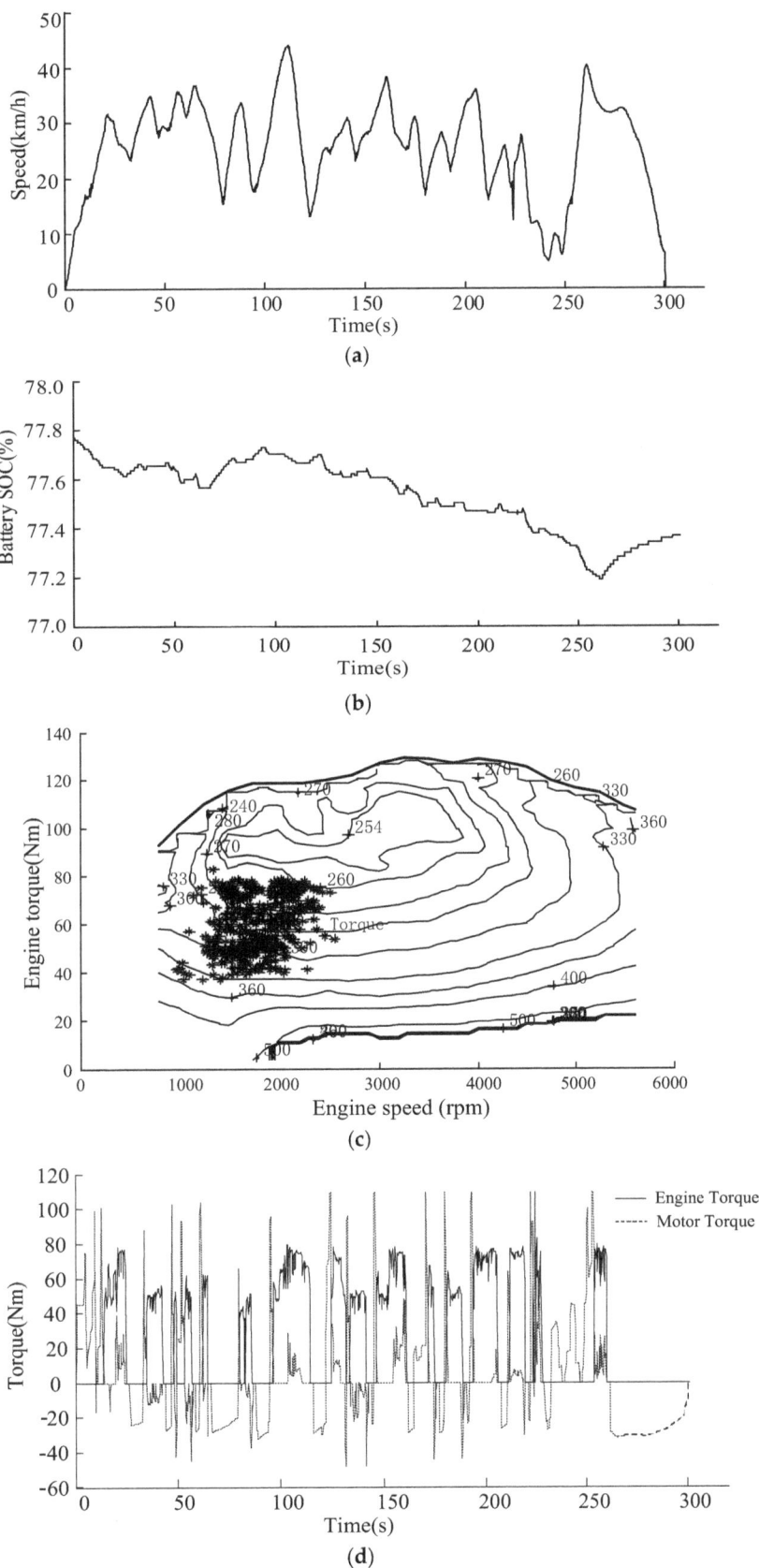

Figure 18. Road test results under the comprehensive driving cycle. (**a**) Time-speed curve of the road test; (**b**) variation of battery *SOC*; (**c**) engine operating points during testing; and (**d**) engine and motor toque distribution during testing.

Table 13. Road test comparison results.

Factor	Rule-Based Control Strategy	Multiple Driving Cycles Optimization	Driving Pattern Recognition
Total Mileage (km)	21.67	22.23	21.24
Fuel Consumption (L)	1.38	1.34	1.25
Corrected Fuel Consumption (L)	1.45	1.40	1.27
Corrected Q_{100} (L/100 km)	6.69	6.30	5.98

7. Conclusions

(1) A new methodology for parameter optimization under multiple driving cycles using SA-PSO algorithm is proposed to the simultaneous optimization for parameters of power system and control strategy. It's beneficial to achieve the best fuel consumption without impairing the dynamic performance.

(2) The EMS of HEVs based on driving pattern recognition, which optimizes the control parameters in real-time, is proposed after the parameter optimization under multiple driving cycle conditions. The proposed dynamic control strategy for HEVs based on parameter optimization under multiple driving cycles and driving pattern recognition has been simulated using Matlab/Simulink platform under the comprehensive driving cycle. Basically, the problem that the optimization based on a certain driving cycle cannot keep the battery *SOC* balance in other cycles has been solved in this paper.

(3) The simulation results show that compared with the original EMS, the former strategy reduces the fuel consumption by 4.36% and the latter one reduces the fuel consumption by 11.68%. The results validate the fact that the fuel consumption of EMS based on driving pattern recognition is greatly improved compared with that of the rule-based control strategy and more effective than that of multiple driving cycles. Meanwhile, the variation of battery *SOC* with the EMS based on driving pattern recognition is more reasonable than that of the optimization based on multiple driving cycles. It will serve as a guideline for calibrating the key parameters for road test.

(4) The proposed dynamic control strategy for HEVs based on driving pattern recognition is validated on a prototype HEV by a road test. The test results show that the EMS developed in this paper can effectively distribute the engine torque and motor torque, and significantly improve the fuel consumption of the vehicle. Furthermore, the battery *SOC* fluctuates smoothly and the battery *SOC* balance is well maintained during the test process. It will serve a reference role in dynamic control strategy for HEVs in real world.

Acknowledgments: The work presented in this paper is funded by the National Natural Science Foundation (No. 51305468), China Postdoctoral Science Foundation (No. 2016M602925XB), and the Fundamental Research Funds for the Central Universities (No. CDJZR14110005) and the Key Laboratory of Advanced Manufacture Technology for Automobile Parts, Ministry of Education (No. 2016KLMT06).

Author Contributions: Yonggang Liu provided algorithms and designed the experiments; Zhenzhen Lei wrote the paper and completed the simulation for case studies; Dong Cheng and Qingbo Xie performed the experiments and analyzed the data; Datong Qin and Yi Zhang conceived the structure and research direction of the paper.

Conflicts of Interest: The authors declare no conflict of interest.

References

1. Fang, L.; Qin, S.; Xu, G.; Li, T.; Zhu, K. Simultaneous optimization for hybrid electric vehicle parameters based on multi-objective genetic algorithms. *Energies* **2011**, *4*, 532–544. [CrossRef]
2. Montazeri-Gh, M.A. Application of genetic algorithm for optimization of control strategy in parallel hybrid electric vehicles. *J. Frankl. Inst.* **2006**, *343*, 420–435. [CrossRef]
3. Panday, A.; Bansal, H.O. Energy Management strategy for hybrid electric vehicles using genetic algorithm. *Renew. Sustain. Energy* **2016**, *8*, 741–746. [CrossRef]

4. Wang, J.; Wang, Q.; Wang, P.; Han, B. The optimization of control parameters for hybrid electric vehicles based on genetic algorithm. In Proceedings of the SAE World Congress, Detroit, MI, USA, 8–10 April 2014.

5. Varesi, K.; Radan, A. A Novel GA Based Technique for Optimizing Both the Design and Control Parameters in Parallel Passenger Hybrid Cars. *Int. Rev. Electr. Eng.* **2011**, *63*, 1279–1286.

6. Wang, J.; Wang, Q.; Zeng, X.; Zhou, N.; Li, L. *The Algorithmic Research of Multi-Operating Mode Energy Management System*; SAE Technical Paper 2013-01-0988; SAE International: Warrendale, PA, USA, 2013.

7. Gao, S.A.; Wang, X.M.; He, H.W.; Guo, H.Q.; Tang, H.L. Powertrain matching based on driving cycle for fuel cell hybrid electricvehicle. *Mech. Mater.* **2013**, *288*, 142–147. [CrossRef]

8. Wu, J.; Zhang, C.H.; Cui, N.X. PSO Algorithn-based parameter optimization for HEV powertrain and its control strategy. *Int. J. Automot. Technol.* **2008**, *38*, 53–59. [CrossRef]

9. Sun, F.; Xiong, R.; He, H. A systematic state-of-charge estimation framework for multi-cell battery pack in electric vehicles using bias correction technique. *Appl. Energy* **2016**, *162*, 1399–1409. [CrossRef]

10. Chen, Z.; Xiong, R.; Wang, K.; Jiao, B. Optimal energy management strategy of a plug-in hybrid electric vehicle based on a particle swarm optimization algorithm. *Energies* **2015**, *8*, 3661–3678. [CrossRef]

11. Fang, L.; Qin, S. Concurrent optimization for parameters of powertrain and control system of hybrid electric vehicle based on multi-objective genetic algorithms. In Proceedings of the 2006 SICE-ICASE International Joint Conference, Busan, Korea, 18–21 October 2006.

12. Li, L.; Zhang, Y.; Yang, C.; Jiao, X.; Zhang, L.; Song, J. Hybrid genetic algorithm-based optimization of powertrain and control parameters of plug-in hybrid electric bus. *J. Frankl. Inst.* **2015**, *352*, 776–801. [CrossRef]

13. Hao, J.; Yu, Z.; Zhao, Z.; Shen, P.; Zhan, X. optimization of key parameters of energy management strategy for hybrid electric vehicle using DIRECT algorithm. *Energies* **2016**, *9*, 997. [CrossRef]

14. Deng, Y.W.; Chen, K.L. Simulated Annealing Particle Swarm Algorithm Based Parameters Optimization for Hybrid Electric Vehicles. *Autom. Eng.* **2012**, *34*, 580–584.

15. Roy, H.K.; McGordon, A.; Jennings, P.A. A generalized powertrain design optimization methodology to reduce fuel economy variability in hybrid electric vehicles. *IEEE Trans. Veh. Technol.* **2014**, *63*, 1055–1070.

16. Zhang, S.; Xiong, R. Adaptive energy management of a plug-in hybrid electric vehicle based on driving pattern recognition and dynamic programming. *Appl. Energy* **2015**, *155*, 68–78. [CrossRef]

17. Chen, Z.; Xiong, R.; Cao, J. Particle swarm optimization-based optimal power management of plug-in hybrid electric vehicles considering uncertain driving conditions. *Energy* **2016**, *96*, 197–208. [CrossRef]

18. Jeon, S.I.; Jo, S.; Park, Y.; Lee, J. Multi-mode driving control of a parallel hybrid electric vehicle using driving pattern recognition. *J. Dyn. Syst. Meas. Control Trans. ASME* **2002**, *124*, 141–149. [CrossRef]

19. Lin, C.C.; Jeon, S.; Peng, H. Driving pattern recognition for control of hybrid electric trucks. *Veh. Syst. Dyn.* **2004**, *42*, 41–58. [CrossRef]

20. Tian, Y.; Zhang, X.; Zhang, L.; Zhang, X. Intelligent energy management based on driving cycle identification using fuzzy neural network. In Proceedings of the Second International Symposium on Computational Intelligence and Design, Changsha, China, 12–14 December 2009; pp. 501–504.

21. Dayeni, M.K.; Soleymani, M. Intelligent energy management of a fuel cell vehicle based on traffic condition recognition. *Clean Technol. Environ. Policy* **2016**, *18*, 1945–1960. [CrossRef]

22. Wang, J.; Wang, Q.N.; Zeng, X.H.; Wang, P.Y.; Wang, J.N. Driving cycle recognition neural network algorithm based on the sliding time window for hybrid electric vehicles. *Int. J. Automot. Technol.* **2015**, *16*, 685–695. [CrossRef]

23. Park, J.; Chen, Z.; Kiliaris, L.; Kuang, M.L.; Masrur, M.A.; Phillips, A.M.; Murphey, Y.L. Intelligent vehicle power control based on machine learning of optimal control parameters and prediction of road type and traffic congestion. *IEEE Trans. Veh. Technol.* **2009**, *58*, 4741–4756. [CrossRef]

24. Huang, X.; Tan, Y.; He, X. An Intelligent Multifeature Statistical Approach for the Discrimination of Driving Conditions of a Hybrid Electric Vehicle. *IEEE Trans. Intell. Transp. Syst.* **2010**, *12*, 1–13. [CrossRef]

25. Zhang, S.; Xiong, R.; Cao, J.Y. Battery durability and longevity based power management for plug-in hybrid electric vehicle with hybrid energy storage system. *Appl. Energy* **2016**, *179*, 316–328. [CrossRef]

26. Alonso, E.; Ruiz, J.; Astruc, D. Power Management Optimization of an Experimental Fuel Cell/Battery/Supercapacitor Hybrid System. *Energies* **2015**, *8*, 6302–6327.

27. Meintz, A.; Ferdowsi, M. Control strategy optimization for a parallel hybrid electric vehicle. In Proceedings of the 2008 IEEE Vehicle Power and Propulsion Conference (VPPC '08), Harbin, China, 3–5 September 2008.

28.   Wu, L.; Wang, Y.; Yuan, X.; Chen, Z. Multiobjective optimization of HEV fuel economy and emissions using the self-adaptive differential evolution algorithm. *IEEE Trans. Veh. Technol.* **2011**, *60*, 2458–2470. [CrossRef]

29.   Metropolis, N.; Rosenbluth, A.W.; Rosenbluth, M.N.; Teller, M.; Teller, E. Equation of state calculations by very fast computing machines. *J. Chem. Phys.* **1953**, *21*, 1087. [CrossRef]

30.   Shieh, H.L.; Kuo, C.C.; Chiang, C.M. Modified particle swarm optimization algorithm with simulated annealing behavior and its numerical verification. *Appl. Math. Comput.* **2011**, *218*, 4365–4383. [CrossRef]

31.   Johnson, V.H.; Wipke, K.B.; Rausen, D.J. HEV Control Strategy for Real Time Optimization on Fuel Economy and Emission. In Proceedings of the 2000 Future Car Congress, Arlington, VA, USA, 2–6 April 2000.

32.   Pisu, P.; Rizzoni, G. A comparative study of supervisory control strategies for hybrid electric vehicles. *IEEE Trans. Control Syst. Technol.* **2007**, *15*, 506–518. [CrossRef]

33.   Clark, N.; Xie, W.; Gautam, M.; Lyons, D.W.; Norton, P.; Balon, T. Hybrid Diesel-Electric Heavy Duty Bus Emissions: Benefits of Regeneration and Need for State of Charge Correction. In Proceedings of the 2000 International Fall Fuels and Lubricants Meeting and Exposition, Baltimore, MD, USA, 16–19 October 2000.

6

Improved Battery Parameter Estimation Method Considering Operating Scenarios for HEV/EV Applications

Jufeng Yang [1,2], **Bing Xia** [2,3], **Yunlong Shang** [2,4], **Wenxin Huang** [1,*] and **Chris Mi** [2,*]

[1] Department of Electrical Engineering, Nanjing University of Aeronautics and Astronautics, Nanjing 211106, China; jufeng.yang@mail.sdsu.edu

[2] Department of Electrical and Computer Engineering, San Diego State University, San Diego, CA 92182, USA; bixia@eng.ucsd.edu (B.X.); shangyunlong@mail.sdu.edu.cn (Y.S.)

[3] Department of Electrical and Computer Engineering, University of California San Diego, San Diego, CA 92093, USA

[4] School of Control Science and Engineering, Shandong University, Jinan 250061, China

[*] Correspondence: huangwx@nuaa.edu.cn (W.H.); cmi@sdsu.edu (C.M.);

Academic Editor: Rui Xiong

Abstract: This paper presents an improved battery parameter estimation method based on typical operating scenarios in hybrid electric vehicles and pure electric vehicles. Compared with the conventional estimation methods, the proposed method takes both the constant-current charging and the dynamic driving scenarios into account, and two separate sets of model parameters are estimated through different parts of the pulse-rest test. The model parameters for the constant-charging scenario are estimated from the data in the pulse-charging periods, while the model parameters for the dynamic driving scenario are estimated from the data in the rest periods, and the length of the fitted dataset is determined by the spectrum analysis of the load current. In addition, the unsaturated phenomenon caused by the long-term resistor-capacitor (RC) network is analyzed, and the initial voltage expressions of the RC networks in the fitting functions are improved to ensure a higher model fidelity. Simulation and experiment results validated the feasibility of the developed estimation method.

Keywords: lithium-ion battery; operating scenario; equivalent circuit modeling; parameter estimation

1. Introduction

Lithium-ion batteries have been widely used in the energy storage systems of hybrid electric vehicles (HEVs) and pure electric vehicles (EVs) because of their low self-discharge rate, high energy and power densities. To ensure the safe and reliable operation of lithium-ion batteries, the battery management system (BMS) is of significant importance. The main task of a BMS includes monitoring of critical states, fault diagnosis and thermal management [1–7].

1.1. Review of the Literature

The performance of a BMS is highly dependent on the accurate description of battery characteristics. Hence, a proper battery model, which can not only correctly characterize the electrochemical reaction processes, but also be easily implemented in embedded microcontrollers, is necessary for a high-performance BMS. There are two common forms of battery models available in the literature: the electrochemical model and the equivalent circuit model (ECM). The electrochemical model expresses the fundamental electrochemical reactions by complex nonlinear partial differential algebraic equations (PDAEs) [8]. It can accurately capture the characteristics of the battery, but requires extensive computational power to obtain the solutions of the equations. Hence, such models are suitable for the battery design rather than the system level simulation. In contrast, the ECM abstracts away the detailed internal electrochemical reactions and characterizes them solely by simple electrical components; thus, it is ideal for circuit simulation software and implementation in embedded microcontrollers. The accuracy of the ECM is highly dependent on the model structure and model parameters. Theoretically, a higher order ECM can represent a wider bandwidth of the battery application and can generate more accurate voltage estimation results. However, the high order ECM can not only increase the computational burden, but also reduce the numerical stability for the further battery states' estimation [9,10]. Hence, considering a tradeoff among the model fidelity, the computational burden and the numerical stability, the second order ECM is employed in this paper [11–18]. The common structure of the second order ECM is illustrated in the top subfigure of Figure 1, where the open circuit voltage (OCV), which is a function of state of charge (SoC), stands for the open circuit voltage, R_{in} is the internal resistance, which represents the conduction and charge transfer processes [19–21], and two resistor-capacitor (RC) networks approximately describe the diffusion process. Among them, the short-term RC network models the fast dynamics diffusion process (Part A in the bottom subfigure of Figure 1), and the long-term RC network represents the slow dynamics diffusion process (Part B in the bottom subfigure of Figure 1). The above model parameters can be identified either through the time-domain or the frequency-domain parameter extraction experiments. For the time-domain parameter estimation methods, model parameters are usually identified through fitting the voltage response from the parameter extraction experiment with the exponential-based functions. The electrochemical impedance spectroscopy (EIS) test is the commonly-used frequency-domain parameter extraction experiment. Compared to the time-domain test process, one limitation of the EIS test is that the amplitude of the current excitation is so low that the battery can be considered as equalized during the whole test process, which seldom happens in HEV/EV applications. In order to overcome the above drawback, references [22–24] propose superimposing the direct current (DC) offset over the EIS signals to determine the current dependency of impedance parameters. However, since significant time is required for the EIS test, the battery SoC changes significantly during the test procedure if the amplitude of the superimposed current is improper. This can reduce the parameter estimation accuracy and make this method practically not applicable at moderate and high current rates [25,26]. Based on the aforementioned analysis, the second order ECM with parameters estimated by the time-domain analysis is discussed in this paper.

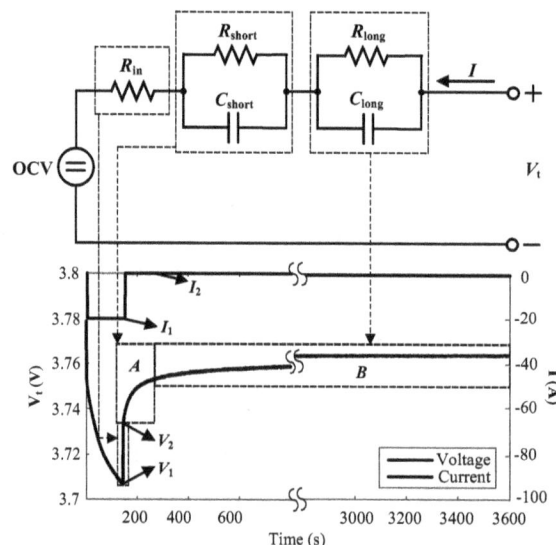

Figure 1. The second order equivalent circuit model (ECM). OCV, open circuit voltage.

Generally speaking, batteries usually operate in two scenarios in automotive applications: The constant-current (CC) charging scenario and the dynamic driving scenario [27]. Usually, the motions of lithium ions under the continuous external excitation (representing the CC charging scenario) and the discontinuous external excitation (representing the dynamic driving scenario) show different characteristics, and this difference is related to the diffusivity of ions. In other words, the model parameters, especially the RC network parameters, show diverse values under different operating scenarios [21,28]. Therefore, battery parameters should be identified separately according to the actual operating scenarios. Abundant research work has been conducted to seek the accurate ECM for the specific operating scenario. For the charging scenario, a universal model based on a simple mathematical equation with constant parameters is proposed [29–31]. The mathematical equations include one polynomial component and one or two exponential functions, and relevant parameters can be obtained by fitting collected charging profiles. Verification results in related literature show that the overall model output profiles match well with the experimental data, but there still exists obvious estimation errors during certain periods (at the beginning of the plateau region and the last charging region). This is mainly caused by the constant parameters during the whole charging process since the actual model parameters, such as time constants, may vary greatly at different SoC regions [32]. The works in [32–34] estimate the model parameters through the data in the rest periods of the pulse-rest test at different SoC points, and the estimated model parameters can be shown as functions of SoC. However, the charging concentration process under continuous excitation is different from the charging recovery process under the rest period [19,35]; thus, the estimated model parameters may not accurately represent the charging characteristics of the battery. For the dynamic driving scenario, many modeling approaches have been reported on the basis of the pulse discharge analysis. In [36–38], model parameters are obtained by simple algebraic operations. This is straightforward, but large estimation errors exist. A more accurate method is to fit the voltage response of the whole rest period with an exponential function [39–41]. The limitation of this method is its poor dynamic performance. In order to improve the battery model accuracy, Hu and Wang in [42] propose a two time-scale identification algorithm to separate the identifications of slow and fast battery dynamics. This method shows better frequency response matching without increasing computational complexity. Xiong in [17] uses the bias correction method to ensure the battery model prediction performance. This approach shows excellent performance and high accuracy against uncertain operating scenarios and battery packs. Instead of the conventional pulse-rest test, [43,44] propose two types of application-oriented parameter extraction tests, leading to a fast dynamics battery model with high fidelity. One major limitation of this kind of method is that the parameter extraction test corresponds to a specific operating scenario. If the actual load profiles show obviously different bandwidths under different working conditions, the parameter extraction test should be re-implemented. One solution to overcome this drawback is to conduct as many parameter extraction tests as possible to cover the typical load characteristics, but this requires an extensive amount of time and effort.

1.2. Contributions of This Paper

Based on the battery parameter estimation methods discussed above, it can be concluded that seldom does work in the previous literature discuss a battery model considering both the CC charging and dynamic driving scenarios. Hence, the focus of this paper is to propose a battery parameter estimation method, which is applicable to common operating scenarios in HEV/EV applications. The main contributions are: (1) both the constant-current charging and the dynamic driving scenarios are taken into consideration, and two separate sets of model parameters are estimated through different parts of the pulse-rest test; (2) the model parameters for the constant-current charging scenario are estimated from the data in the pulse-charging periods; (3) the model parameters for the dynamic driving scenario are estimated from the data in the rest periods, and the length of the fitted dataset is determined by the spectrum analysis of the load current; (4) the unsaturated phenomenon caused by the long-term RC network is analyzed, and the initial voltage expressions of the RC networks in the fitting functions are improved to ensure a higher model fidelity; (5) both the simulation and experiment results agree with the analysis and demonstrate the improvement of the proposed battery parameter estimation method over the existing ones.

2. Parameter Extraction Procedure

2.1. Parameter Extraction Test Design

It can be seen from Figure 1 that the second order ECM contains one *OCV-SoC* relationship and five impedance parameters (R_{in}, R_{short}, C_{short}, R_{long} and C_{long}), which need to be estimated. Theoretically, all of the impedance parameters mentioned above should be multivariable functions of *SoC*, the C-rate of the load current (C is the amplitude of the current with which the battery can be fully discharged in 1 h), temperature and cycle numbers [39,45]. These functions not only make the parameter extraction process complex and time consuming, but also increase the computational burden of the BMS. Hence, within certain error tolerance, some relationships can be simplified or ignored. Usually, aging periods are generally in the range of months to years. While for the system-level simulations of automotive applications, the time periods of interest are typically in the range of seconds to hours or days in special cases [43,45]. Hence, the long-term aging effect is usually ignored in the parameter estimation process and handled separately in most cases [39,46].

In this paper, all of the model parameters are estimated through the discharging/charging pulse-rest test at room temperature (22 °C–25 °C). A lithium-ion polymer battery with nickel-manganese-cobalt-based cathode and graphite-based anode is under test. Its specifications are given in Table 1, and the detailed experimental steps are described as follows.

Table 1. Specification of the tested battery.

Charge Capacity	40.99 Ah
Discharge capacity	40.89 Ah
Nominal voltage	3.7 V
Charge cutoff voltage	4.2 V
Discharge cutoff voltage	2.7 V

The discharging pulse-rest test starts with a fully-charged battery. In each cycle of the test, the battery is discharged at a 2% *SoC* step with C/2 constant current, then followed by a rest period. This cycle is repeated until the battery is fully discharged. Data points (including current, voltage, charging capacity and discharging capacity) are collected with the sampling frequency of 1 Hz. The relevant voltage and current profiles of the discharging pulse-rest test during the 66%–64% *SoC* interval are plotted in the bottom subfigure of Figure 1. The charging pulse-rest test is conducted similarly, that is it begins with a fully-discharged battery, then charged at a 2% *SoC* step with C/2 constant current and followed by a rest period. In order to eliminate the polarization voltage, the *OCV* values are extracted at the end of each rest period. Too short a rest time leads to a large *OCV* estimation error, whereas too long a rest time makes the whole test time consuming. It has been shown previously that for the lithium-ion polymer batteries, electrochemical reactions are negligible after a 2-h rest period [47,48]. Therefore, the rest time in this paper is predetermined as 2 h.

2.2. Parameter Estimation Algorithm

The electrical behavior of the ECM is expressed as the following state space formalism:

$$\begin{bmatrix} dV_{RC,short}/dt \\ dV_{RC,long}/dt \end{bmatrix} = \begin{bmatrix} -1/R_{short}C_{short} & 0 \\ 0 & -1/R_{long}C_{long} \end{bmatrix} \begin{bmatrix} V_{RC,short} \\ V_{RC,long} \end{bmatrix} + \begin{bmatrix} 1/C_{short} \\ 1/C_{long} \end{bmatrix} I \quad (1)$$

$$V_t = OCV(SoC) + IR_{in} + V_{RC,short} + V_{RC,long} \quad (2)$$

where Equation (1) is the state equation and Equation (2) is the output equation, $V_{RC,short}$ and $V_{RC,long}$ represent the voltages across the short-term and the long-term RC networks, respectively, $OCV(SoC)$ is an eighth-order polynomial equation as a function of SoC, V_t is the battery terminal voltage and the positive current I represents charging. R_{in} represents the internal resistance; R_{short} and R_{long} denote the diffusion resistances; and C_{short} and C_{long} represent the diffusion capacitances. Among them, R_{in} can be directly obtained from each pulse-rest cycle through Equation (3); the corresponding four variables (V_1, V_2, I_1 and I_2) are marked in the bottom subfigure of Figure 1, and the variation of identified R_{in} with SoC is shown in Figure 2. SoC can be calculated through Equation (4), in which C_{ap} denotes the capacity of the battery in Ah.

$$R_{in} = \frac{V_2 - V_1}{I_2 - I_1} \tag{3}$$

$$SoC = SoC(0) + \frac{1}{3600C_{ap}} \int_0^t I(\tau)d\tau \tag{4}$$

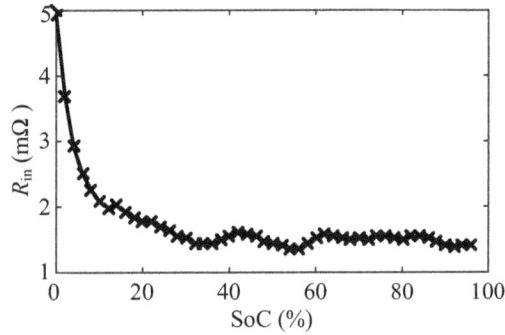

Figure 2. R_{in} variation with different state of charge (SoC).

For the CC operating scenario ($I \neq 0$), the analytical solutions of Equation (1) are derived as:

$$\begin{cases} V_{RC,short}(t) = V_{RC,short}(0)e^{-\frac{t}{\tau_{short}}} + IR_{short}(1 - e^{-\frac{t}{\tau_{short}}}) \\ V_{RC,long}(t) = V_{RC,long}(0)e^{-\frac{t}{\tau_{long}}} + IR_{long}(1 - e^{-\frac{t}{\tau_{long}}}) \end{cases} \tag{5}$$

where $V_{RC,short}(0)$ and $V_{RC,long}(0)$ are the initial voltages of corresponding RC networks and $\tau_{short} = R_{short}C_{short}$, $\tau_{long} = R_{long}C_{long}$, which represent the short-term and the long-term time constants, respectively.

Substituting Equation (5) into Equation (2), the output equation is rewritten as:

$$V_t(t) = OCV(SoC) + IR_{in} + V_{RC,short}(0)e^{-\frac{t}{\tau_{short}}} + V_{RC,short}(0)e^{-\frac{t}{\tau_{long}}} + IR_{short}(1 - e^{-\frac{t}{\tau_{short}}}) + IR_{long}(1 - e^{-\frac{t}{\tau_{long}}}) \tag{6}$$

During the rest period, where there is no current excitation ($I = 0$), Equation (6) can be simplified to:

$$V_t(t) = OCV(SoC) + V_{RC,short}(0)e^{-\frac{t}{\tau_{short}}} + V_{RC,long}(0)e^{-\frac{t}{\tau_{long}}} \tag{7}$$

With the knowledge of R_{in} and charging/discharging OCV-SoC relationships, RC network parameters (R_{short}, C_{short}, R_{long} and C_{long}) can be obtained through fitting the experimental data with relevant exponential functions, as

$$\begin{cases} y = IR_{short}(1 - e^{-\frac{t}{\tau_{short}}}) + IR_{long}(1 - e^{-\frac{t}{\tau_{long}}}) & I \neq 0 \\ y = V_{RC,short}(0)e^{-\frac{t}{\tau_{short}}} + V_{RC,long}(0)e^{-\frac{t}{\tau_{long}}} & I = 0 \end{cases} \tag{8}$$

where $y = V_t - OCV(SoC) - IR_{in}$. Since there only exists 2% SoC variation during each pulse-charging/discharging period, it is reasonable to make an assumption that the RC network parameters keep constant during this period. In addition, considering that the battery has converged to the steady state after a 2-h rest, $V_{RC,short}(0)$ and $V_{RC,long}(0)$ are set as zero at the beginning of the pulse-charging/discharging period.

Based on the above analysis, the RC network parameters can be estimated through fitting the experimental dataset with Equation (8). The cost function of the curve fitting method J is to minimize the sum of squared errors between the estimation results and the measured data, subjected to the following constraints:

$$\begin{cases} J = \min\limits_{r,\tau} \sum\limits_{k=1}^{n} [V_t^m(t_k) - V_t^e(r,\tau,t_k)]^2 \\ s.t.\ R_{short},\ \tau_{short},\ R_{long},\ \tau_{long} > 0 \end{cases} \tag{9}$$

where t_k is the input time sequence, n is the length of the fitted experimental dataset, $r = [R_{short}, R_{long}]$, $\tau = [\tau_{short}, \tau_{long}]$, V_t^e is the model estimated voltage and V_t^m is the voltage measurements from the pulse-rest test.

3. RC Network Parameters Estimation

Based on the Introduction in Section 1, the RC network parameters show diverse values under different operating scenarios. In HEV/EV applications, batteries usually work in two typical scenarios: the CC charging scenario and the dynamic driving scenario. In the CC charging scenario, continuous external charging currents are applied to the batteries, and the transport of ions is mainly driven by the electric field. While for the dynamical driving scenario, especially for the urban driving condition, the load current has the characteristics of discontinuous amplitude values and a wide-spread frequency spectrum. In this case, besides the electric field, the gradient in concentration is also largely responsible for the transport of ions within batteries [45]. Therefore, the RC network parameters employed in different operating scenarios should be identified through different identification approaches.

3.1. RC Network Parameters for the CC Charging Scenario

The polarization voltage (V_P) is adopted to illustrate the variation of RC network parameters under the CC excitation. According to the aforementioned battery output equation, V_P can be obtained as:

$$V_P = V_{RC,short} + V_{RC,long} = V_t - OCV(SoC) - IR_{in} \tag{10}$$

The V_P-SoC profile during the C/2 rate CC charging process is shown in Figure 3. Since in the HEV/EV application, batteries seldom work in the extremely low or high $SoCs$, the voltage profile from 10%–90% SoC is covered. It can be observed from Figure 3 that the polarization voltage increases dramatically in Stage I (10%–18% SoC), then it declines slowly and shows a concave shape curve in Stage II, with the local minimum value at around 30% SoC. During Stage III (40%–70% SoC), the polarization voltage becomes relatively stable. After that (70%–90% SoC), the polarization voltage rises sharply.

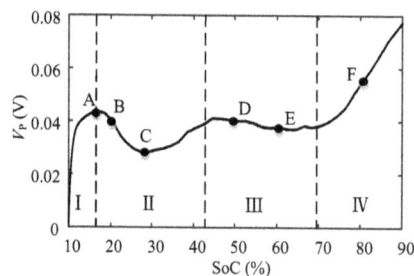

Figure 3. V_P versus SoC under constant-current (CC) charging.

The variation of the polarization voltage during the above *SoC* range is closely related to the internal electrochemical reaction process during charging. In the initial *SoC* region, a relatively large amount of energy is needed to form the nucleation on the surfaces of the electrodes; thus, the polarization voltage increases quickly. Once the nuclei are formed, the following lithium ions' removal process needs less energy. This explains the concave shape voltage curve occurring from 18% *SoC* to 40% *SoC*. While in the last charging stage, the lithium-ion concentration increases in the negative materials. Hence, a large amount of energy is needed to insert the lithium ions, which leads to the obvious growth of the polarization voltage in the high *SoC* region. The detailed explanation for the electrochemical reaction mechanism occurring during the CC charging process can be found in [28,32].

As mentioned in Section 2, the model parameters are estimated through fitting the measured data either from the pulse-charging period or the rest period. In order to select the proper experimental datasets that can better describe the charging characteristic of the battery, the profiles of the polarization voltage during the pulse-charging and the following rest periods, which are also calculated from Equation (10), are compared in Figure 4. Figure 4a shows the polarization voltage under the pulse-charging excitation, and Figure 4b plots the absolute values of the polarization voltage during the following rest. It can be seen from both figures that the shape of the polarization voltage curve strongly depends on the *SoC*. In Figure 4a, it is obvious that the final value of the polarization voltage obtained from 26%–28% *SoC* is the lowest, which is similar to point C in Figure 3. In addition, the final values of the voltage curves obtained from 18%–20% *SoC* and 50%–52% *SoC* are almost coincident with each other, which approximately matches the corresponding parts (point B and point D) in Figure 3. Meanwhile, the relations among the final voltage values collected from 14%–16% *SoC*, 60%–62% *SoC* and 80%–82% *SoC* are also identical to the relations among point A, point E and point F in Figure 3, respectively. Hence, it can be summarized from Figure 4a that the final values of the polarization voltage obtained from different pulse-charging periods are approximately consistent with the corresponding points in Figure 3. While in Figure 4b, the variation trend of the predicted stable voltage values differs greatly compared to the results in Figure 4a. This is because in the pulse-charging period, the ion migration is driven by external electric potential. While in the rest period, the transport of ions is mainly dominated by diffusion, owing to the concentration gradient. The detailed explanation of the electrochemical reactions occurring under different load current has been discussed in [21,45].

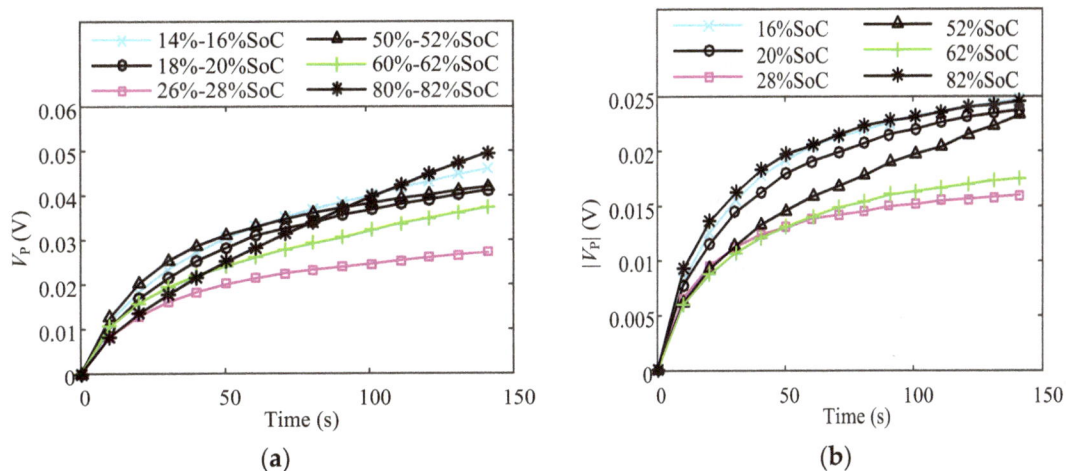

Figure 4. (a) The profiles of V_P at different *SoC* intervals during the pulse-charging period; (b) the profiles of $|V_P|$ at different *SoC* points during the rest period.

Consequently, it can be concluded that the voltage response during the pulse-charging period can better describe the characteristic of the CC charging process because of the similar current excitation.

3.2. RC Network Parameters for the Dynamic Driving Scenario

3.2.1. Typical Dynamic Driving Scenarios

For the dynamic driving scenario, especially for the urban driving scenario, vehicles accelerate and brake frequently, which cause the long lasting load current to seldom exist. There are two typical kinds of standard urban driving cycles, namely the urban dynamometer driving schedule (UDDS) and the worldwide harmonized light vehicles test procedure (WLTP), which are the American and European certification cycles, respectively. The load current profiles and the load current amplitude distributions of the two driving cycles are plotted in Figure 5. It can be observed from Figure 5a,b that both of the dynamic current profiles vary frequently over the test span. Meanwhile, from Figure 5c,d, it can be concluded that: (1) the discharging current accounts for a much larger portion, compared to the charging current during the regenerative process; (2) among the load currents, the low C-rate discharging current, particularly around zero-value amplitudes, accounts for a larger portion in both tests. Hence, the voltage response during the rest period can be employed to estimate the RC network parameters for the dynamic driving scenario.

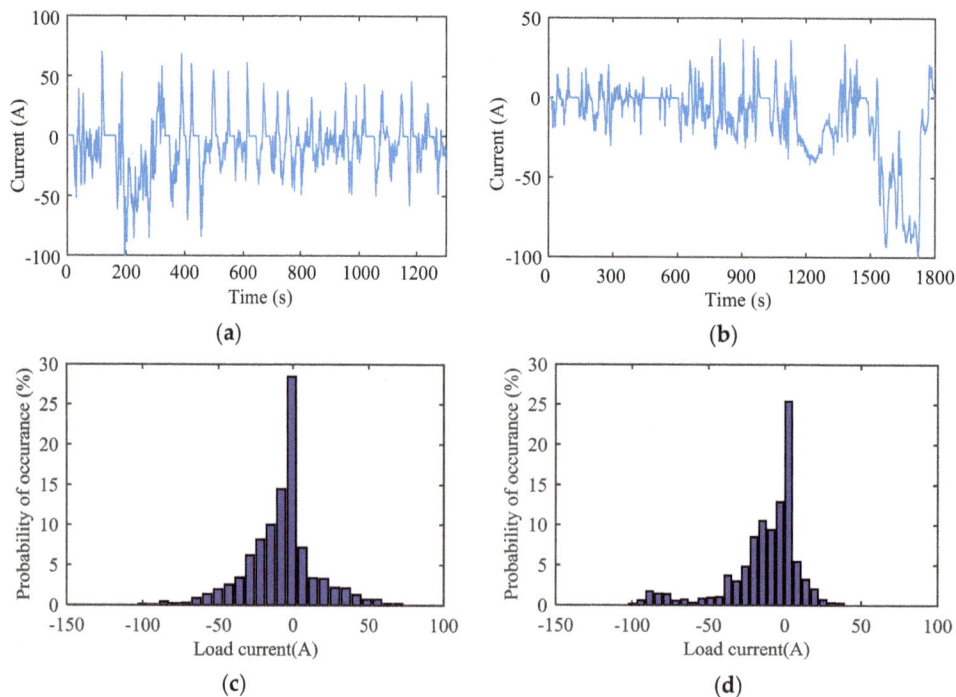

Figure 5. (**a**) The load current profile of the urban dynamometer driving schedule (UDDS) test; (**b**) the load current profile of the worldwide harmonized light vehicles test procedure (WLTP) test; (**c**) the load current amplitude distribution of the UDDS test; (**d**) the load current amplitude distribution of the WLTP test.

3.2.2. Determination of the Length of the Fitted Experimental Dataset

The diffusion process, which is caused by the gradient in concentration, plays a major role in the low C-rate load current and rest cases. Since the electrochemical reactions occurring during the diffusion process are very complex, these reactions can be accurately modeled as infinite series-connected RC networks with a wide range of time constants ($\tau_1, \tau_2, \ldots, \tau_j$). Usually, the values of time constants depend on the electrode thickness and the structure of the battery to a great extent, and typical time constants are in the range of seconds to minutes [45]. The second order RC network can only approximate the diffusion process by two parts: the fast dynamics part (the short-term RC network with τ_{short}) and the slow dynamics part (the long-term RC network with τ_{long}).

In general, the values of the two time constants are closely related to the length of the fitted experimental data Δt. When only the initial segment of the voltage response is employed in parameter estimation, such as Part A in the bottom subfigure of Figure 1, the voltages across the shorter-term RC networks have a larger degree of variability, which means that the shorter-term RC networks have a greater impact on the initial segment of the voltage response. This in turn leads to the smaller estimated time constants and subsequently ignores the slower dynamics diffusion process. On the contrary, after the initial phase of the rest period, such as Part B in the bottom subfigure of Figure 1, the voltages across the shorter-term RC networks have converged to zero; thus, the voltage variation caused by the shorter-term RC networks is negligible. Instead, the voltages across the longer term RC networks make a remarkable contribution to the total voltage response. Subsequently, it can be inferred that the measured data show a slower varying characteristic, which represent the slower dynamics diffusion process and can be modeled by the RC networks with larger time constants. Hence, if the whole voltage response of the long time rest period is adopted, data with slower varying values will account for a large portion, which will lead to the relatively larger estimated time constants. However, too large time constants will make the model output voltage severely lag behind the actual response and result in a poor dynamic performance.

In order to further illustrate the above analysis, a third order RC network circuit is simulated in MATLAB; two equivalent time constants (τ'_{short} and τ'_{long}) are estimated from the different value of Δt. In the simulation, the resistances of the three RC networks are all set as 1 mΩ, and the time constants are predetermined as $\tau_1 = 40$ s, $\tau_2 = 200$ s and $\tau_3 = 2000$ s ($\tau_3 \gg \tau_2 > \tau_1$). The applied excitation consists of a 400-s pulse-discharging current and a 2-h rest period, and the amplitude of the current is 20 A. Time constants estimated by different lengths of the voltage response are given in Table 2. It can be clearly seen from Table 2 that both τ'_{short} and τ'_{long} decrease simultaneously with the reduced value of Δt, which is consistent with the previous analysis. Hence, to obtain the appropriate values of the time constants, Δt should be predetermined properly, which is illustrated in detail as follows.

Table 2. Equivalent time constant estimation results with different values of Δt.

Δt (s)	7200	3600	1800	1400	1200	1000	900	850	800
τ'_{short} (s)	88.67	67.18	48.53	45.10	43.74	42.59	42.08	41.83	41.63
τ'_{long} (s)	971.0	484.3	284.4	256.7	245.3	235.3	230.9	228.8	226.8
k [1]	4.049×10^{-12}	4.395×10^{-5}	0.1448	0.8759	2.154	5.299	8.311	10.41	13.03

[1] k represents the degree of resistor-capacitor (RC) voltage variability; the detailed expression can referred to in Equation (13).

During Δt, the derivative of Equation (13) with respect to τ_i during the rest period is expressed as:

$$\left| \frac{dV_{RC,i}}{d\tau_i} \right| = \frac{\Delta t |V_{RC,i}(0)|}{\tau_i^2} e^{-\frac{\Delta t}{\tau_i}} \tag{11}$$

where $V_{RC,i}$ is the voltage across the i-th RC network, $i \in \{1,2,3,\ldots,j\}$, $V_{RC,i}(0)$ is the corresponding initial voltage, R_i is the resistance of the i-th RC network and τ_i is the time constant of the i-th RC network, which is subject to $\tau_1 < \tau_2 < \ldots < \tau_j$.

After the pulse-discharging period, $|V_{RC,i}(0)|$ can be expressed as:

$$|V_{RC,i}(0)| = |I|R_i(1 - e^{-\frac{D}{\tau_i}}) \tag{12}$$

where D denotes the length of the pulse-discharging period.

For the two well-separated time constants τ_i and τ_{i+m} ($\tau_{i+m} \geq 10\tau_i$ and $0 < m < j - i$), the voltage across the shorter term RC network $V_{RC,i}$ has a larger degree of variability when satisfying the following requirement:

$$\frac{|dV_{RC,i}/d\tau_i|}{|dV_{RC,i+m}/d\tau_{i+m}|} = k \qquad (13)$$

where the constant k denotes the degree of variability, and it is subject to $k > 1$.

Substituting Equations (11) and (12) into Equation (13), the value of Δt can be derived as:

$$\Delta t = \ln\left[\frac{R_i(1-e^{-\frac{D}{\tau_i}})\tau_{i+m}^2}{kR_{i+m}(1-e^{-\frac{D}{\tau_{i+m}}})\tau_i^2}\right]\frac{\tau_i\tau_{i+m}}{\tau_{i+m}-\tau_i} \qquad (14)$$

In Equation (14), since the values of R_i and R_{i+m} are nearly of the same order of magnitude [39,43,46], the value of R_i/R_{i+m} can be neglected when compared to the value of τ_{i+m}^2/τ_i^2; thus, Δt can be simplified as:

$$\Delta t = \ln\left[\frac{(1-e^{-\frac{D}{\tau_i}})\tau_{i+m}^2}{k(1-e^{-\frac{D}{\tau_{i+m}}})\tau_i^2}\right]\frac{\tau_i\tau_{i+m}}{\tau_{i+m}-\tau_i} \qquad (15)$$

Equation (15) shows that k and τ_i should be determined before calculating Δt. In the aforementioned simulation, the value of k for τ_2 and τ_3 can be obtained directly from Equation (13), as shown in Table 2. This indicates that when k is larger than one, the estimated τ'_{short} and τ'_{long} are closer to τ_1 and τ_2. This is because the voltage across the RC network with τ_3 has a lower degree of variability, compared to those with τ_1 and τ_2. It can be observed from Table 2 that τ'_{short} and τ'_{long} are nearly stable when k is larger than 10. Hence, k is selected as 10 throughout the paper.

In order to set a proper τ_i in Equation (15), the discrete Fourier analysis of the load current is employed to determine the lower bandwidth limitation of the ECM. The current spectrums of UDDS and WLTP tests are shown in Figure 6. It can be observed in Figure 6a,b that there exists a large DC component (Points A and C) due to the nonzero mean value of the two current profiles. Since the characteristics of the DC component cannot be modeled by the RC circuit, they are neglected when determining the length of the fitted dataset. The major low frequency components for the two profiles are around 0.00146 Hz (point B) and 0.00138 Hz (the mean value from point D to point E), respectively. Hence, the mean value of the long-term time constant is selected as 704 s. In order to exclude the voltage variation caused by the larger time constants (larger than $10\tau_i$), the prior 1-h measured battery voltage dataset is employed to estimate the RC parameters.

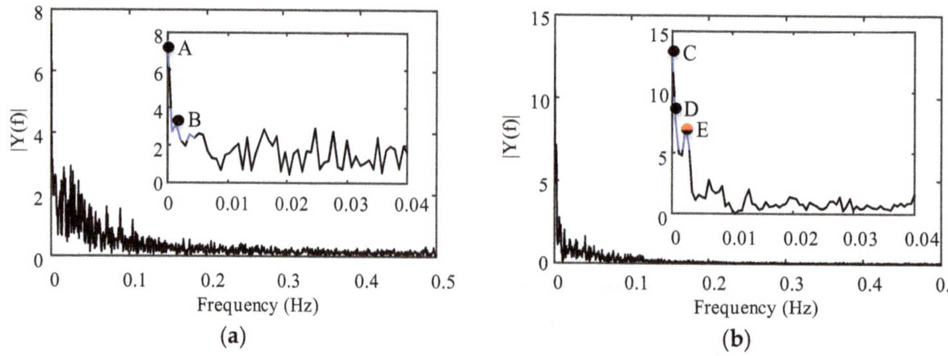

Figure 6. The spectral analysis of the load current: (**a**) the urban dynamometer driving schedule (UDDS) test; (**b**) the worldwide harmonized light vehicles test procedure (WLTP) test.

3.2.3. Improved Fitting Function

From Equations (6) and (7), it can be observed that only the initial values $V_{RC,short}(0)$, $V_{RC,long}(0)$ and time constants τ_{short}, τ_{long} can be obtained directly from the fitting results; thus, we should do the further computations to obtain the resistances and capacitances of RC networks.

In [37,39–41], two initial voltages across the RC networks are predetermined as IR_{short} and IR_{long} respectively, from which the resistances of the RC networks can be derived under the knowledge of the current value. In [49], the capacitances of the RC networks are firstly obtained from the initial voltage values. Both of the above two methods have an assumption that the capacitors of the RC networks have already converged to the steady state at the end of the pulse-discharging period.

Usually, in the parameter extraction test, in order to obtain as much data as possible at different *SoC* intervals, the length of the pulse-charging/discharging period is usually set as several minutes (resulting in 2% *SoC* variation in this paper), while the rest time is usually set as one or more hours (such as 2 h in this paper) to get an accurate *OCV* value. For the short-term RC network, the voltage can easily converge to the equilibrium state during the pulse-discharging process, which is shown in Figure 7. In other words, there is no current flowing through the capacitor branch of the short-term RC network during the last stage of the pulse-discharging period; thus, $V_{RC,short}(0)$ at the beginning of the rest period can be expressed as:

$$V_{RC,short}(0) = IR_{short} \qquad (16)$$

However, for the long-term RC network, the voltage varies continuously due to a relatively large time constant, as illustrated in Figure 7. The voltage across the long-term RC network has not reached the equilibrium state at the end of the pulse-discharging period; thus, there always exists a significant proportion of the load current $I(1 - e^{-D/\tau_{long}})$ flowing through the corresponding capacitor. Consequently, $V_{RC,long}(0)$ at the beginning of the rest period should be written as:

$$V_{RC,long}(0) = IR_{long}\left(1 - e^{-\frac{D}{\tau_{long}}}\right) \qquad (17)$$

where I is the value of the pulse-discharging current. Since the *SoC* variation in each test cycle is set as 2% in this paper, it can be assumed that the model parameters keep constant during the pulse-discharging period.

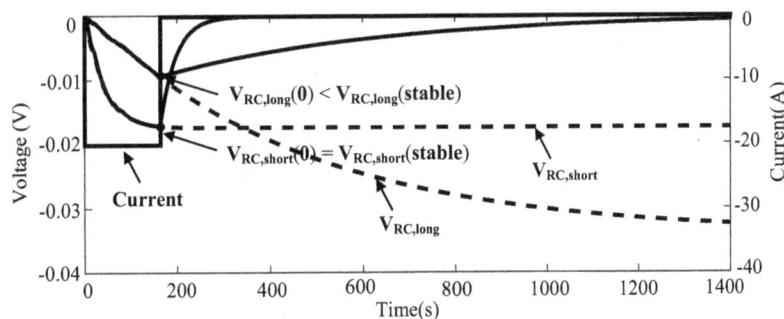

Figure 7. The voltage curve of RC networks during one cycle of the discharging pulse-rest test.

4. Experimental Results and Discussions

4.1. RC Network Parameter Estimation Results

Based on the aforementioned analysis in Section 3.1, for the case of the CC charging scenario, the charging pulse-rest test is implemented firstly. The parameters are estimated from the voltage response of the pulse-charging period, and the estimation results are shown in Figure 8. Figure 8a plots two estimated time constants; it can be seen that the general order of the magnitude of the short-term time constant is 10 s; it fluctuates greatly when the *SoC* changes, especially in the middle *SoC* region, while the order of the magnitude of the long-term time constant is 100 s; it is relatively flat during the whole *SoC* region. Figure 8b plots two estimated resistances; it can be observed that in the middle *SoC* range, the short-term resistance has a larger value, which means that the voltage across the short-term RC network accounts for more weight during this period. Hence, it can be observed from

Figures 3 and 8b that the variation tendencies of the polarization voltage and the short-term resistance are similar during the middle *SoC* range. At the end of the charging process, the short-term resistance decreases and stabilizes around a very small value, while the long-term resistance increases almost linearly after 60% *SoC*, leading to a similar variation tendency of the polarization voltage, compared to the corresponding part in Figure 3. Hence, it can be concluded that the long-term diffusion process plays a major role in this stage.

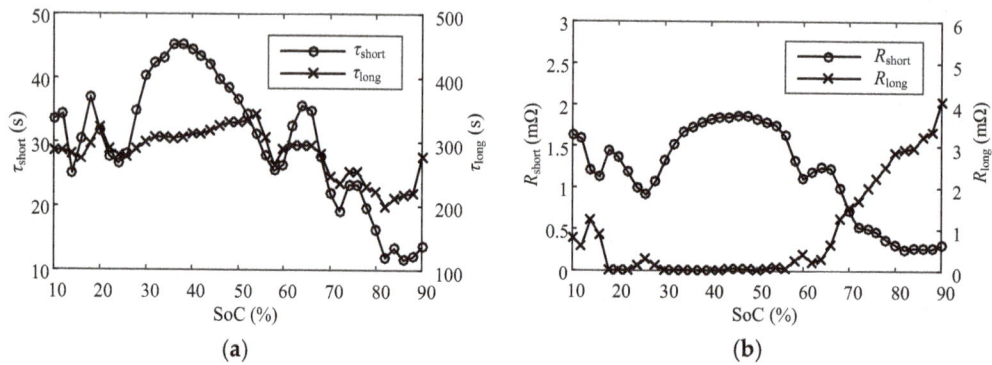

Figure 8. Parameter estimation results for the constant-current (CC) charging scenario: (a) time constant; (b) resistance.

For the case of the dynamic driving scenario, the discharging pulse-rest test is implemented, and the data from the rest periods are adopted in the parameter estimation. According to the analysis in Section 3.2.2, different time constants will be obtained from the fitted experimental datasets in different lengths. Firstly, in order to compare the best fit performances for the measured datasets in different lengths, the measured battery terminal voltage response at 60% *SoC* during a 2-h rest period is adopted, and the curve fitting results are shown in Figure 9. It can be observed from Figure 9a that the fitting result of the whole measured voltage response shows a better performance during most of the rest period, especially in the equilibrium state. Whereas for the performance of the first 200 s, the fitting result through the prior 0.5-h measured voltage response yields less errors, which is illustrated in Figure 9b. Parameter estimation results in Figure 10 show the time constants estimated from the measured voltage dataset in different lengths, ranging from 30 min–2 h with a 30-min interval. It can be observed that the time constants, both for the long term and the short term, increase simultaneously when the length of the fitted dataset increases. In addition, by comparing Figure 10 with Figure 8a, it can be concluded that the time constants applied in the CC charging scenario and the dynamic driving scenario show different variation tendencies. Hence, it is essential to adopt different sets of model parameters for different operating scenarios.

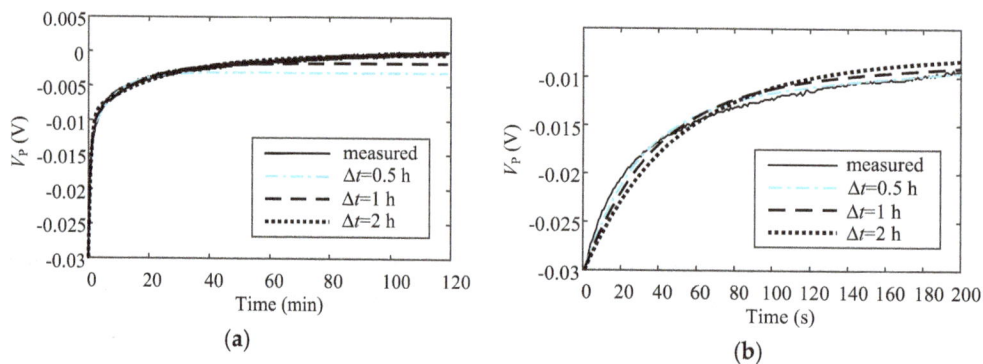

Figure 9. Curve fitting results of V_P during the rest period of the discharging pulse-rest test at 60% *SoC*: (a) the overall result; (b) a close look at the transient part at the beginning.

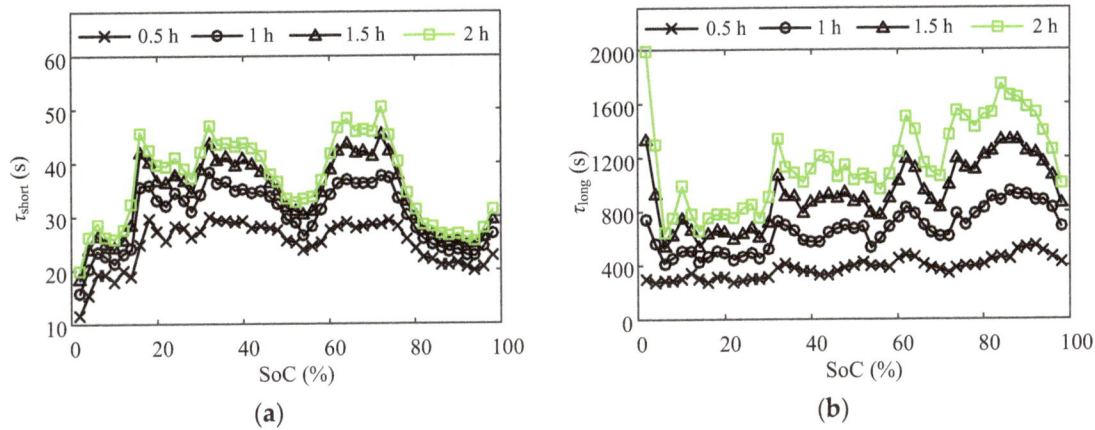

Figure 10. Time constant estimation results with different lengths of the experimental dataset: (a) τ_{short}; (b) τ_{long}.

After determining the length of the fitted experimental dataset, we can subsequently obtain the resistances. Figure 11 shows the R_{long} estimation results by the conventional fitting function and the improved fitting function. It can be concluded from Figure 11 that the R_{long} estimated by the conventional fitting function is generally less than the one estimated by the improved fitting function, because it neglects the $(1 - e^{-D/\tau_{long}})$ part. In order to demonstrate the advantage of the improved fitting function, data from the 20th cycle of the discharging pulse-rest test are adopted. In this cycle, SoC changes from 62% to 60% during the pulse-discharging period, then keeps the value of 60% during the following rest period. The current profile of the 20th discharging pulse-rest test is applied on the ECM MATLAB/SIMULINK model as an excitation. Figure 12a,b shows the model output voltage responses with two sets of estimated model parameters. It can be seen that the model with parameters estimated by the proposed fitting function outputs better estimation results. The lower voltage error is mainly contributed by the higher voltage drop across the long-term RC network, as plotted in Figure 12c. In addition, the root mean square errors (RMSEs) between the measured voltage and the model output voltage at different SoCs are given in Table 3. It can also be seen that the model parameters estimated by the proposed fitting function show a better performance for a wide range of SoC.

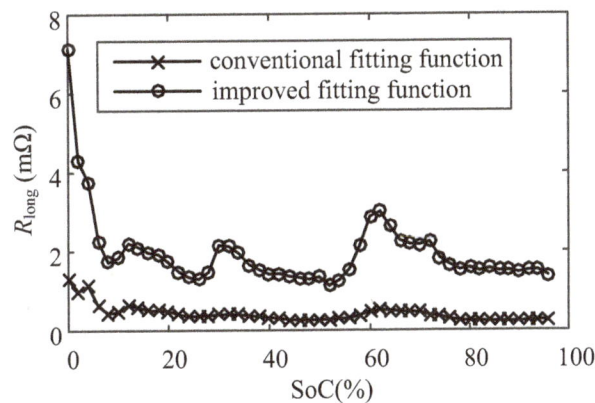

Figure 11. R_{long} estimation results.

Table 3. Comparison of RMSE at different SoC.

SoC (%)		10	20	30	40	50	60	70	80	90
RMSE (mV)	Conventional fitting function	1.802	1.714	2.167	1.540	1.268	2.803	2.416	1.558	1.444
	Improved fitting function	0.7658	0.7582	0.9707	0.7643	0.5000	1.202	1.242	0.7104	0.6482

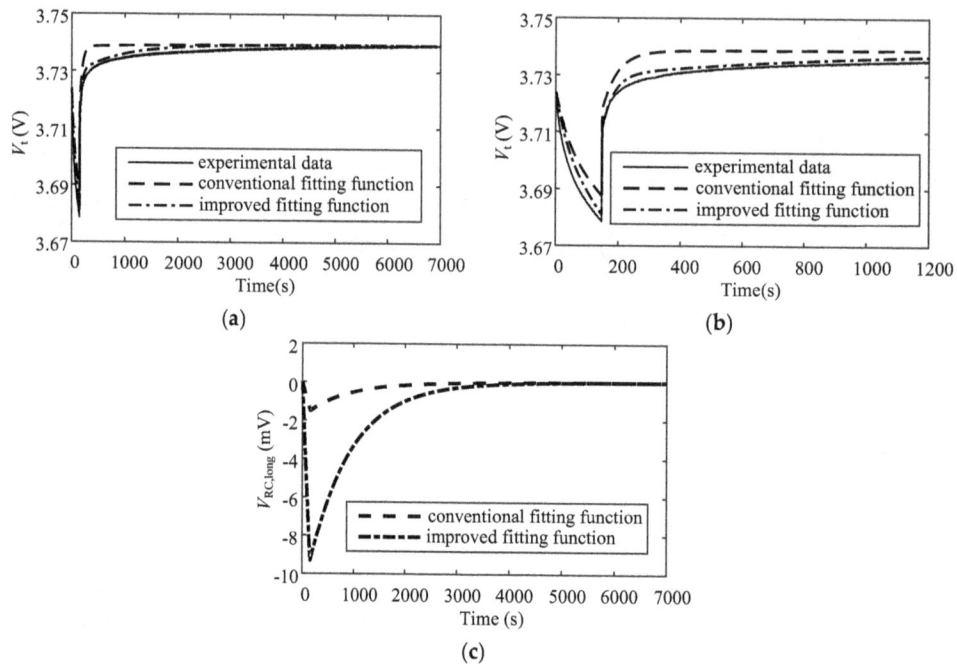

Figure 12. Voltage curves of one cycle of the discharging pulse-rest test (62%–60%): (**a**) the overview; (**b**) a close look; (**c**) the voltage across the long-term RC network.

4.2. Model Verification

In this paper, the CC charging test and the consecutive UDDS test, which respectively represent two typical operating scenarios in HEV/EV applications, are conducted separately to verify the effectiveness of the model. For the charging condition, the battery is charged from 10%–90% SoC. The typical charging current in practice varies from C/8 to 2C [50], and a C/2 rate current is employed in the charging test. The consecutive UDDS test starts from 90% SoC to 20% SoC, with a 10-min rest period in between to simulate a short parking time. In the real application, a specific set of parameters can be selected by the characteristics of the measured load current. For example, if the values of the current are approximately constant over a certain time interval, parameters estimated from the data in the pulse-charging periods are employed. On the other hand, parameters estimated from the data in the rest periods are employed when the load current shows the characteristics of high dynamics over a certain time interval.

Firstly, for the CC charging scenario, three model outputs and measured battery terminal voltage curves are plotted in Figure 13, and the corresponding RMSEs are given in Table 4. It can be observed that during the whole charging process, the model with parameters estimated from the data in pulse-charging periods outputs a voltage curve matching the measured curve better because of considering the continuous external electric driving forces. However, parameters estimated from the data in the rest periods result in relatively larger errors, especially in the high SoC region. In addition, during most part of the charging period, the model with parameters used in the dynamic driving scenarios outputs a voltage higher than the experimental voltage. Comparing the corresponding curves in Figures 8b and 11, it can be deduced that the higher estimated voltage is mainly caused by the larger value of estimated R_{long}, especially during the middle range of the SoC region.

Table 4. RMSE of model voltage estimation under the CC charging test.

Modeling Methods	Dynamic Condition	Rest-Period	Pulse-Period
RMSE (mV)	18.41	19.76	5.448

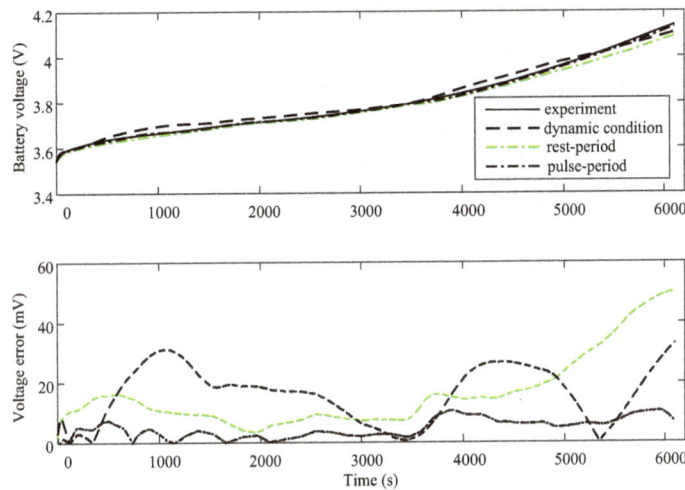

Figure 13. Verification results of different parameter estimation methods under the CC charging test.

In order to verify the robustness of the proposed parameter estimation method, the CC charging voltage profiles at different initial SoC are plotted in Figure 14. This shows that the estimated voltage curves match well with the measurement voltage curves, despite the different initial SoC.

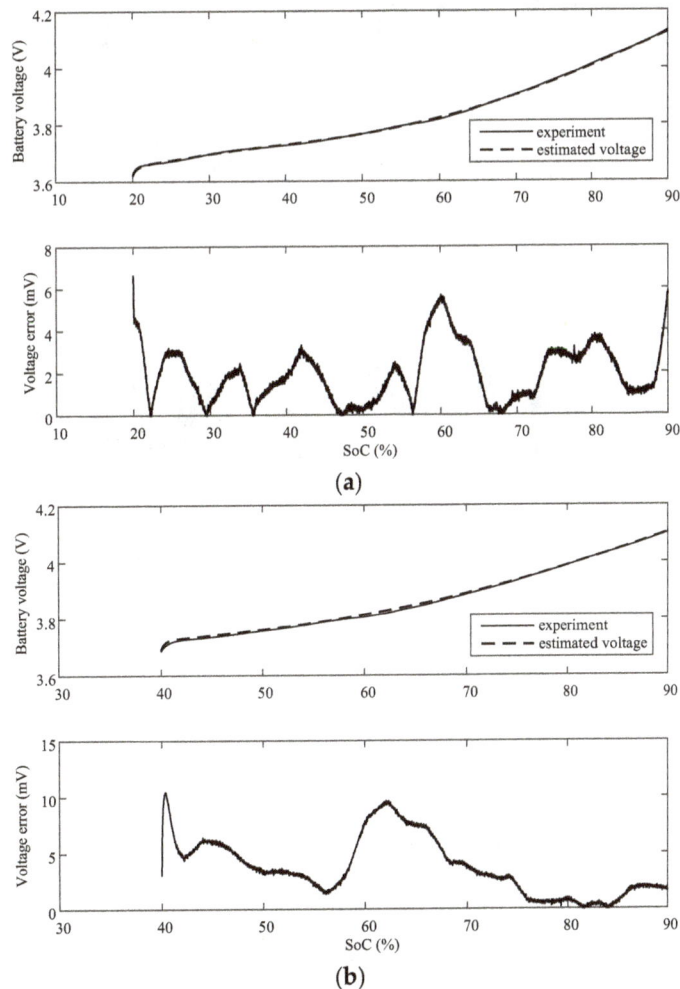

(a)

(b)

Figure 14. CC charging voltage profiles at different initial SoC: (a) initial $SoC = 20\%$; (b) initial $SoC = 40\%$.

Secondly, in order to demonstrate the improvement of the proposed battery modelling approach during the dynamic driving scenario, the model and experimental voltage outputs in the consecutive UDDS validation are plotted in Figure 15a, the corresponding calculated *SoC* profile is shown in Figure 15b, and the detailed figure from 10,000 s to 12,000 s is plotted in Figure 15c. The RMSE of the aforementioned estimation methods during the whole consecutive UDDS test are also shown in Table 5. Figure 15b shows that the consecutive UDDS test is started from 90% *SoC*, and terminated when the value of *SoC* drops below 20%. It can be observed from Figure 15c that parameters estimated by the improved fitting function generally demonstrate a better performance, especially during the dynamic period (ranging from 10,000 s to 11,400 s), because considering the unsaturated phenomenon of the long-term RC network. It can also be concluded that the model containing parameters estimated by the prior 1-h experimental data from the rest period gives voltage output with the least error, especially during the short-time rest period. In addition, it can be seen from Figure 5a that there exists a relatively long-time and high C-rate discharging current in the UDDS cycle approximately ranging from 150 s to 300 s. Since larger time constants are obtained from the data of the whole rest period, this causes the corresponding voltage output not to recover fast after a relatively long-time discharging current, which leads to an offset of voltage errors in comparison to the voltage error caused by the proposed approach.

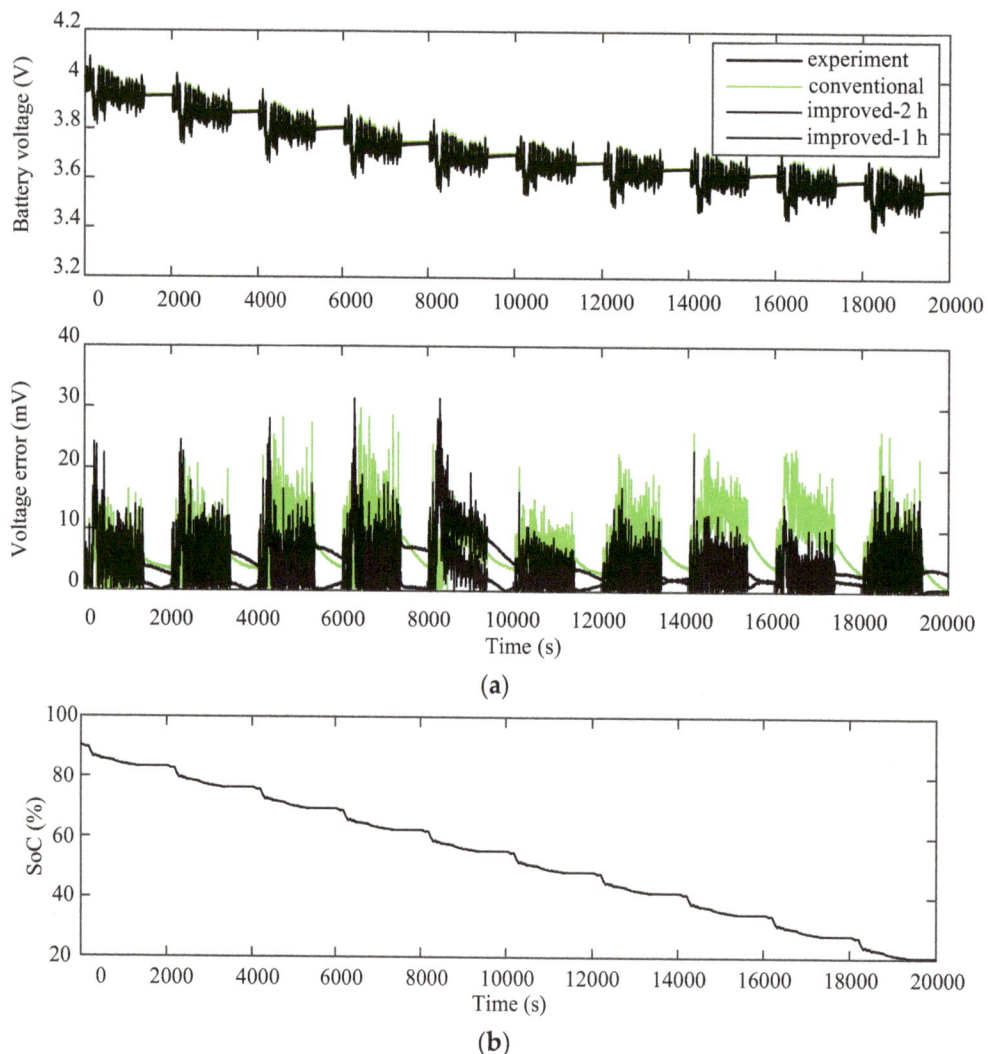

(a)

(b)

Figure 15. *Cont.*

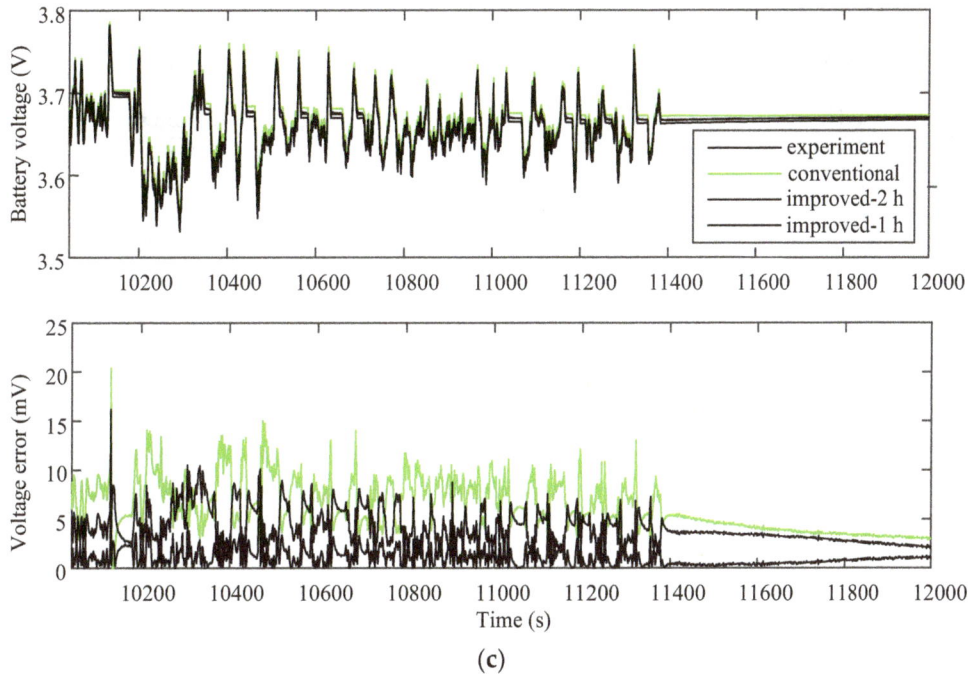

Figure 15. Verification results of different parameter estimation methods under the UDDS tests: (**a**) The overall look; (**b**) The calculated *SoC* profile (**c**) The close look.

Table 5. RMSE of the model voltage estimation under the urban dynamometer driving schedule (UDDS) test.

Modeling Methods	Conventional	Improved-2 h	Improved-1 h
RMSE (mV)	8.504	6.329	4.244

5. Conclusions

In this paper, an advanced battery parameter estimation method based on two general operating scenarios in HEV/EV applications is proposed. Firstly, the second order ECM is employed, and the model parameter extraction process is described in detail. Considering the typical operating scenarios in HEV/EV applications, namely the CC charging scenario and the dynamic driving scenario, two sets of model parameters are extracted from the charging/discharging pulse-rest tests. Specifically, voltage responses of the pulse-charging phases are selected to estimate model parameters applied in the CC charging scenario. For the dynamic driving scenario, the model parameters are identified through the measured data from the rest period. Instead of employing the data from the whole rest period, only the prior portion of the collected data is selected, and the length of the fitted data is determined by the frequency spectrum analysis of the load current under two typical urban driving conditions. In addition, an unsaturated phenomenon caused by the long-term RC network is analyzed in detail, and subsequently, an improved fitting equation with more accurate initial voltage expression of the RC network is adopted. Finally, verification tests simulating the CC charging scenario and the dynamic driving scenario are conducted, respectively, and comparisons between the conventional and the proposed battery parameter estimation methods are given. Experimental results show that in both cases, the voltage profiles predicted from the proposed model show a better conformity to the experimental data.

It is important to note that the proposed battery parameter estimation method for the dynamic driving scenario only considers the typical urban driving conditions at room temperature. However, the characteristics of the load current under the other special conditions (such as the highway driving condition and the extremely cold condition) will be obviously different. For the future work, the influence caused by different C-rates of the current profiles, bandwidths of the current profiles and temperature effects will be considered, and the parameter extraction test will be modified accordingly.

Acknowledgments: The authors would like to acknowledge the funding support from the China Scholarship Council (CSC); the U.S. DOE Graduate Automotive Technology Education (GATE) Center of Excellence; and Nanjing Golden Dragon Bus Co., Ltd.

Author Contributions: Jufeng Yang handled the technical modeling, drafted and revised the manuscript. Bing Xia revised the manuscript. Jufeng Yang and Bing Xia designed the experiments and analyzed the data. Yunlong Shang participated in the experiment. Wenxin Huang revised the manuscript. Chris Mi contributed the experiment platform, gave great suggestions and polished the manuscript.

Conflicts of Interest: The authors declare no conflict of interest.

References

1. Xiong, R.; Sun, F.; Chen, Z.; He, H. A data-driven multi-scale extended kalman filtering based parameter and state estimation approach of lithium-ion olymer battery in electric vehicles. *Appl. Energy* **2014**, *113*, 463–476. [CrossRef]

2. Zou, Z.; Xu, J.; Mi, C.; Cao, B.; Chen, Z. Evaluation of model based state of charge estimation methods for lithium-ion batteries. *Energies* **2014**, *7*, 5065–5082. [CrossRef]

3. Shang, Y.; Zhang, C.; Cui, N.; Guerrero, J.M. A cell-to-cell battery equalizer with zero-current switching and zero-voltage gap based on quasi-resonant lc converter and boost converter. *IEEE Trans. Power Electron.* **2015**, *30*, 3731–3747. [CrossRef]

4. Xia, B.; Mi, C. A fault-tolerant voltage measurement method for series connected battery packs. *J. Power Sources* **2016**, *308*, 83–96. [CrossRef]

5. Sun, F.; Xiong, R.; He, H. A systematic state-of-charge estimation framework for multi-cell battery pack in electric vehicles using bias correction technique. *Appl. Energy* **2016**, *162*, 1399–1409. [CrossRef]

6. Xia, B.; Shang, Y.; Nguyen, T.; Mi, C. A correlation based fault detection method for short circuits in battery packs. *J. Power Sources* **2017**, *337*, 1–10. [CrossRef]

7. Salameh, M.; Schweitzer, B.; Sveum, P.; Al-Hallaj, S.; Krishnamurthy, M. Online temperature estimation for phase change composite-18650 lithium ion cells based battery pack. In Proceedings of the 2016 IEEE Applied Power Electronics Conference and Exposition (APEC), Long Beach, CA, USA, 20–24 March 2016; pp. 3128–3133.

8. Seaman, A.; Dao, T.-S.; McPhee, J. A survey of mathematics-based equivalent-circuit and electrochemical battery models for hybrid and electric vehicle simulation. *J. Power Sources* **2014**, *256*, 410–423. [CrossRef]

9. Zou, Y.; Hu, X.; Ma, H.; Li, S.E. Combined state of charge and state of health estimation over lithium-ion battery cell cycle lifespan for electric vehicles. *J. Power Sources* **2015**, *273*, 793–803. [CrossRef]

10. Wei, Z.; Tseng, K.J.; Wai, N.; Lim, T.M.; Skyllas-Kazacos, M. Adaptive estimation of state of charge and capacity with online identified battery model for vanadium redox flow battery. *J. Power Sources* **2016**, *332*, 389–398. [CrossRef]

11. He, H.; Xiong, R.; Guo, H.; Li, S. Comparison study on the battery models used for the energy management of batteries in electric vehicles. *Energy Convers. Manag.* **2012**, *64*, 113–121. [CrossRef]

12. He, H.; Zhang, X.; Xiong, R.; Xu, Y.; Guo, H. Online model-based estimation of state-of-charge and open-circuit voltage of lithium-ion batteries in electric vehicles. *Energy* **2012**, *39*, 310–318. [CrossRef]

13. Nejad, S.; Gladwin, D.; Stone, D. A systematic review of lumped-parameter equivalent circuit models for real-time estimation of lithium-ion battery states. *J. Power Sources* **2016**, *316*, 183–196. [CrossRef]

14. Xia, B.; Zhao, X.; De Callafon, R.; Garnier, H.; Nguyen, T.; Mi, C. Accurate lithium-ion battery parameter estimation with continuous-time system identification methods. *Appl. Energy* **2016**, *179*, 426–436. [CrossRef]

15. Pérez, G.; Garmendia, M.; Reynaud, J.F.; Crego, J.; Viscarret, U. Enhanced closed loop state of charge estimator for lithium-ion batteries based on extended kalman filter. *Appl. Energy* **2015**, *155*, 834–845. [CrossRef]

16. Chen, Z.; Fu, Y.; Mi, C.C. State of charge estimation of lithium-ion batteries in electric drive vehicles using extended kalman filtering. *IEEE Trans. Veh. Technol.* **2013**, *62*, 1020–1030. [CrossRef]

17. Sun, F.; Xiong, R. A novel dual-scale cell state-of-charge estimation approach for series-connected battery pack used in electric vehicles. *J. Power Sources* **2015**, *274*, 582–594. [CrossRef]

18. Li, K.; Tseng, K.J. An equivalent circuit model for state of energy estimation of lithium-ion battery. In Proceedings of the 2016 IEEE Applied Power Electronics Conference and Exposition (APEC), Long Beach, CA, USA, 20–24 March 2016; pp. 3422–3430.

19. Fuller, T.F.; Doyle, M.; Newman, J. Relaxation phenomena in lithium-ion-insertion cells. *J. Electrochem. Soc.* **1994**, *141*, 982–990. [CrossRef]

20. Smith, K.A. Electrochemical Modeling, Estimation and Control of Lithium Ion Batteries. Ph.D. Thesis, The Pennsylvania State University, State College, PA, USA, 2006.

21. Park, M.; Zhang, X.; Chung, M.; Less, G.B.; Sastry, A.M. A review of conduction phenomena in Li-ion batteries. *J. Power Sources* **2010**, *195*, 7904–7929. [CrossRef]

22. Karden, E.; Buller, S.; De Doncker, R.W. A method for measurement and interpretation of impedance spectra for industrial batteries. *J. Power Sources* **2000**, *85*, 72–78. [CrossRef]

23. Thele, M.; Bohlen, O.; Sauer, D.U.; Karden, E. Development of a voltage-behavior model for nimh batteries using an impedance-based modeling concept. *J. Power Sources* **2008**, *175*, 635–643. [CrossRef]

24. Buller, S.; Thele, M.; De Doncker, R.; Karden, E. Impedance-based simulation models of supercapacitors and Li-ion batteries for power electronic applications. *IEEE Trans. Ind. Appl.* **2005**, *41*, 742–747. [CrossRef]

25. Waag, W.; Käbitz, S.; Sauer, D.U. Experimental investigation of the lithium-ion battery impedance characteristic at various conditions and aging states and its influence on the application. *Appl. Energy* **2013**, *102*, 885–897. [CrossRef]

26. Howey, D.A.; Mitcheson, P.D.; Yufit, V.; Offer, G.J.; Brandon, N.P. Online measurement of battery impedance using motor controller excitation. *IEEE Trans. Veh. Technol.* **2014**, *63*, 2557–2566. [CrossRef]

27. Zheng, Y.; Lu, L.; Han, X.; Li, J.; Ouyang, M. Lifepo 4 battery pack capacity estimation for electric vehicles based on charging cell voltage curve transformation. *J. Power Sources* **2013**, *226*, 33–41. [CrossRef]

28. Nakayama, M.; Iizuka, K.; Shiiba, H.; Baba, S.; Nogami, M. Asymmetry in anodic and cathodic polarization profile for LiFePO4 positive electrode in rechargeable Li ion battery. *J. Ceram. Soc. Jpn.* **2011**, *119*, 692–696. [CrossRef]

29. Musio, M.; Damiano, A. A simplified charging battery model for smart electric vehicles applications. In Proceedings of the 2014 IEEE International Energy Conference (ENERGYCON), Dubrovnik, Croatia, 13–16 May 2014; pp. 1357–1364.

30. Tsang, K.; Sun, L.; Chan, W. Identification and modelling of lithium ion battery. *Energy Convers. Manag.* **2010**, *51*, 2857–2862. [CrossRef]

31. Yao, L.W.; Aziz, J.; Kong, P.Y.; Idris, N.; Alsofyani, I. Modeling of lithium titanate battery for charger design. In Proceedings of the 2014 IEEE Australasian Universities Power Engineering Conference (AUPEC), Perth, Australia, 28 Sepember–1 October 2014; pp. 1–5.

32. Jiang, J.; Liu, Q.; Zhang, C.; Zhang, W. Evaluation of acceptable charging current of power Li-ion batteries based on polarization characteristics. *IEEE Trans. Ind. Electron.* **2014**, *61*, 6844–6851. [CrossRef]

33. Kim, N.; Ahn, J.-H.; Kim, D.-H.; Lee, B.-K. Adaptive loss reduction charging strategy considering variation of internal impedance of lithium-ion polymer batteries in electric vehicle charging systems. In Proceedings of the 2016 IEEE Applied Power Electronics Conference and Exposition (APEC), Long Beach, CA, USA, 20–24 March 2016; pp. 1273–1279.

34. Chen, Z.; Xia, B.; Mi, C.C.; Xiong, R. Loss-minimization-based charging strategy for lithium-ion battery. *IEEE Trans. Ind. Appl.* **2015**, *51*, 4121–4129. [CrossRef]

35. Rao, R.; Vrudhula, S.; Rakhmatov, D.N. Battery modeling for energy aware system design. *Computer* **2003**, *36*, 77–87.

36. Fleischer, C.; Waag, W.; Heyn, H.-M.; Sauer, D.U. On-line adaptive battery impedance parameter and state estimation considering physical principles in reduced order equivalent circuit battery models: Part 1. Requirements, critical review of methods and modeling. *J. Power Sources* **2014**, *260*, 276–291. [CrossRef]

37. Schweighofer, B.; Raab, K.M.; Brasseur, G. Modeling of high power automotive batteries by the use of an automated test system. *IEEE Trans. Instrum. Meas.* **2003**, *52*, 1087–1091. [CrossRef]

38. Castano, S.; Gauchia, L.; Voncila, E.; Sanz, J. Dynamical modeling procedure of a Li-ion battery pack suitable for real-time applications. *Energy Convers. Manag.* **2015**, *92*, 396–405. [CrossRef]

39. Chen, M.; Rincon-Mora, G.A. Accurate electrical battery model capable of predicting runtime and iv performance. *IEEE Trans. Energy Convers.* **2006**, *21*, 504–511. [CrossRef]

40. Baronti, F.; Fantechi, G.; Leonardi, E.; Roncella, R.; Saletti, R. Enhanced model for lithium-polymer cells including temperature effects. In Proceedings of the IECON 2010—36th Annual Conference on IEEE Industrial Electronics Society, Glendale, AZ, USA, 7–10 November 2010; pp. 2329–2333.

41. Lam, L.; Bauer, P.; Kelder, E. A practical circuit-based model for li-ion battery cells in electric vehicle applications. In Proceedings of the 2011 IEEE 33rd International Telecommunications Energy Conference (INTELEC), Amsterdam, The Netherlands, 9–13 October 2011; pp. 1–9.

42. Hu, Y.; Wang, Y.-Y. Two time-scaled battery model identification with application to battery state estimation. *IEEE Trans. Control Syst. Technol.* **2015**, *23*, 1180–1188. [CrossRef]

43. Li, J.; Mazzola, M.S. Accurate battery pack modeling for automotive applications. *J. Power Sources* **2013**, *237*, 215–228. [CrossRef]

44. Widanage, W.; Barai, A.; Chouchelamane, G.; Uddin, K.; McGordon, A.; Marco, J.; Jennings, P. Design and use of multisine signals for Li-ion battery equivalent circuit modelling. Part 1: Signal design. *J. Power Sources* **2016**, *324*, 70–78. [CrossRef]

45. Jossen, A. Fundamentals of battery dynamics. *J. Power Sources* **2006**, *154*, 530–538. [CrossRef]

46. Hentunen, A.; Lehmuspelto, T.; Suomela, J. Time-domain parameter extraction method for thévenin-equivalent circuit battery models. *IEEE Trans. Energy Convers.* **2014**, *29*, 558–566. [CrossRef]

47. Petzl, M.; Danzer, M.A. Advancements in *OCV* measurement and analysis for lithium-ion batteries. *IEEE Trans. Energy Convers.* **2013**, *28*, 675–681. [CrossRef]

48. Barai, A.; Widanage, W.D.; Marco, J.; McGordon, A.; Jennings, P. A study of the open circuit voltage characterization technique and hysteresis assessment of lithium-ion cells. *J. Power Sources* **2015**, *295*, 99–107. [CrossRef]

49. Hariharan, K.S.; Kumar, V.S. A nonlinear equivalent circuit model for lithium ion cells. *J. Power Sources* **2013**, *222*, 210–217. [CrossRef]

50. Gong, X.; Xiong, R.; Mi, C.C. A data-driven bias-correction-method-based lithium-ion battery modeling approach for electric vehicle applications. *IEEE Trans. Ind. Appl.* **2016**, *52*, 1759–1765.

7

Non-Linear Behavioral Modeling for DC-DC Converters and Dynamic Analysis of Distributed Energy Systems

Xiancheng Zheng, Husan Ali *, Xiaohua Wu, Haider Zaman and Shahbaz Khan

Department of Electrical Engineering, School of Automation, Northwestern Polytechnical University, Xi'an 710000, China; zxcer@nwpu.edu.cn (X.Z.); wxh@nwpu.edu.cn (X.W.); hdrzaman@hotmail.com (H.Z.); muhd_shahbaz@yahoo.com (S.K.)
* Correspondence: engr.husan@gmail.com

Academic Editor: João P. S. Catalão

Abstract: In modern distributed energy systems (DES), focus is shifting from the conventional centralized approach towards distributed architectures. However, modeling and analysis of these systems is more complex, as it involves the interface of multiple energy sources with many different type of loads through power electronics converters. The integration of power electronics converters allows distributed renewable energy sources to become part of modern electronics power distribution systems (EPDS). It will also facilitate the ongoing research towards DC-based DES which is mostly composed of commercial DC-DC converters whose internal structure and parameters are unknown. For the system level analysis, the behavioral modeling technique is the only choice. Since most power electronics converters are non-linear systems and linear models can't model their dynamics to a desired level of accuracy, hence non-linear modeling is required for accurate modeling. The non-linear modeling approach presented here aims to develop behavioral models that can predict the response of the system over the entire operating range. In this work, either a lookup table or a polytopic structure-based modeling technique is used. The technique is further applied to cascade and parallel connected converters, being two DES scenarios. First the procedure is verified via application to switching models in a simulation and then validated for commercial converters via experiments. The results show that the developed behavioral models accurately predict both the transient and steady state response.

Keywords: distributed energy system (DES); electronic power distribution; power electronics converter; behavioral modeling; non-linear

1. Introduction

Modern distributed energy systems (DES) are comprised of multiple converters in which the loads are supplied by several low power converters, distributed throughout the system [1–3]. These systems include more than one source of energy, energy storage elements and several active and passive loads, either DC, AC or both. This implies the use of power electronics converters, required for power distribution over the entire system, hence also called electronic power distribution systems (EPDS). Today's AC distribution systems are swiftly being replaced by DC distribution systems in many areas, motivated by the extensive use of electronic loads and integration of renewable energy sources into existing systems [3].

In the literature, DC-based energy distribution systems have been discussed for more electric aircraft (MEA) [4], electric/hybrid electric vehicles (HEVs) [5,6], all electric ships (AESs) [7], telecom applications [8,9] and for commercial and residential services [10,11]. The active nature of power electronics converters results in complex dynamic behavior during interconnection [12]. It makes the system level analysis of interconnected converters much more complicated, hence simulation tools are required to analyze and predict the behavior of complete systems [13,14]. Modeling and simulation are essential steps during the design stage of the complete system. The requirement to model power converters for system level analysis was first discussed in [15], and subsequently work has been done in this direction [2,16]. However conventional modeling techniques rely on the availability of information about the internal structure, i.e., topology, control, etc., of power electronics converters.

Modeling of power converters can be broadly categorized into white-box [2,17,18] and *black-box* approaches [19–21]. When all the necessary data to model the system's behavior is available, then in such cases a white-box modeling approach is useful, while black-box or behavioral modeling refer to the modeling technique in which models for power converters and passive modules e.g., electromagnetic induction (EMI) filters are built without any available information about their internal design and components. The models of power electronics converters with minimum or no detail about the system are used to analyze the input-output behavior of the system. Such models can be easily interconnected with each other in various configurations, such as cascade, parallel, series and stacking form to build distributed energy systems.

The two port network-based linear behavioral modeling approach doesn't require any details about the internal design and structure of the converter to be known. Linear behavioral modeling is well suited for converters whose behavior is linear over the entire operating range, e.g., un-regulated buck converters, but when applied to non-linear converters, it fails to give accurate results for large signal perturbations (load current or input voltage step). In fact it causes the operating point to move away from the region at which the converter was linearized, so non-linear behavioral modeling must be employed for power converters with strong non-linearities.

One type of non-linear behavioral modeling is based upon the series connection of a linear model with a non-linear function. If the non-linear function precedes the linear model, it is called a Hammerstein model and it is called a Wiener model when vice-versa. In [22] a hybrid Wiener-Hammerstein structure was developed using the data provided in the datasheets along with the transient response of the converter. It is based on the cascade combination of a non-linear static network and a linear dynamic network which cover the steady state and transient behavior of the converter, respectively. The non-linear hybrid terminal behavioral model proposed in [23], is based upon a Hammerstein approach, but it is limited to the converters that feature only static non-linearity and are dynamically linear. This approach models the static behavior, but is unable to accurately predict the dynamic behavior.

In this paper a look up table or polytopic structure-based non-linear behavioral modeling approach is applied to power converters exhibiting non-linear behavior. It is investigated first if each of the four g-parameters behave in a linear or non-linear way, as determined from the transient as well as the frequency response data. For the non-linear case, if the lookup table-based approach can't handle the non-linearity, then the more complex polytopic structure-based approach is used. When a polytopic structure is used to model each dynamic system, it results in a highly accurate model, but at the cost of increased complexity and computational time, so there a trade-off must be made between accuracy and simplicity.

In modern distributed energy systems power electronics converters are often connected in various configurations, i.e., cascade, series, parallel and stacking. The non-linear behavioral modeling is further extended to analyze cascade and parallel configurations for system level analysis. First the behavioral models are built for individual converters and then the models are interconnected for the dynamic analysis of the complete system.

The behavioral modeling approach is first verified via simulation using the MATLAB/Simulink Software package (MathWorks, Natick, MA, USA) [24]. Then the approach is validated experimentally for commercial DC-DC converters. Both in the case of verification via simulation and validation via experiment, the results from the actual converters are compared with the behavioral models. The close agreement of the results demonstrates that the non-linear behavioral modeling approach is able to predict the transient as well as the steady responses of the modeled systems with high accuracy.

The paper is organized as follows: Section 2 explains the two port network-based behavioral modeling of DC-DC converters. Section 3 covers the non-linear behavioral modeling. Section 4 explains the modeling of a distributed energy system. Finally Section 5 gives the conclusion of the work presented.

2. Behavioral Modeling of DC-DC Converters

The two port network-based modeling technique was first applied to DC-DC converters in [15,25], while the first identification procedure was proposed in [21]. It is based upon the measurement of frequency responses via small signal perturbations, obtained using a network analyzer. Another method, which doesn't require expensive equipment is based upon the step change in transient response, the time domain data is then used for the identification of frequency responses [26].

The g-parameters-based two port network is a hardware-oriented behavioral modeling approach, which doesn't require any knowledge about the internal design of the converter. Hence there is no difference in the modeling methodology for various type of converters, i.e., buck, boost etc. The complete model is based upon the measurement and identification of four linear time invariant (LTI) models as transfer functions in the Laplace domain, i.e., output impedance (Z_o), back current gain (H_i), audiosusceptibility (G_o) and input admittance (Y_i). The two-port network shown in Figure 1 represents un-terminated model, so the dynamic system based upon it should model only the internal dynamics of the converter. To achieve this the measurement setup should have almost no interaction either with the source or load. This is possible if the converter is fed from a low output impedance voltage source and connected to an electronic load in constant current sink mode [27]. Using such a decoupling procedure the influence of external elements such as filters and other converters can be removed from the measurements.

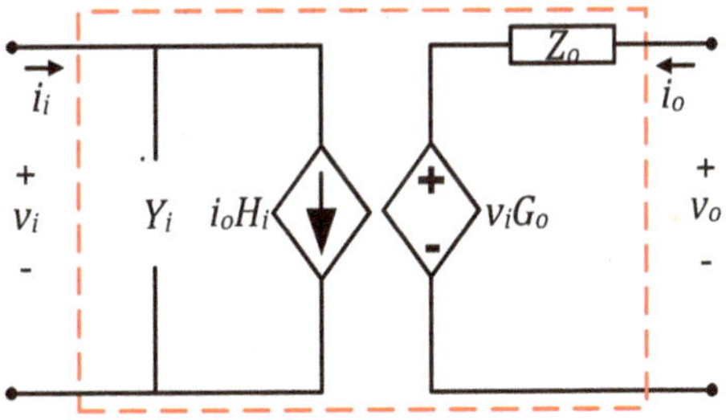

Figure 1. G-parameters based two port network model for DC-DC converter.

The g-parameter set required to be measured is given in Equation (1):

$$Y_i = \frac{i_i}{v_i}\Big|_{i_o=0} \quad H_i = \frac{i_i}{i_o}\Big|_{v_i=0}$$
$$G_o = \frac{v_o}{v_i}\Big|_{i_o=0} \quad Z_o = \frac{v_o}{i_o}\Big|_{v_i=0} \tag{1}$$

The input variables of the two port network are the input voltage and output current (v_i, i_o) while the output variables are the output voltage and input current (v_o, i_i).

The problem with linear two port network-based modeling is that it is assumed that converters are mildly non-linear [1]. For linear or mildly non-linear dynamic relations a single LTI model is sufficient. In practice most of the power electronic converters are non-linear systems, so some non-linear behavioral modeling technique must be employed. The non-linear behavioral modeling technique employed here differs from the linear behavioral modeling shown in Figure 1 in the sense that single LTI models are replaced by a lookup table or a polytopic structure.

The modification of the linear behavioral model requires some sort of iteration of the linearization process applied at different operating points. In the end all these linear models can be unified into a single non-linear structure called polytopic modeling. A polytopic structure is built by the combination of several local LTI models. In it, small perturbations are applied on the input variables at different operating points. The use of a number of local linear models for a non-linear system have been used in several other applications, e.g., the fuzzy logic-based systems called Takagi-Sugeno models [28,29], neural network-based systems [30,31], mechanics [32,33], robotics [34], aeronautics [35] and electromechanical systems [36]. The application of polytopic modeling to power electronic systems is discussed in [37,38].

In case of polytopic structure-based non-linear behavioral modeling, the system is modeled by a set of linear models which are valid for different operating points. These local linear models are then combined by interpolation to cover the entire operating range. When the operating point changes, the model makes a smooth transition from one local linear model to another. The width of the space for each local model is chosen such as the system's behavior is linear within a sub-space. Linear models are measured and identified for each sub-space. Then with the change in operating condition, the shift from one operating point to another is made by using some validity function.

For a general non-linear system:

$$\dot{x} = f(x, u)$$
$$y = h(x, u)$$
(2)

where x, u and y represent the state, input and output variables respectively. The mathematical representation of the polytopic structure is:

$$\dot{x} = \sum_{i=1}^{n} w_i(x, u)(A_i x + B_i u)$$
$$y = \sum_{i=1}^{n} w_i(x, u)(C_i x + D_i u)$$
(3)

$w_i(\cdot)$ is called validity or interpolation function, describing the region where the local models are valid, and it may be a function of either x or u. To avoid switching and have a smooth transition from one local model to another, the validity function should satisfy the condition, i.e.,:

$$0 \leq w_i(u_i) \leq 1$$
(4)

Furthermore it is also necessary that the sum of weighting functions at any point within the entire operating space should be equal to one:

$$\sum_{i=1}^{n} w_i(u_i) = 1$$
(5)

The weighting functions are placed in the center of each sub-space, such that:

$$w_i(u) = 1, \text{ if } u \in u_i$$
$$w_i(u) = 0, \text{ if } u \notin u_i$$
(6)

A few of the commonly used weighting functions are triangular [28], sigmoid, double sigmoid and trapezoidal functions. The triangular function used in this work is defined by Equation (7):

$$f(x; a, b, c) = \begin{cases} \frac{x-a}{b-a}, & a \leq x \leq b \\ \frac{c-x}{c-b}, & b \leq x \leq c \\ 0, & x \leq a, c \leq x \end{cases}$$
(7)

where the parameters a and c represent the end points of the triangle and b shows the center.

The partitioning of the operating region into sub-spaces mainly depends upon the non-linearity shown by the converter over the entire range [39]. The center of the sub-space is the point at which the frequency response function is measured. If the number of local linear models obtained are unable to predict the response accurately over a certain region, then more local models should be obtained by further reducing the width of the sub-space.

The output of local linear models with input u and weighting function w_i can be calculated by weighting the validity function values and adding them together:

$$Y = \sum_{i=1}^{n} w_i(u) f(u_i) \tag{8}$$

When the local linear models are dependent on two variables, then 2-d polytopic modeling is employed in which several 1-d polytopic models are grouped together, but increase in the number of variables for polytopic modeling cause an exponential increase in the number of local models.

In order to identify the dynamic models from the measured data, parametric and non-parametric methods can be used [40]. Parametric methods are applied to data obtained either from frequency or transient response measurements, from which transfer functions or state-space models are obtained. Among the available model structures, a few commonly used ones are: ARX, ARMAX, OE, BJ and the state-space model, and the most appropriate one should be selected [40–42]. A proper model structure is important for accurate identification. Also, the order of the model plays its role in terms of fitting accuracy. Normally there is a trade-off between accuracy and simplicity of the model. Once the transfer functions are identified, behavioral model is constructed and implemented in MATLAB/Simulink.

3. Non-Linear Behavioral Modeling

3.1. Verification via Simulation

The aim of this section is to explain the non-linear behavioral modeling procedure and verify it by simulation. A switch model of a regulated DC-DC buck converter (30/10 V, 100 W) is simulated using MATLAB/Simulink. The dynamic systems as transfer functions are subsequently identified from step transient response data. Depending upon the behavior exhibited by a dynamic relation, either a linear or a non-linear model is used. Finally the response of the behavioral and switching model is compared.

The step change in the output current is applied to identify output impedance and back current gain. Also step changes in the input voltage are used to identify input admittance and audiosusceptibility. The step tests are performed such that only one input signal is perturbed at a time, while the other is kept constant. The step transient response data is used for the identification of corresponding frequency responses, where the midpoint of the step change is considered as the operating point. The transient as well as corresponding frequency responses are given for each case.

To evaluate output impedance, the output voltage response is analyzed for step changes in the load current. In first case the input voltage is kept constant and a load current step with different values is applied as shown in Figure 2a,b. In the second case the input voltage is changed for the same load current step shown in Figure 2c,d. It can be seen that if the input voltage is kept constant for each load step, the response is same, thus a single LTI model is sufficient. However in the second case when the same load step is applied at different input voltage values, both the transient as well as the frequency response vary, indicating non-linearity.

Figure 3 shows a 1-d polytopic model for output impedance, where the local linear models are evaluated for different values of input voltage.

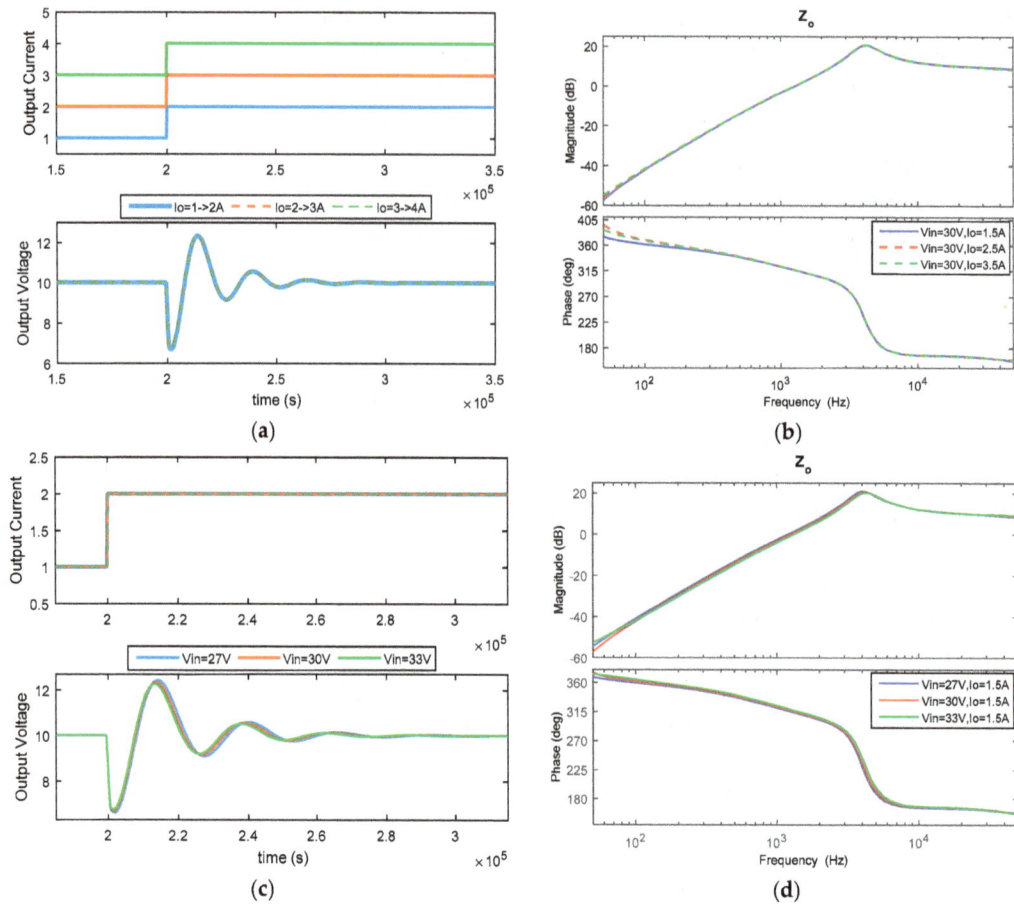

Figure 2. Transient and frequency responses as a function of (**a,b**) load current; and (**c,d**) input voltage.

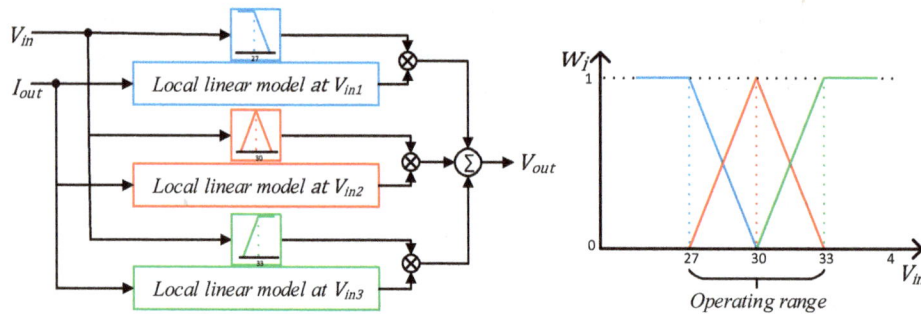

Figure 3. 1-d polytopic model.

To evaluate back current gain, the input current response is analyzed for step change in load current. In first case the input voltage is kept constant and load current step is applied at different values as shown in Figure 4a,b. In the second case the input voltage is changed for the same load current step shown in Figure 4c,d. It can be seen that in either case, both transient as well as the frequency response vary, indicating 2-d non-linearity. It should be noted that for step change from 2–3 A and 3–4 A the input current values also increases, but the increase is subtracted to match the current value at 1 A, so that comparison for all the responses can be made easily. A similar approach is also used in the figures ahead where there is an increase in input current.

Figure 5 shows a 2-d polytopic model for back current gain, where first the local linear models are evaluated for different values of input voltage V_{in}, and then for each value of V_{in} models are evaluated for different values of I_{out}.

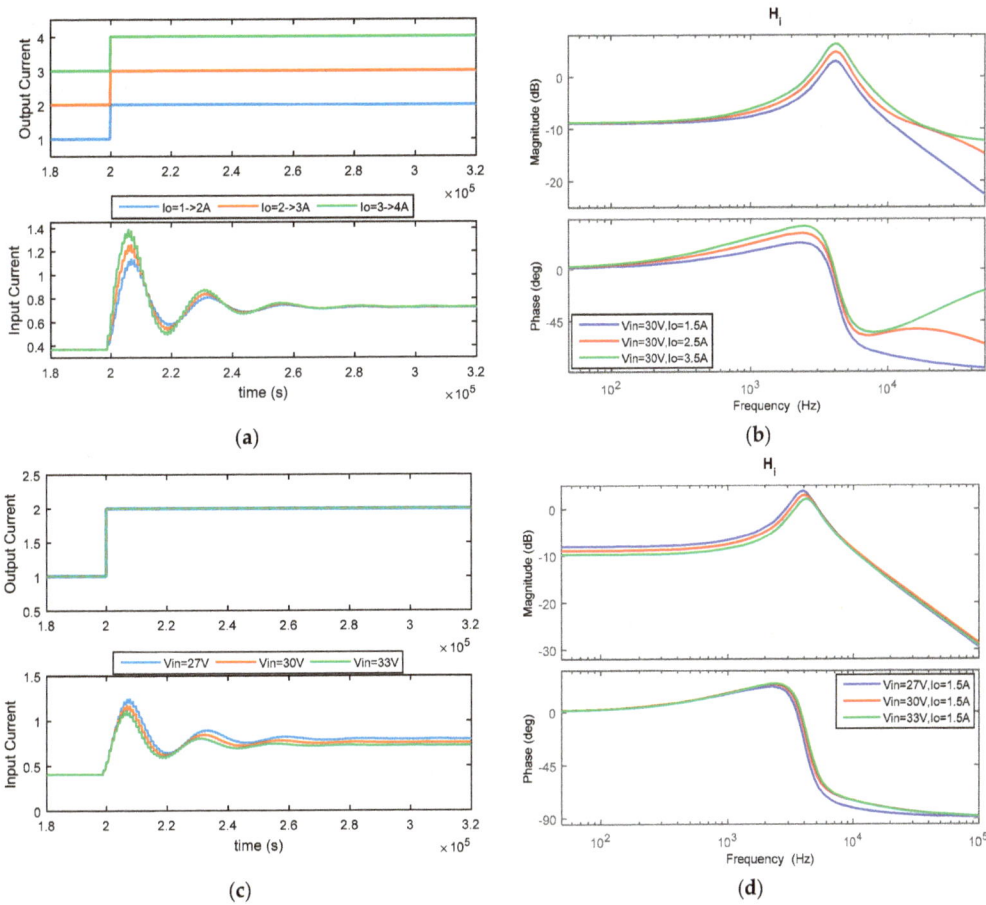

Figure 4. Transient and frequency responses as a function of (**a**,**b**) load current; and (**c**,**d**) input voltage.

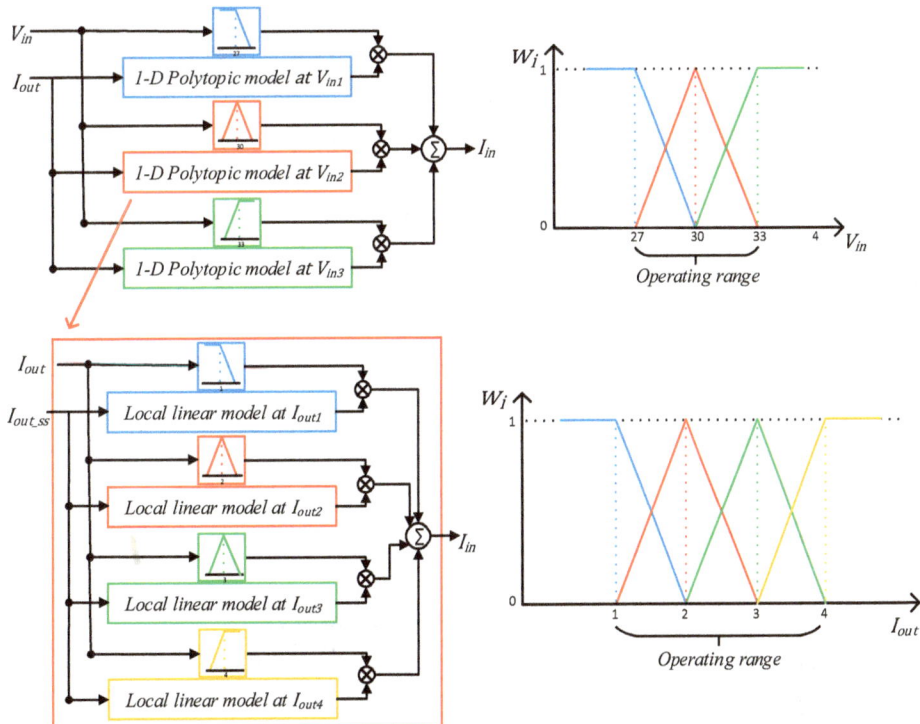

Figure 5. 2-d polytopic model.

To evaluate the audiosusceptibility, the output voltage response is analyzed for step changes in input voltage. In the first case the load current is kept constant and different values of the input voltage step are applied as shown in Figure 6a,b. In the second case the load current is changed for the same input voltage step as shown in Figure 6c,d.

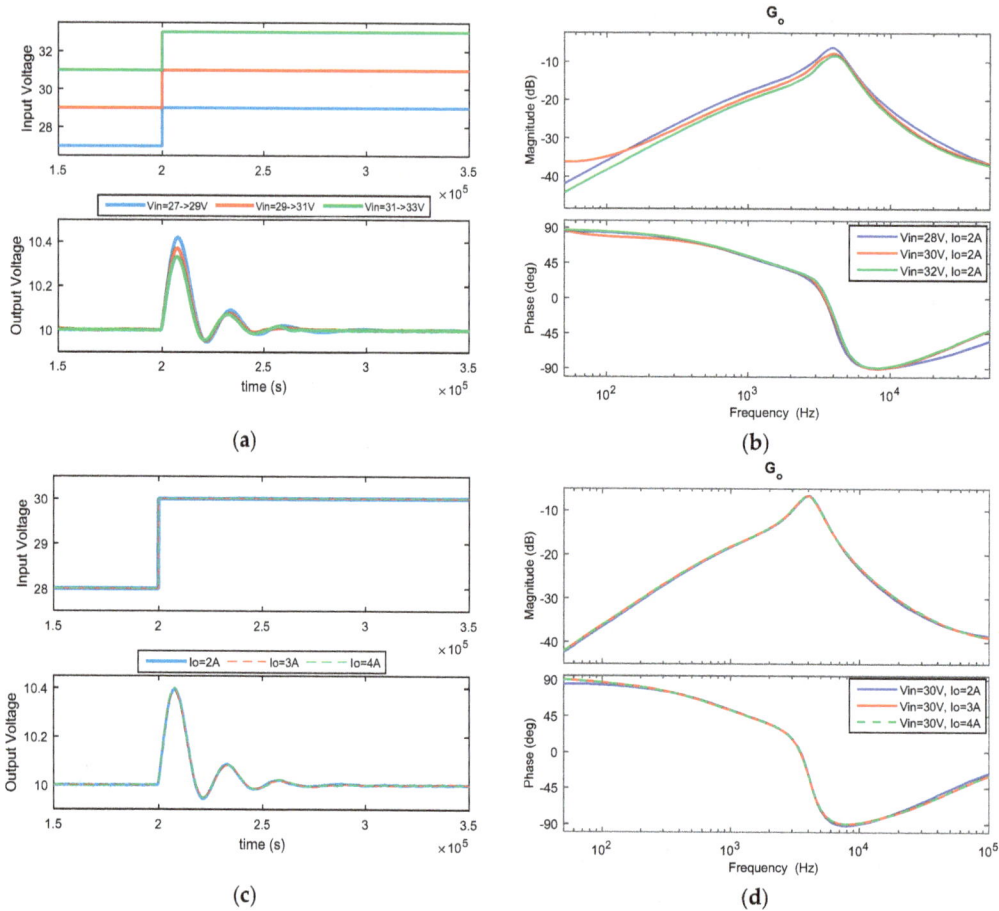

Figure 6. Transient and frequency responses as a function of (**a,b**) input voltage; and (**c,d**) load current.

It can be seen that if the load current is kept constant for each voltage step, both transients as well as frequency responses vary with a constant gain, so a lookup table-based approach is used, where the input voltage serves as an input that looks for a constant multiplier to be multiplied with the LTI model, i.e., a transfer function measured at the nominal operating point. However in the second case when the same voltage step is applied at different values of load current, the response is the same. Hence a single LTI model is sufficient.

Figure 7 shows the representation of 1-d lookup table structure.

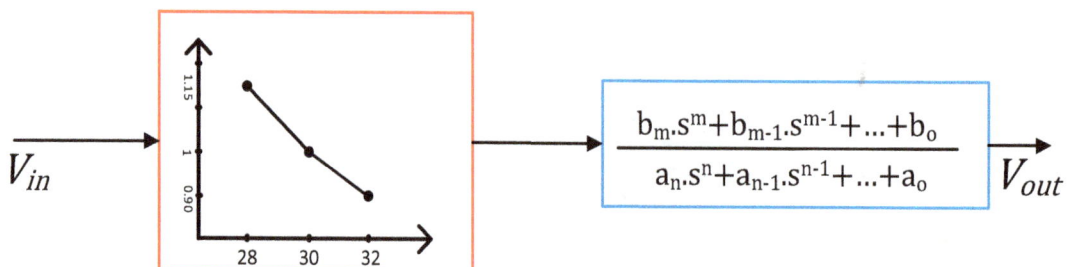

Figure 7. 1-d lookup table model.

To evaluate the input admittance, the input current response is analyzed for step changes in input voltage. In the first case the load current is kept constant and different value input voltage steps are applied as shown in Figure 8a,b. In the second case the load current is changed for the same input voltage step as shown in Figure 8c,d.

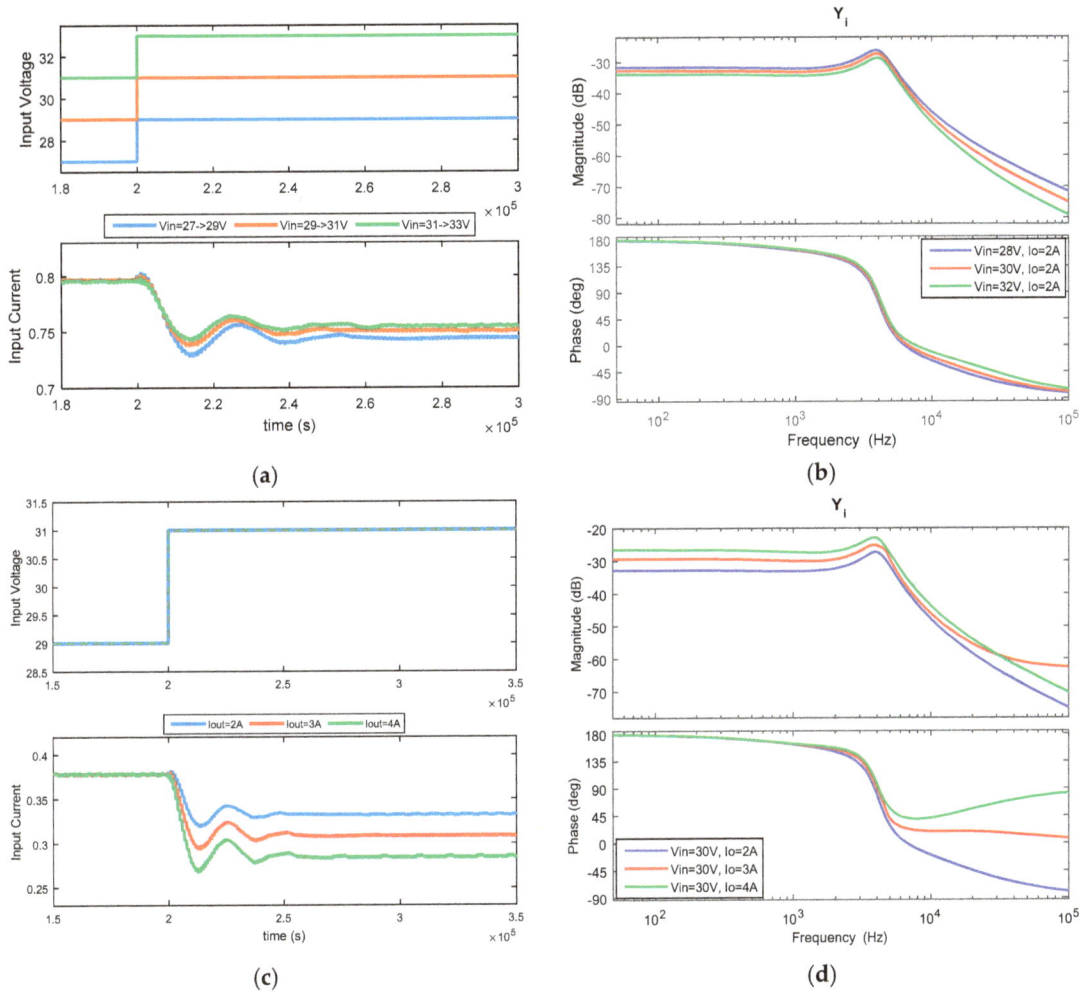

Figure 8. Transient and frequency responses as a function (**a,b**) input voltage; and (**c,d**) load current.

It can be seen that in both cases the responses are related by a linear gain, so a 2-d lookup table is used to model this non-linear dynamic relation.

Figure 9 shows the 2-d lookup table structure.

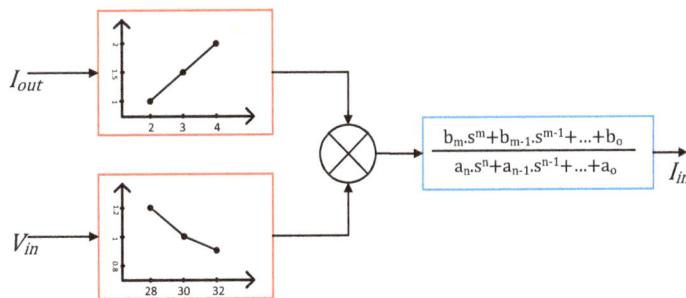

Figure 9. 2-d lookup table model.

To verify the modeling done for four g-parameters a step change in load current is applied. The model is verified with data that is different from that used to build the model [40]. The response of the switch model for each of the output variables is compared against the behavioral model and their close match suggests the effectiveness of the model, as shown in Figure 10. It is thus verified that the developed behavioral model has the capability to accurately represent the dynamic behavior of the converter over a wide operating region.

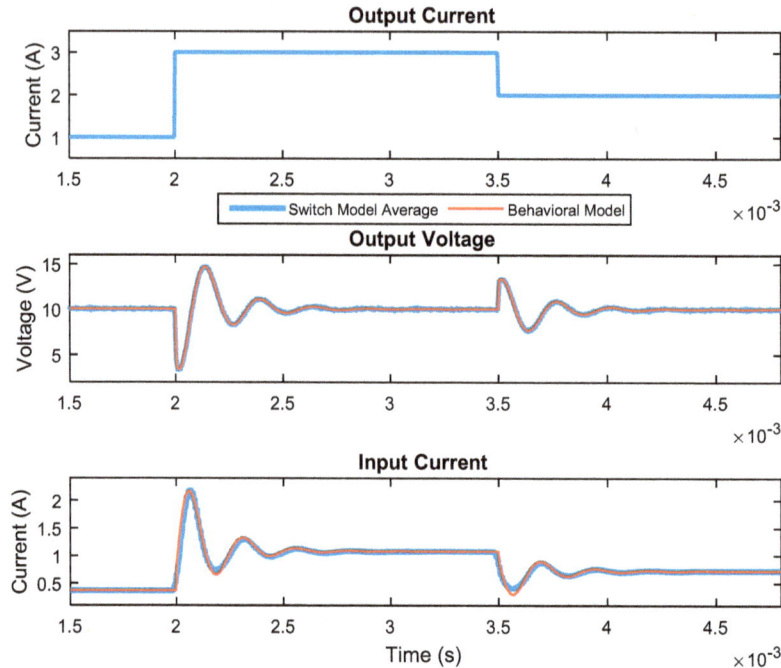

Figure 10. Non-linear behavioral model verification.

For further verification of the model, root mean square deviation (*RMSD*) values are calculated for the above waveforms. The switch model's response is compared with the behavioral model's response for output voltage and input current:

$$RMSD\left(v_o^{\text{swt}}, v_o^{\text{beh model}}\right) = \sqrt{\frac{\sum_{x=1}^{n}\left(v_{o,x}^{\text{swt}} - v_{o,x}^{\text{beh model}}\right)^2}{n}} = 0.0767$$

$$RMSD\left(i_i^{\text{swt}}, i_i^{\text{beh model}}\right) = \sqrt{\frac{\sum_{x=1}^{n}\left(i_{i,x}^{\text{swt}} - i_{i,x}^{\text{beh model}}\right)^2}{n}} = 0.0455$$

(9)

3.2. Validation via Experiment

In this section the non-linear behavioral modeling procedure developed in the previous section is experimentally validated for a commercial DC-DC converter, i.e., a SD-100B-12 (30/12 V, 100 W) [43]. Figure 11 shows the identified frequency responses for Z_o, H_i and Y_i, while in case of G_o the output voltage remains unperturbed to change in input voltage or load current so its value is introduced as a constant in the model.

In order to validate the performance of the behavioral model, step changes in the load current are applied. The input and output signals are recorded using an oscilloscope and then imported into MATLAB. The actual load current from the experiment is applied to the behavioral model and its output is compared with that of the experiment. In Figure 12 it can be clearly seen that the response of the behavioral model closely matches that of the actual converter.

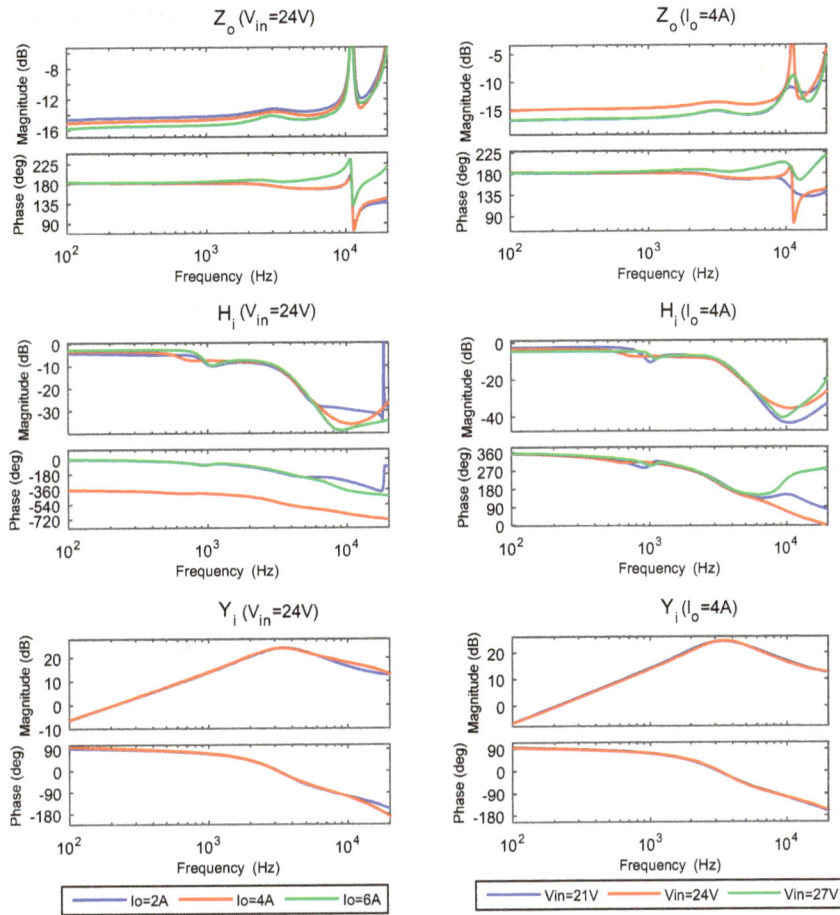

Figure 11. Frequency responses for Z_o, H_i and Y_i of commercial converter.

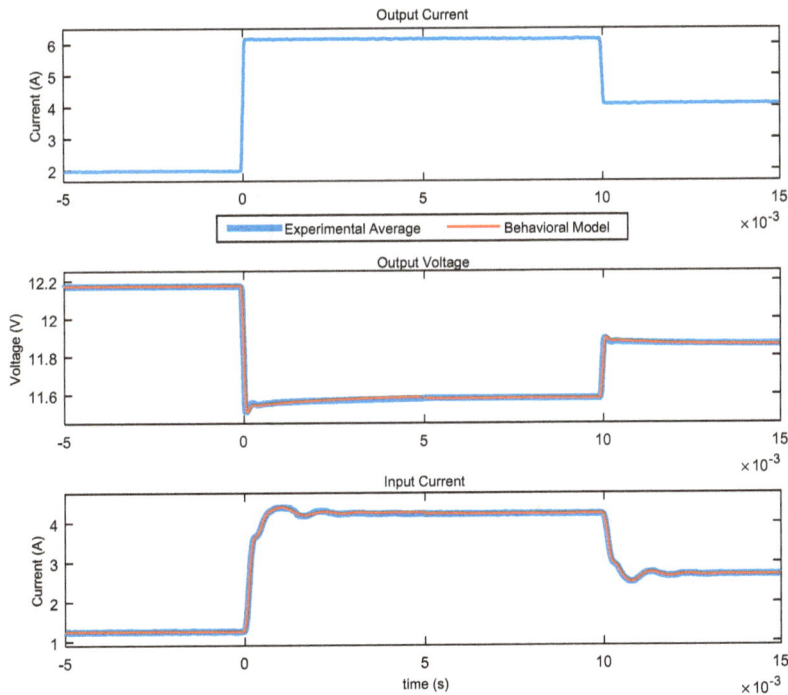

Figure 12. Experimental validation for non-linear behavioral modeling.

The *RMSD* values are also calculated for the experimentally obtained output voltage and input current and compared with its corresponding behavioral model's response:

$$RMSD\left(v_o^{\text{expt}}, v_o^{\text{beh model}}\right) = 0.0158$$

$$RMSD\left(i_i^{\text{expt}}, i_i^{\text{beh model}}\right) = 0.0184$$

(10)

The results show that the non-linear behavioral model is able to predict not only the steady state value, but also the transient response, i.e., overshoot and natural frequency of oscillations both in the case of simulation and experiment.

4. Modeling of Distributed Energy Systems

Behavioral models based upon two port networks are suitable for the system level design and analysis of larger distributed energy systems. The two terminal nature of these models makes it an appropriate choice for cascade and parallel connected converters. Such configurations are commonly found in modern EPDS [8,9], so it is necessary to build models for the analysis of such interconnected systems.

The transient dynamics of the interconnected converters are not only dependent upon the converters themselves, but also on the elements to which they are connected [44,45], so it is investigated whether the models developed for converters working in standalone mode remain valid when they become part of a distributed system. Here two commonly used configurations, i.e., cascade and parallel, are analyzed.

4.1. Parallel Connected Converters

The advantage of converters connected in parallel configuration allows for online replacement of any converter which stops working, thus the system keeps running uninterruptedly. The parallel operation of DC-DC converters may also be employed for loads which demand high current. It is also often used in distributed systems to improve the reliability of $N+1$ power converters and also reduce stress on each converter. Figure 13 shows two parallel connected converters, represented in terms of their g-parameters.

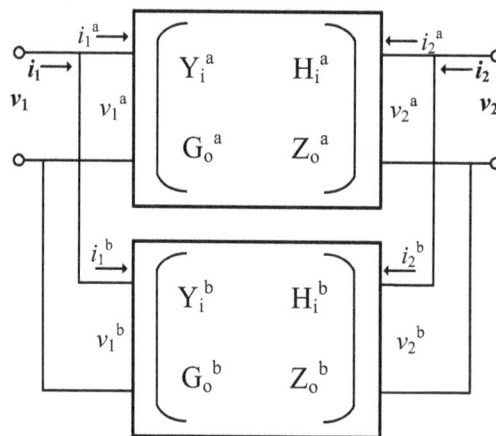

Figure 13. Parallel connected converters as two port network.

First the frequency responses of each converter are measured individually. The matrix representation of both is given below:

$$\text{For Converter } A, \quad \begin{bmatrix} I_1^a \\ V_2^a \end{bmatrix} = \begin{bmatrix} Y_i^a & H_i^a \\ G_o^a & Z_o^a \end{bmatrix} \begin{bmatrix} V_1^a \\ I_2^a \end{bmatrix}$$

(11)

$$\text{For Converter } B, \quad \begin{bmatrix} I_1^b \\ V_2^b \end{bmatrix} = \begin{bmatrix} Y_i^b & H_i^b \\ G_o^b & Z_o^b \end{bmatrix} \begin{bmatrix} V_1^b \\ I_2^b \end{bmatrix} \tag{12}$$

When the two systems are connected in parallel, they can be represented as a single system:

$$\begin{bmatrix} I_1 \\ V_2 \end{bmatrix} = \begin{bmatrix} Y_i & H_i \\ G_o & Z_o \end{bmatrix} \begin{bmatrix} V_1 \\ I_2 \end{bmatrix} \tag{13}$$

As per the procedure discussed in [46], the g-parameters of the overall system are:

$$\begin{bmatrix} Y_i & H_i \\ G_o & Z_o \end{bmatrix} = \begin{bmatrix} \left(Y_i^a + Y_i^b \right) - \dfrac{\left(H_i^a - H_i^b \right)\left(G_o^a - G_o^b \right)}{Z_o^a + Z_o^b} & \dfrac{H_i^a Z_o^b + Z_o^a H_i^b}{Z_o^a + Z_o^b} \\ \dfrac{G_o^a Z_o^b + Z_o^a G_o^b}{Z_o^a + Z_o^b} & \dfrac{Z_o^a Z_o^b}{Z_o^a + Z_o^b} \end{bmatrix} \tag{14}$$

4.1.1. Model Verification via Simulation

For verification via simulation two un-regulated buck converters are connected in parallel. First the two converters are individually operated at the following same operating point:

$$\begin{bmatrix} V_1^a & I_1^a & V_2^a & I_2^a \end{bmatrix} = \begin{bmatrix} 20\text{ V} & 4.6\text{ A} & 9\text{ V} & 10\text{ A} \end{bmatrix} \\ \begin{bmatrix} V_1^b & I_1^b & V_2^b & I_2^b \end{bmatrix} = \begin{bmatrix} 20\text{ V} & 4.6\text{ A} & 9\text{ V} & 10\text{ A} \end{bmatrix} \tag{15}$$

Figure 14 shows the g-parameters measured for each converter and for the parallel system. Since both the converters have identical parameters and have the same operating point, they have similar frequency responses, as shown by the "blue" and "green" colored plots.

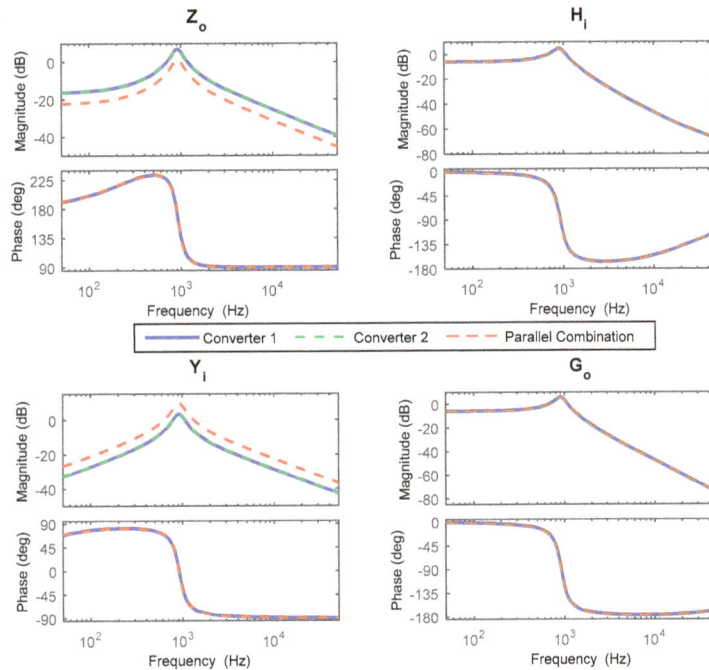

Figure 14. Frequency responses for individual and parallel connected converters.

To verify the behavioral model of the overall parallel system, step change in load current is applied at the input and Figure 15 shows the output voltage and input current response comparison for the switch and behavioral model.

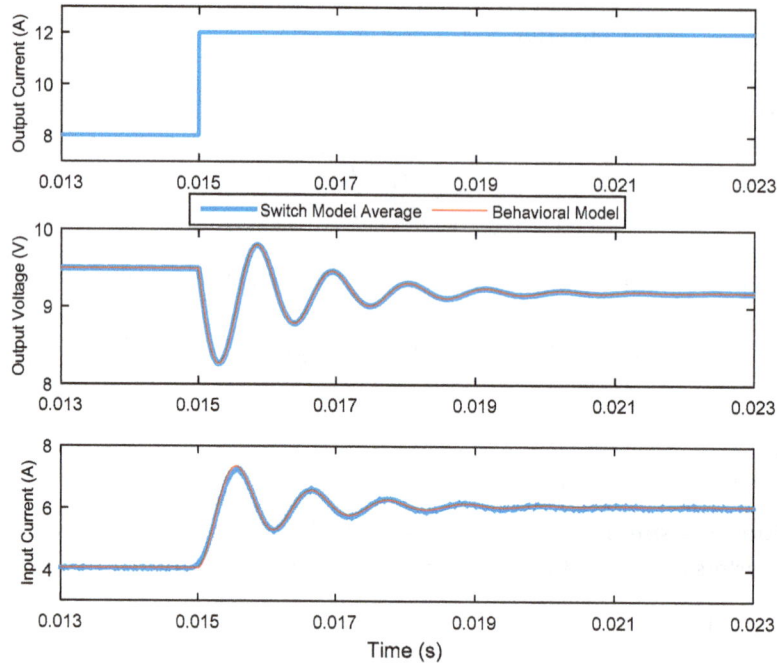

Figure 15. Parallel configuration's modeling verification.

4.1.2. Model Validation via Experiment

The experimental setup is based upon two commercial DC-DC converters, i.e., an SD-100B-12 (30/12V, 100W, MEAN WELL, New Taipei City, Taiwan) and an SD-200B-12 (30/12V 200W), connected to an electronic load. The passive current sharing method is employed and a diode is used with each converter for output decoupling. Due to the passive current sharing, the current drawn from each converter depends upon the output impedance of the two. Hence each converter must be modeled separately.

Using the same procedure as employed in the simulation, first the two converters are individually operated at certain operating point. The g-parameters are measured for each converter and then Equation (14) is used to obtain the equivalent frequency response for the parallel system, which is then used to construct the behavioral model of the parallel connected converters:

$$
\begin{bmatrix} V_1^a & I_1^a & V_2^a & I_2^a \\ V_1^b & I_1^b & V_2^b & I_2^b \end{bmatrix} = \begin{bmatrix} 24\,\text{V} & 0.9\,\text{A} & 12\,\text{V} & 1.1\,\text{A} \\ 24\,\text{V} & 1.4\,\text{A} & 12\,\text{V} & 1.9\,\text{A} \end{bmatrix} \tag{16}
$$

To validate the behavioral model of the parallel system, a step change in load current is applied and the actual input as well as the output signals from the experiment are recorded. The experimental step load current signal is applied to the behavioral model setup constructed in MATLAB and the actual output from experiment is compared with the output of behavioral model, shown in Figure 16.

Both in the case of simulation as well as the experiments the results match pretty well, suggesting that the behavioral model developed for converters obtained in standalone mode remains valid when they are connected in parallel configuration.

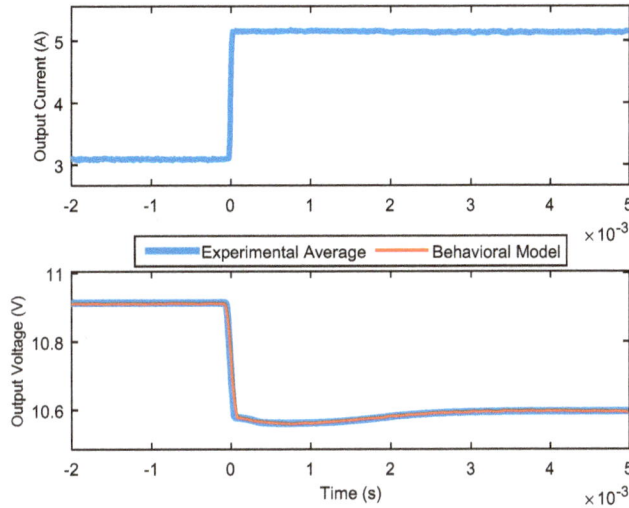

Figure 16. Parallel configuration's modeling validation.

4.2. Cascade Connected Converters

Figure 17 shows two cascade connected converters, represented in terms of their g-parameters.

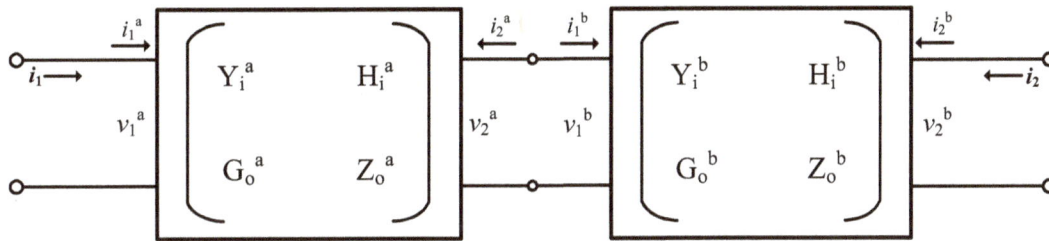

Figure 17. Cascade connected converters as two port network.

First the frequency responses of each converter are measured individually. The matrix representation of both is given below:

For Converter A,
$$\begin{bmatrix} I_1^a \\ V_2^a \end{bmatrix} = \begin{bmatrix} Y_i^a & H_i^a \\ G_o^a & Z_o^a \end{bmatrix} \begin{bmatrix} V_1^a \\ I_2^a \end{bmatrix} \tag{17}$$

For Converter B,
$$\begin{bmatrix} I_1^b \\ V_2^b \end{bmatrix} = \begin{bmatrix} Y_i^b & H_i^b \\ G_o^b & Z_o^b \end{bmatrix} \begin{bmatrix} V_1^b \\ I_2^b \end{bmatrix} \tag{18}$$

When the two systems are connected in cascade, they can be represented as a single system:

$$\begin{bmatrix} I_1 \\ V_2 \end{bmatrix} = \begin{bmatrix} Y_i & H_i \\ G_o & Z_o \end{bmatrix} \begin{bmatrix} V_1 \\ I_2 \end{bmatrix} \tag{19}$$

The equivalent g-parameters of cascade connected converters are represented as [46]:

$$\begin{bmatrix} Y_i & H_i \\ G_o & Z_o \end{bmatrix} = \begin{bmatrix} Y_i^a - \frac{H_i^a G_o^a Y_i^b}{1+Z_o^a Y_i^b} & -\frac{H_i^a H_i^b}{1+Z_o^a Y_i^b} \\ \frac{G_o^a G_o^b}{1+Z_o^a Y_i^b} & Z_o^b - \frac{Z_o^a H_i^b G_o^b}{1+Z_o^a Y_i^b} \end{bmatrix} \tag{20}$$

4.2.1. Model Verification via Simulation

For verification via simulation a regulated boost converter and an un-regulated buck converter are connected in cascade. First the two converters are individually operated at the following operating points:

$$\begin{bmatrix} V_1^a & I_1^a & V_2^a & I_2^a \end{bmatrix} = \begin{bmatrix} 24\,\text{V} & 5.84\,\text{A} & 28\,\text{V} & 4.99\,\text{A} \end{bmatrix}$$
$$\begin{bmatrix} V_1^b & I_1^b & V_2^b & I_2^b \end{bmatrix} = \begin{bmatrix} 28\,\text{V} & 4.99\,\text{A} & 14\,\text{V} & 10\,\text{A} \end{bmatrix} \tag{21}$$

Figure 18 shows the g-parameters measured for each converter and for cascade system.

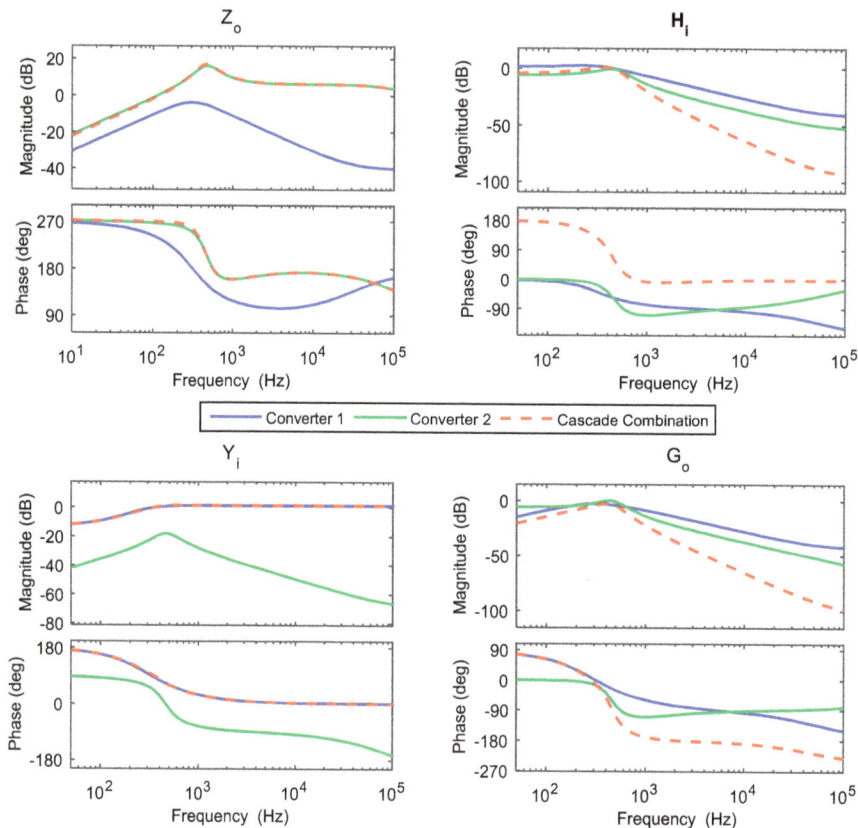

Figure 18. Frequency responses for individual and cascade connected converters.

To verify the behavioral model of the overall cascade system, a step change in load current is applied and responses for output voltage and input current are compared. Figure 19 shows the equivalent behavioral model where the transfer functions used are computed using Equation (20).

However, when the behavioral model's response is compared with the actual switch model response, there is slight mismatch between the results. After some further research it is found that the dynamic response to a step change in load current of an un-regulated buck converter is different when it becomes part of a cascade network compared to standalone mode of operation. Figure 20 shows the step load change response of the un-regulated buck converter in isolated vs. cascade mode.

Therefore, in order to include the effect of these small dynamic changes the frequency responses are measured while the two converters are connected in cascade configuration. Then a modified behavioral model is built as shown in Figure 21, based upon each converter's frequency responses.

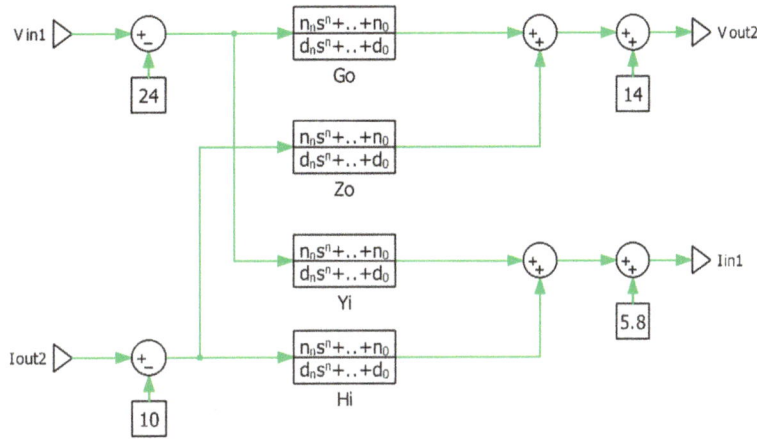

Figure 19. Model for cascade connected converters.

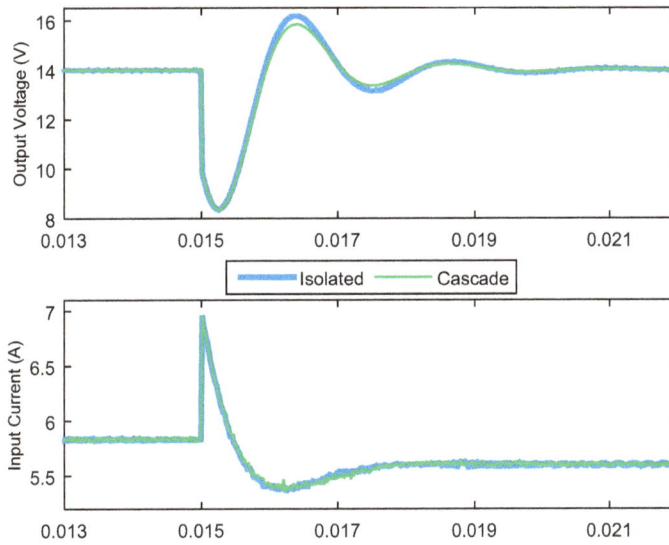

Figure 20. Step load change response for isolated vs. cascade mode.

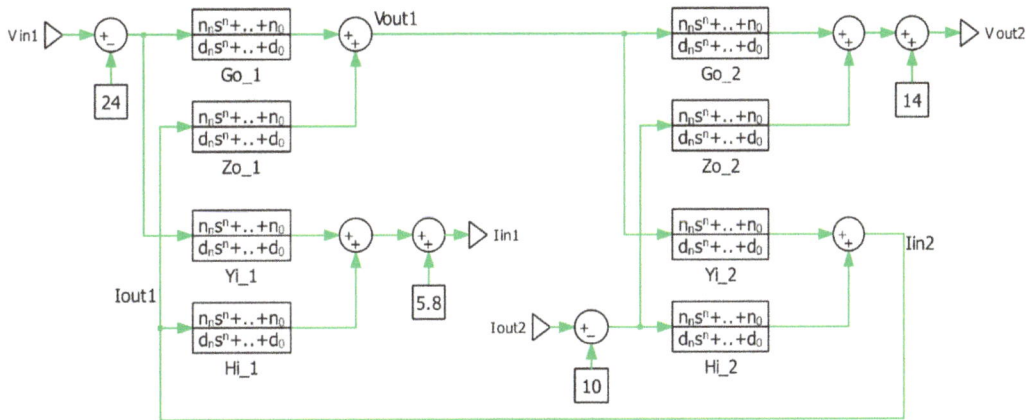

Figure 21. Modified model for cascade connected converters.

Now the result of step change in load current for the switch model is compared with the response of original and modified behavioral model. Figure 22 shows that there is slight mismatch for the switch and original model, while the modified model results are in good agreement with the actual switch model results.

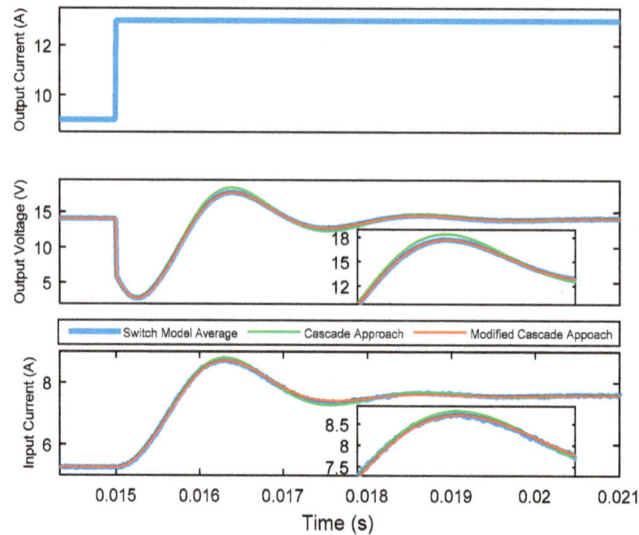

Figure 22. Cascade configuration's modeling verification.

4.2.2. Model Validation via Experiment

The experimental setup is based upon two commercial DC-DC converters, i.e., an SD-100B-24 (30/24 V, 100 W) and an SD-100B-12 (24/12 V 100 W), connected to an electronic load.

A modified behavioral modeling approach is used for experimental validation as well. The g-parameters are measured for each converter, while connected in cascade mode and then Equation (20) is used to obtain the equivalent frequency response for the overall cascade system, which is then used to construct the behavioral model of the cascade connected converters. The measurements are done at the following operating points:

$$
\begin{bmatrix} V_1^a & I_1^a & V_2^a & I_2^a \end{bmatrix} = \begin{bmatrix} 30\text{ V} & 2.1\text{ A} & 24\text{ V} & 2.5\text{ A} \end{bmatrix}
$$
$$
\begin{bmatrix} V_1^b & I_1^b & V_2^b & I_2^b \end{bmatrix} = \begin{bmatrix} 24\text{ V} & 2.5\text{ A} & 12\text{ V} & 4.8\text{ A} \end{bmatrix} \tag{22}
$$

To validate the behavioral model of the cascade system, step changes are applied in the load current and the actual input as well as the output signals from the experiment are recorded. The experimental step load current signal is applied to the behavioral model constructed in MATLAB and the output signals from experiment are compared with that of the behavioral model, shown in Figure 23.

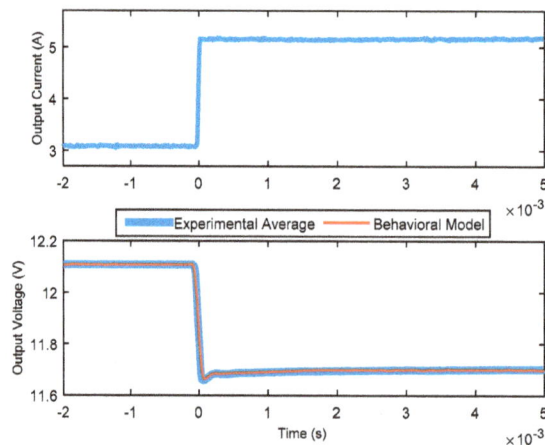

Figure 23. Cascade configuration's modeling validation.

The results from the actual experiment match with those of the behavioral model, validating the procedure described for the analysis of overall cascade and parallel connected converters.

5. Conclusions

The integration of several different energy sources thorough power electronics converters to become electronic power distribution system also requires simulation tools for the design and analysis of such systems. As most of the power converters are provided by different vendors, this means less data is available for their modeling. The two-port network-based behavioral modeling approach provides the solution as it relies upon experimental data to model the input-output behavior of the system. It also enables interconnection of different types of power electronics devices for system level analysis from the design perspective. The switching action of power electronics devices causes them to behave in a non-linear way and linear modeling techniques fail to model the entire operating range accurately. A non-linear modeling methodology is presented based upon either a lookup table or a polytopic structure for different dynamic relations. The concept is further extended and applied to two types of distributed energy system, i.e., cascade and parallel configuration. All the work done is first verified by simulation in MATLAB/Simulink and then validated experimentally for commercial converters. Both in the cases of simulation and experiments, the results from the actual system match well with the developed behavioral model, thus proving the effectiveness of the presented work.

Author Contributions: Xiancheng Zheng and Husan Ali contributed equally to the research described in this work. Xiancheng Zheng and Husan Ali conceived and designed the experiments. Husan Ali performed the experiments, analyzed the data and wrote the paper. Xiaohua Wu and Haider Zaman provided significant comments and technical feedback throughout the research. Xiaohua Wu and Shahbaz Khan reviewed and improved the paper.

Conflicts of Interest: The authors declare no conflict of interest.

References

1. Boroyevich, D.; Burgos, R.; Arnedo, L.; Wang, F. Synthesis and integration of future electronic power distribution systems. In Proceedings of the Power Conversion Conference (PCC'07), Nagoya, Japan, 2–5 April 2007.

2. Franz, G.A.; Ludwig, G.W.; Steigerwald, R.L. Modelling and simulation of distributed power systems. In Proceedings of 21st Annual IEEE Power Electronics Specialists Conference (PESC'90), San Antonio, TX, USA, 11–14 June 1990.

3. Dragicevic, T.; Vasquez, J.C.; Guerrero, J.M.; Skrlec, D. Advanced LVDC electrical power architectures and microgrids: A step toward a new generation of power distribution networks. *IEEE Electr. Mag.* **2014**, 2, 54–65. [CrossRef]

4. Izquierdo, D.; Azcona, R.; Cerro, F.J.L.; Fernandez, C.; Delicado, B. Electrical power distribution system (HV270DC), for application in more electric aircraft. In Proceedings of the 25th Annual IEEE Applied Power Electronics Conference and Exposition (APEC), Palm Springs, CA, USA, 21–25 February 2010.

5. Jayabalan, R.; Fahimi, B.; Koenig, A.; Pekarek, S. Applications of power electronics-based systems in vehicular technology: State-of-the art and future trends. In Proceedings of the 35th Annual IEEE Power Electronics Specialists Conference (PESC 04), Aachen, Germany, 20–25 June 2004.

6. Emadi, A.; Lee, Y.J.; Rajashekara, K. Power electronics and motor drives in electric, hybrid electric, and plug-in hybrid electric vehicles. *IEEE Trans. Ind. Electron.* **2008**, 55, 2237–2245. [CrossRef]

7. Ciezki, J.G.; Ashton, R.W. Selection and stability issues associated with a navy shipboard DC zonal electric distribution system. *IEEE Trans. Power Deliv.* **2000**, 15, 665–669. [CrossRef]

8. Schulz, W. ETSI standards and guides for efficient powering of telecommunication and datacom. In Proceedings of the IEEE 29th International Telecommunications Energy Conference (INTELEC), Rome, Italy, 30 September–4 October 2007.

9. Miftakhutdinov, R. Power distribution architecture for tele- and data communication system based on new generation intermediate bus converter. In Proceedings of the IEEE 30th International Telecommunications Energy Conference (INTELEC), San Diego, CA, USA, 14–18 September 2008.

10. Salomonsson, D.; Sannino, A. Low-voltage DC distribution system for commercial power systems with sensitive electronic loads. *IEEE Trans. Power Deliv.* **2007**, *22*, 1620–1627. [CrossRef]

11. Wu, T.-F.; Chen, Y.-K.; Yu, G.-R.; Chang, Y.-C. Design and development of DC-distributed system with grid connection for residential applications. In Proceedings of the IEEE 8th International Conference on Power Electronics and ECCE Asia (ICPE&ECCE), Jeju, Korea, 30 May–3 June 2011.

12. Sanz, M.; Valdivia, V.; Zumel, P.; Moral, D.L.; Fernández, C.; Lázaro, A.; Barrado, A. Analysis of the Stability of Power Electronics Systems: A Practical Approach. In Proceedings of the 29th Annual IEEE Applied Power Electronics Conference and Exposition (APEC), Charlotte, NC, USA, 16–20 March 2014.

13. Feng, X.; Liu, J.; Lee, F.C. Impedance specifications for stable DC distributed power systems. *IEEE Trans. Power Electron.* **2000**, *17*, 157–162. [CrossRef]

14. Liu, J.; Feng, X.; Lee, F.C.; Boroyevich, D. Stability margin monitoring for DC distributed power systems via perturbation approaches. *IEEE Trans. Power Electron.* **2003**, *18*, 1254–1261.

15. Cho, B.H.; Lee, F.C.Y. Modeling and analysis of spacecraft power systems. *IEEE Trans. Power Electron.* **1988**, *3*, 44–54. [CrossRef]

16. Karimi, K.J.; Booker, A.; Mong, A. Modeling, simulation, and verification of large DC power electronics systems. In Proceedings of the 27th Annual IEEE Power Electronics Specialists Conference (PESC'96), Baveno, Italy, 23–27 June 1996.

17. Tam, K.-S.; Yang, L. Functional models for space power electronic circuits. *IEEE Trans. Aerosp. Electron. Syst.* **1995**, *31*, 288–296.

18. Wang, R.; Liu, J.; Wang, H. Universal approach to modeling current mode controlled converters in distributed power systems for large-signal subsystem interactions investigation. In Proceedings of the 22nd Annual IEEE Applied Power Electronics Conference (APEC 2007), Anaheim, CA, USA, 25 February–1 March 2007.

19. Prieto, R.; Laguna-Ruiz, L.; Oliver, J.A.; Cobos, J.A. Parameterization of DC/DC converter models for system level simulation. In Proceedings of the European Conference on Power Electronics and Applications, Aalborg, Denmark, 2–5 September 2007.

20. Oliver, J.A.; Prieto, R.; Romero, V.; Cobos, J.A. Behavioral modeling of DC-DC converters for large-signal simulation of distributed power systems. In Proceedings of the 21st Annual IEEE Applied Power Electronics Conference and Exposition (APEC'06), Dallas, TX, USA, 19–23 March 2006.

21. Arnedo, L.; Burgos, R.; Wang, F.; Boroyevich, D. Black-box terminal characterization modeling of DC-to-DC Converters. In Proceedings of the 22nd Annual IEEE Applied Power Electronics Conference (APEC 2007), Anaheim, CA, USA, 25 February–1 March 2007.

22. Oliver, J.; Prieto, R.; Cobos, J.; Garcia, O.; Alou, P. Hybrid Wiener-Hammerstein structure for grey-box modeling of DC-DC converters. In Proceedings of the 24th Annual IEEE Applied Power Electronics Conference and Exposition (APEC), Washington, DC, USA, 15–19 February 2009.

23. Cvetkovic, I.; Boroyevich, D.; Mattavelli, P.; Lee, F.C.; Dong, D. Nonlinear, hybrid terminal behavioral modeling of a dc-based nanogrid system. In Proceedings of the 26th Annual IEEE Applied Power Electronics Conference and Exposition (APEC), Orlando, FL, USA, 6–11 March 2011.

24. MathWorks. Available online: http://www.mathworks.com (accessed on 5 April 2016).

25. Maranesi, P.G.; Tavazzi, V.; Varoli, V. Two-port characterization of PWM voltage regulators at low frequencies. *IEEE Trans. Ind. Electron.* **1988**, *35*, 444–450. [CrossRef]

26. Valdivia, V.; Barrado, A.; Lazaro, A.; Zumel, P.; Raga, C. Easy modeling and identification procedure for "black box" behavioral models of power electronics converters with reduced order based on transient response analysis. In Proceedings of the 24th Annual IEEE Applied Power Electronics Conference and Exposition (APEC), Washington, DC, USA, 15–19 February 2009.

27. Arnedo, L.; Boroyevich, D.; Burgos, R.; Wang, F. Un-terminated frequency response measurements and model order reduction for black box terminal characterization models. In Proceedings of the 23rd Annual IEEE Applied Power Electronics Conference and Exposition (APEC), Austin, TX, USA, 24–28 February 2008.

28. Takagi, T.; Sugeno, M. Fuzzy identification of systems and its applications to modeling and control. *IEEE Trans. Syst. Man Cybern.* **1985**, *1*, 116–132. [CrossRef]

29. Zhang, H.; Liu, D. *Fuzzy Modeling and Fuzzy Control*, 1st ed.; Birkhäuser Basel: Boston, MA, USA, 2006.

30. Murray-Smith, R.; Kenneth, H. Local model architectures for nonlinear modelling and control. In *Neural Network Engineering in Dynamic Control Systems*; Hunt, K.J., Irwin, G.R., Warwick, K., Eds.; Springer: London, UK, 1995; pp. 61–82.

31. Lin, F.-J.; Huang, M.-S.; Hung, Y.-C.; Kuan, C.-H.; Wang, S.-L.; Lee, Y.-D. Takagi-sugeno-kang type probabilistic fuzzy neural network control for grid-connected LiFePO$_4$ battery storage system. *IET Power Electron.* **2013**, *6*, 1029–1040. [CrossRef]

32. Fujimori, A.; Ljung, L. A polytopic modeling of aircraft by using system identification. In Proceedings of the International Conference on Control and Automation, Budapest, Hungary, 27–29 June 2005.

33. Grof, P.; Petres, Z.; Gyeviki, J. Polytopic model reconstruction of a pneumatic positioning system. In Proceedings of the 5th International Symposium on Applied Computational Intelligence and Informatics, Timisoara, Romania, 28–29 May 2009.

34. Chen, J.; Li, R.; Cao, C. Convex polytopic modeling for flexible joints industrial robot using TP-model transformation. In Proceedings of the IEEE International Conference on Information and Automation (ICIA), Hailar, China, 28–30 July 2014.

35. Huang, Y.; Sun, C.; Qian, C.; Zhang, J.; Wang, L. Polytopic LPV modeling and gain-scheduled switching control for a flexible air-breathing hypersonic vehicle. *J. Syst. Eng. Electron.* **2013**, *24*, 118–127. [CrossRef]

36. Ren, L.; Irwin, G.W.; Flynn, D. Nonlinear identification and control of a turbo generator an on-line scheduled multiple model controller approach. *IEEE Trans. Energy Convers.* **2005**, *20*, 237–245. [CrossRef]

37. Sudhoff, S.D.; Glover, S.F.; Zak, S.H.; Pekarek, S.D.; Zivi, E.J.; Delisle, D.E.; Clayton, D. Stability analysis methodologies for dc power distribution systems. In Proceedings of the 13th International Ship Control Systems Symposium (SCSS), Orlando, FL, USA, 7–9 April 2003.

38. Arnedo, L.; Boroyevich, D.; Burgos, R.; Wang, F. Polytopic Black-Box modeling of DC-DC converters. In Proceedings of the 39th Annual IEEE Power Electronics Specialists Conference (PESC), Rhodes, Greece, 15–19 June 2008.

39. Nelles, O. Axes-oblique partitioning strategies for local model networks. In Proceedings of the IEEE International Symposium on Intelligent Control, Munich, Germany, 4–6 October 2006.

40. Ljung, L. *System Identification—Theory for the User*, 2nd ed.; Prentice Hall: Upper Saddle River, NJ, USA, 1999.

41. Ljung, L. Integrated frequency-time domain tools for system identification. In Proceedings of the American Control Conference, Boston, MA, USA, 30 June–2 July 2004.

42. Ljung, L.; Glad, T. *Modeling of Dynamic Systems*; Prentice Hall: Upper Saddle River, NJ, USA, 1994.

43. MeanWell. Available online: http://www.meanwell.com/webapp/product/search.aspx?prod=SD-100 (accessed on 20 July 2016).

44. Hankaniemi, M.; Suntio, T.; Sippola, M.; Oyj, E. Characterization of regulated converters to ensure stability and performance in distributed power supply systems. In Proceedings of the 27th International Telecommunications Conference (INTELEC), Berlin, Germany, 18–22 September 2005.

45. Hankaniemi, M.; Karppanen, M.; Suntio, T. Load-imposed instability and performance degradation in a regulated converter. *IEE Proc. Electr. Power Appl.* **2006**, *153*, 781–786. [CrossRef]

46. Guotian, Y. The g parameters of series, parallel, cascade connection in two-port networks. *J. Xuzhou Norm. Univ.* **1993**, *11*, 29–33.

8

Performance Analysis of Data-Driven and Model-Based Control Strategies Applied to a Thermal Unit Model

Cihan Turhan [1], Silvio Simani [2,*], Ivan Zajic [3] and Gulden Gokcen Akkurt [4]

[1]   Mechanical Engineering, Izmir Institute of Technology, Gulbahce Campus, Urla, 35430 Izmir, Turkey; cihanturhan@iyte.edu.tr

[2]   Dipartimento di Ingegneria, Università degli Studi di Ferrara. Via Saragat 1E, 44122 Ferrara (FE), Italy

[3]   Control Theory and Applications Centre, Coventry University, Coventry CV1 5FB, UK; zajici@uni.coventry.ac.uk

[4]   Energy Engineering Program, Izmir Institute of Technology, Gulbahce Campus, Urla, 35430 Izmir, Turkey; guldengokcen@iyte.edu.tr

*   Correspondence: silvio.simani@unife.it

Academic Editor: Lei Feng

Abstract: The paper presents the design and the implementation of different advanced control strategies that are applied to a nonlinear model of a thermal unit. A data-driven grey-box identification approach provided the physically–meaningful nonlinear continuous-time model, which represents the benchmark exploited in this work. The control problem of this thermal unit is important, since it constitutes the key element of passive air conditioning systems. The advanced control schemes analysed in this paper are used to regulate the outflow air temperature of the thermal unit by exploiting the inflow air speed, whilst the inflow air temperature is considered as an external disturbance. The reliability and robustness issues of the suggested control methodologies are verified with a Monte Carlo (MC) analysis for simulating modelling uncertainty, disturbance and measurement errors. The achieved results serve to demonstrate the effectiveness and the viable application of the suggested control solutions to air conditioning systems. The benchmark model represents one of the key issues of this study, which is exploited for benchmarking different model-based and data-driven advanced control methodologies through extensive simulations. Moreover, this work highlights the main features of the proposed control schemes, while providing practitioners and heating, ventilating and air conditioning engineers with tools to design robust control strategies for air conditioning systems.

Keywords: modelling and simulation for control; advanced control design; model-based and data-driven approaches; artificial intelligence; thermal unit nonlinear system

1. Introduction

The energy cost used in buildings for the developed countries in Europe was very high, up to 50% of which was due to air conditioning systems. On the other hand, water heating represented 13%, whilst lighting and electric appliances contributed to a further 12% in 2015 [1,2]. Air conditioning modules in buildings yield to heating/cooling tools, which can require also the regulation of advanced thermal comfort parameters represented by the air relative humidity and its temperature. The thermal unit (TU) module is fundamental in these systems, as it supplies the properly-treated air to the buildings. Therefore, the dynamic behaviour of TU modules represents the key point for decreasing the energy consumption and achieving the thermal comfort in buildings [3].

The TU module can be described as a nonlinear time-invariant multivariable dynamic process, which is affected by disturbance and uncertainty terms when analysed for control applications [4,5]. Different control schemes exploited classic regulation approaches, such as on-off control laws and proportional, integral and differential (PID) standard compensators [6–8]. These control schemes are simple with low-cost implementations, but sometimes unable to achieve accurate solutions. Moreover, the TU module can include nonlinear functions [9], such as products between air temperature and mass flow rate, which can require advanced control strategies to achieve more complex thermal comfort indices and lower energy consumption. To overcome these problems, control strategies relying on artificial intelligence (AI) tools, namely artificial neural networks (ANN), fuzzy logic (FL), adaptive neuro-fuzzy inference systems (ANFIS) and model predictive controllers (MPC) have been proposed to obtain more advanced comfort issues in building applications [10–13]. As an example, an FL control scheme was proposed in [14], where the heat, the humidity and the oxygen particle concentration represented the control variables, while the fresh air inflow and the fan circulation rate were the monitored outputs. It was shown that FL allowed for more accurate and straightforward results when compared to linear control schemes. A different FL controller to regulate the air conditioning system temperature was proposed in [15], which was able to easily manage the system nonlinearity.

Other contributions considered different ANN tools that are able to enhance the design of suitable controllers used in air conditioning applications [3,12]. As an example, ANN controllers were proposed in [12] for an air conditioning system and compared with a standard PID regulator. It was shown how these ANN controllers allowed one to achieve a controlled output with a shorter settling time and almost zero overshoot. The main advantage of ANN controllers is represented by their interesting features of automatic learning, easy adaptation and straightforward generalisation. However, more efficient solutions were proposed and based on the ANFIS tool [15–18]. In particular, in [15], ANFIS was successfully exploited as an alternative control strategy with heating, ventilation and air conditioning (HVAC) systems to achieve accurate tracking errors.

Other works proposed MPC schemes for the temperature control of buildings [2,19–21]. As an example, in [2], it was shown that the MPC scheme was able to achieve both thermal comfort and energy saving features. Similarly, [19] suggested a more efficient control method when compared with traditional weather-compensated control schemes. Moreover, [22] addressed an interesting overview of MPC methodologies for HVAC systems.

Note that recent studies considered the achievement of thermal comfort and energy efficiency issues using AI tools. However, the performances obtained by these AI-based methods were not analysed in detail and compared via extensive simulations as proposed in this paper using the Monte Carlo (MC) tool. Moreover, this work illustrates the design and the implementation of different control schemes with application to a nonlinear TU dynamic module proposed by the same authors in [9]. Note also that the same authors presented some preliminary results in [23–25], but the analysis of the achievable properties and the robustness features of the proposed solutions with their reliability characteristics have been described in detail in this paper.

Another key issue of the present study consists of illustrating the viable application of the suggested control schemes to real air conditioning systems. This point is fundamental for enhancing practitioners and HVAC young engineers to acquire the fundamentals and the basic design tools for effective HVAC controller development and application. To this aim, the suggested simulations have been synthesised in the MATLAB® and Simulink® environments and exploiting their standard toolboxes or free software tools. Note that some control strategies proposed in this work were already successfully applied to nonlinear models of energy conversion systems as shown, e.g., in [26–28].

Finally, it is worth observing that several research papers have already dealt with this issue in the past (see, e.g., [22,29]), even if they are limited to MPC solutions. However, this work recalls, analyses and implement different control solutions when applied to the thermal unit model already developed by the authors in [9]. Therefore, the key contribution of the work consists of investigating the viability and the reliability features of the proposed solutions with respect to the considered application example. The control performances achieved in simulation are verified and validated by means of the Monte Carlo tool. The proposed control strategies and the validation tools serve to highlight the potential application of the suggested methodologies to real dynamic processes, such as passive and active air conditioning systems.

The remainder of the paper is organised as follows. Section 2 provides an overview of the TU module and its mathematical description. Section 3 illustrates the suggested control schemes exploited in this study, whilst the obtained results are reported in Section 4. The reliability and robustness characteristics of the proposed tools in simulation are discussed in Section 4.1. Finally, Section 5 ends the work by summarising the main achievements of the paper. Open problems and future issues that require further investigations are also suggested.

2. Thermal Unit Mathematical Description

The TU module considered in this study consists of a fundamental block of the whole test-rig proposed for the description and the assessment of the dynamic behaviour of phase change material (PCM) systems used in passive air conditioning plants. Figure 1 represents the complete PCM system facility considered in [9], where the air flow speed and the temperature are the controlled variables that are exploited to perform the presented simulations and experiments. The heating element included into the PCM system is also sketched.

(a)   (b)

Figure 1. (a) The complete phase change material (PCM) system and (b) the heating element.

On the other hand, Figure 2 illustrates how the inflow air is treated by means of the heating element (HE) of the TU. Moreover, the TU module is in a downstream series connection with a cooling unit, which is not reported in Figure 2.

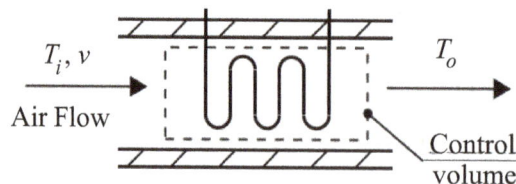

Figure 2. The Thermal Unit (TU) module scheme.

With reference to Figure 2, the measured inlet air temperature, T_i (K), and the air velocity, v (m/s), represent the system inputs considered in this study for describing the dynamic behaviour of the TU module. On the other hand, the system output is the outlet air temperature, T_o (K). The experimental data are acquired from the test-rig of Figure 1, such that the HE sketched in Figure 2 supplies a constant power q (W), with an average value of $q = 830$ (W). As highlighted in Figure 2, the signal measurements of v, T_i and T_o are acquired from the cross-sectional area centre of the supply duct. Note that the input $T_i(t)$ is considered as a disturbance acting on the controlled system.

The mathematical expressions describing the energy balance of the TU module, its air control volume and the energy balance with respect to the adjacent duct walls have the following form:

$$C_h \frac{d\,T_h(t)}{dt} = q(t) - (UA)_h \, (T_h(t) - T_o(t)) \tag{1}$$

$$0 = (UA)_h \, (T_h(t) - T_o(t)) - v(t)\,\rho_a \, A_a \, c_a \, (T_o(t) - T_i(t)) - (UA)_{int} \, (T_o(t) - T_w(t)) \tag{2}$$

$$C_w \frac{d\,T_w(t)}{dt} = (UA)_{int} \, (T_o(t) - T_w(t)) - (UA)_{ext} \, (T_w(t) - T_a(t)) \tag{3}$$

where the variable C_h (J/K) indicates the heating element thermal capacity, C_w (J/K) is the thermal wall capacity (insulated plywood), c_a (J/kg·K) represents the air specific heat capacity, A_a (m^2) is the cross-sectional area of the duct, ρ_a (kg/m^3) denotes the air density, $q(t)$ represents the supplied constant heat gain, whilst U (J/m^2·K) indicates the heat transfer coefficient. Note that in Equations (1)–(3), the term $(UA)_h$ (J/K) indicates the product of the heat transfer coefficient U (J/m^2·K) with the efficient surface area, A (m^2), through which the heat is transmitted and regarding the TU module. On the other hand, $(UA)_{int}$ (J/K) indicates the coefficient with reference to the inner duct wall, whilst $(UA)_{ext}$ (J/K) denotes the same term regarding the outer duct wall.

The variables $T_h(t)$ (K), $T_w(t)$ (K) and $T_a(t)$ (K) denote the average heating element temperature, the wall temperature and outside air temperature, respectively. Note that the air surrounding the heating element is assumed to be perfectly mixed. Therefore, the outlet air temperature $T_o(t)$ is equal to the mean temperature of the whole control volume under the lumped parameter modelling approach (see Equation (1)). Moreover, the thermal capacity of the air passing thought the TU element of Figure 2 is assumed to be very small, so that the heat transfer between the heating element and the air is instantaneous. Under this assumption, the left side of Equation (2) is zero. Finally, in Equation (3), it is assumed that the heat loss occurs only through the walls of the duct.

A standard thermocouple type K has been used to measure the air temperatures. The accuracy is around 1 °C for the whole measurement range. The airflow has been measured using a Hot Wire Thermo-Anemometer with a declared accuracy of 5%. For more details regarding the TU module, which is beyond the scope of this paper, the interested reader is referred to [9].

The authors in [9] showed that the complete dynamic behaviour of the TU module is described by a continuous-time time-invariant nonlinear model consisting of a product of two second-order continuous-time time-invariant transfer functions in the form of Equation (4):

$$T_o(t) = \frac{\hat{\beta}_1 s + \hat{\beta}_2}{s^2 + \hat{\alpha}_1 s + \hat{\alpha}_2} \, (T_i(t)\,v(t)) + \frac{\hat{\eta}_1 s + \hat{\eta}_2}{s^2 + \hat{\alpha}_1 s + \hat{\alpha}_2} \, (T_o(t)\,v(t)) + \hat{\varepsilon} \tag{4}$$

where s denotes a differential operator. The parameters $\hat{\alpha}_i$, $\hat{\beta}_i$, $\hat{\eta}_i$ and $\hat{\varepsilon}$ of the model in the form of Equation (4) were obtained by using a refined instrumental variable method described in [9]. These values are reported in Table 1.

Table 1. Estimated model parameters with their accuracy [9].

Parameter	$\hat{\alpha}_1$	$\hat{\alpha}_2$	$\hat{\beta}_1$	$\hat{\beta}_2$
Value	55.026 ± 2.011	-92.835 ± 3.398	9.9837 ± 0.2512	7.7661 ± 0.2835

Parameter	$\hat{\eta}_1$	$\hat{\eta}_2$	$\hat{\varepsilon}$	
Value	-8.5689 ± 0.2070	-8.3067 ± 0.2984	-122.95 ± 4.426	

The model is considered to be of low complexity yet achieves high simulation performance. The physical meaningfulness of the model provides enhanced insight into the performance and functionality of the system. In return, this information can be used during the system simulation and improved model-based and data-driven control designs for temperature regulation, as shown in the following sections.

3. Control Designs for the TU Module

With reference to the systems sketched in Figure 2 and modelled by the expressions of Equations (1)–(3), the general plant can be described as a Multiple-Input Single-Output (MISO) time-invariant nonlinear model, where the input-output air temperatures and its air flow represent the main input-output variables. Its input-output dynamic behaviour can be described as a nonlinear dynamic function $\mathcal{F}$ in the general form of Equation (5):

$$y(t) = \mathcal{F}(u(t), t) \tag{5}$$

where $y(t)$ is the output variable, i.e., $T_o(t)$, $u(t)$ is the input vector, i.e., $[T_i(t), v(t)]^T$ and t is the time. The control law designed to be applied to the TU module in general determines the control input injected into the controlled plant of Equation (5) in order to track a given reference, or set-point, denoted as $r(t)$.

It is worth observing that the design and the performance of control systems for generic TU processes are strongly determined by the bilinear terms represented in Equation (4), where the nonlinear behaviour is described by the product between the air temperature and its mass flow rate. Under this consideration, in order to enhance the control law designs and their implementation, the inlet air temperature $T_i(t)$ is considered as a measurable disturbance $d(t)$. The input-output data acquired from the test-rig and the TU module in Figures 1 and 2 are represented by the inlet air temperature $T_i(t)$, the air flow $v(t)$ and the outlet air temperature $T_o(t)$.

In the remainder of this section, different control laws and their implementations are summarised. The methods include the standard PID regulator and nonlinear control methodologies relying on AI techniques, such as FL and adaptive schemes, as well as the model predictive control. These control strategies, which are exploited for the the regulation of the outlet air temperature $T_o(t)$, will be applied to the TU system of Figure 2 described by the model of Equation (4).

3.1. Standard PID Controller Design

Several works [5–7,12,30] highlighted that standard PID controllers can be commonly used in general HVAC applications. In fact, it is shown that simple PID controllers are able to achieved interesting results based on the direct and straightforward computation of the tracking error $e(t)$ computed as the difference between the reference and the measured values of the output, respectively, i.e., $e(t) = r(t) - y(t)$. The continuous-time standard PID controller can be represented in the following parallel form [31,32]:

$$u(t) = K_p + K_i \int_0^t e(\tau)d\tau + K_d \frac{de(t)}{dt} \tag{6}$$

where Kp, K_i and K_d are the PID proportional, integral and derivative gains, respectively. Note that the derivative term of the PID controller is usually implemented as the first-order filter whose pole location is defined by the time–constant T_f. The derivation of the PID gains when this standard controller is applied to the TU module of Section 2 will be achieved by means of the auto-tuning approach proposed, e.g., in [32] and implemented in the MATLAB® environment.

3.2. Fuzzy Controller Design

A controller relying on the FL strategy can be described as statements *IF–THEN–ELSE*, as addressed, e.g., in [33,34]. Successful application of FL to HVAC systems was presented, e.g., in [35,36]. This work will show that the derivation of the controller mathematical description can exploit the direct identification of rules in the form of Takagi–Sugeno (TS) prototypes [37]. These models can be derived by exploiting the ANFIS tool already available from the Simulink® toolbox [38].

According to this description, the TS fuzzy prototype relies on a suitable number of rules denoted as R_i, where the consequent terms are deterministic functions in the form of $f_i(.)$. The subscript i indicates the i-th rule, which is usually represented in the form of:

$$R_i: \quad IF \ x \in A_i \quad THEN \ y_i = f_i(x) \tag{7}$$

where $i = 1, 2, \ldots, K$ and K represents a suitable number of rules. In Equation (7), the variable x indicates the antecedent terms, whilst the scalar y_i represents the consequent output. For the i-th rule, the fuzzy set A_i is represented in general by a multivariable membership function $\mu_{A_i}(.)$ described by the relation of Equation (8) [39]:

$$\mu_{A_i}(x): \quad A_i(x) \mapsto [0, 1] \tag{8}$$

The consequent functions $f_i(.)$ can be represented by parametric models, with fixed structure and varying parameters, as addressed by the same authors, e.g., in [40]. The function $f_i(.)$ can be described with a suitable parametrisation in affine form and usually represented in the form of Equation (9):

$$y_i = a_i^T x + b_i \tag{9}$$

where the model parameters are the column vector a_i and the scalar b_i, for a number of rules $i = 1, 2, \ldots, K$. The variable x is a column vector consisting of an appropriate number n of delayed samples of the input and output signals $u(t)$ and $y(t)$ acquired from the controlled process. Under this description, the term $a_i^T x$ represents a linear regression [41].

It is important to note that the prototypes in the form of Equation (7) have interesting approximation properties [42]. In fact, if the consequent functions $f_i(.)$ are represented in the form of Equation (10) [41]:

$$R_i: \quad IF \ x \in A_i \quad THEN \ y_i(t_k) = \sum_{j=1}^{n} \alpha_j^{(i)} y(t_k - Tj) + \sum_{j=1}^{n} \beta_j^{(i)} u(t_k - Tj) + b_i \tag{10}$$

the collection of the systems of Equation (10) can approximate the dynamic behaviour of any process with an accuracy depending on the choice of the structure. t_k is the time sample Tk corresponding to the sampling time T. According to this description, n represents the order of the regression model; the antecedents depend on the column vector $x = x(t_k) = [y(t_k - T), \ldots, y(t_k - Tn), u(t_k - T), \ldots, u(t_k - Tn)]^T$, whilst the consequents are affine with parameter vector $a_i = \left[\alpha_1^{(i)}, \ldots, \alpha_n^{(i)}, \beta_1^{(i)}, \ldots, \beta_n^{(i)}\right]^T$ and scalar b_i.

It is worth noting that the complete behaviour of the discrete-time TS fuzzy prototype of Equation (7), whose output is y, can be expressed in the form of Equation (11):

$$y = \frac{\sum_{i=1}^{K} \mu_{A_i}(x) \, y_i(x)}{\sum_{i=1}^{K} \mu_{A_i}(x)} \tag{11}$$

According to this representation, this work proposes to use the TS fuzzy model as the prototype for providing the mathematical description of the controller exploited for the compensation of the TU module of Section 2. The estimation of the structure of the model of Equation (11) can be obtained by means of the ANFIS tool relying on the following steps [38]:

1. A TS prototype structure with order n, the membership functions $\mu_{A_i}(.)$ and an appropriate number of rules K are assumed;
2. The input and output data sampled from the process under control are exploited by the ANFIS tool for providing the TS model parameters a_i and b_i according to a selected error criterion;
3. By varying the design parameters n and K with a trial and error procedure, the optimal values of the parameters a_i and b_i are obtained in order to achieve the minimisation of the selected error criterion.

This study proposes also a different methodology based on the Fuzzy Modelling and Identification (FMID) toolbox developed in the MATLAB® environment [43]. This tool allows one to obtain in an easy and straightforward way the parameters of the TS fuzzy structure of Equation (11). Moreover, the FMID strategy provides the controller model simply using a data-driven approach scheme addressed in [43]. This approach exploits again the estimation of the rule-based fuzzy model parameters and requires only the input-output data sampled from the controlled process. In particular, the FMID scheme uses the Gustafson–Kessel clustering methodology to partition the input-output data into suitable regions, denoted again as R_i, the so-called clusters [43]. For each i-th cluster, the parameters a_i and b_i of the affine models of Equation (9) with their membership function $\mu_{A_i}(.)$ are derived. The estimation of the TS fuzzy model in the form of Equation (11) is based on the choice of a suitable model structure n and a number of rules K (usually equal to the number of clusters). The selection of these parameters is performed in order to minimise a prescribed cost function usually related with the closed-loop system performance [43].

In this way, the FMID approach estimates the parameters a_i, b_i and the membership functions $\mu_{A_i}(.)$. Moreover, this strategy is exploited again for identifying the mathematical description of the fuzzy controller that minimises a suitable cost function of the tracking error $e(t)$. Note finally that the FL controller in the form of Equation (11) is implemented as discrete-time model that will be connected to the TU process of Equation (4) via suitable Digital-to-Analogue (D/A) and Analogue-to-Digital (A/D) converter devices [31].

3.3. Adaptive Controller Design

This study proposes the derivation of the controller model for the regulation of the TU module of Section 2 by means of an adaptive strategy. This on-line approach relies on the recursive identification of second-order discrete-time in its difference form of Equation (12):

$$y(t_k) = \hat{\beta}_1 u(t_{k-1}) + \hat{\beta}_2 u(t_{k-2}) - \hat{\alpha}_1 y(t_{k-1}) - \hat{\alpha}_2 y(t_{k-2}) \tag{12}$$

where its time-varying parameters $\hat{\alpha}_i$ and $\hat{\beta}_i$ are recursively identified at each sampling time $t_k = kT$, with k the sample index ($k = 1, \ldots, N$, and N the total number of samples) and T the sampling time. This adaptive identification mechanism uses the Recursive Least Squares Method (RLSM) with adaptive directional forgetting as described in [44], since it is already implemented and ready to use in the Simulink® environment [45].

Under this assumption, the adaptive controller design approach exploits a modified Ziegler–Nichols method that is used to achieve the control law in the form of Equation (13) [44]:

$$u(t_k) = q_0 e(t_k) + q_1 e(t_k - T) + q_2 e(t_k - 2T) + (1 - \gamma) u(t_k - T) + \gamma u(t_k - 2T) \tag{13}$$

where $e(t_k)$ is the tracking error at the instant $t_k = kT$ and $u(t_k)$ is the control signal at the sampling time kT. The variables q_0, q_1, q_2 and γ in Equation (13) represent the time-varying controller parameters, which are obtained by solving the Diophantine expressions represented in the form of Equation (14) [44]:

$$\begin{cases} q_0 = \frac{1}{\hat{\beta}_1}\left(d_1 + 1 - \hat{\alpha}_1 - \gamma\right) \\ q_1 = \frac{\hat{\alpha}_2}{\hat{\beta}_2} - q_2\left(\frac{\hat{\beta}_1}{\hat{\beta}_2} - \frac{\hat{\alpha}_1}{\hat{\alpha}_2} + 1\right) \end{cases} \tag{14}$$

where the following relations hold:

$$\begin{cases} \gamma = q_2\frac{\hat{\beta}_2}{\hat{\alpha}_2} \\ q_2 = \frac{\hat{\alpha}_2\left(\left(\hat{\beta}_1+\hat{\beta}_2\right)\left(\hat{\alpha}_1\,\hat{\beta}_2-\hat{\alpha}_2\,\hat{\beta}_1\right)+\hat{\beta}_2\left(\hat{\beta}_1\,d_2-\hat{\beta}_2\,d_1-\hat{\beta}_2\right)\right)}{\left(\hat{\beta}_1+\hat{\beta}_2\right)\left(\hat{\alpha}_1\,\hat{\beta}_1\,\hat{\beta}_2-\hat{\alpha}_2\,\hat{\beta}_1^2-\hat{\beta}_2^2\right)} \end{cases} \tag{15}$$

It is worth noting that the dominant poles of the controlled system can be represented via the characteristic polynomial $P(s)$ in the form of Equation (16):

$$P(s) = s^2 + 2\,\delta\,\omega_n\,s + \omega_n^2 \tag{16}$$

where the variables δ and ω_n indicate the damping factor and the resonant natural frequency, respectively. Therefore, they can be used for computing adaptive controller parameters in Equation (13) since these relations are already available from the Digital Self-Tuning Controller (DSTC) toolbox implemented in the MATLAB® and Simulink® environments [45].

Note finally that the difference equation of Equation (13) represents a discrete-time control law that requires suitable D/A and A/D converters to be applied to the continuous-time TU model of Equation (4) in Section 2. Therefore, with reference to Equation (13), the tracking error $e(t_k)$ is computed as the difference between the sampled reference signal $r(t_k)$ and the sampled controlled output $y(t_k)$.

3.4. Model Predictive Controller Designs

The regulation strategy relying on the model predictive controller (MPC) method exploits the reconstruction of the system output $y(t_k)$ for a number of step-ahead predictions, i.e., the so-called prediction horizon, in order to generate a suitable control sequence $u(t_k)$ [46]. This methodology provides the control law at the current sampling time $t_k = k\,T$, once it has been derived and optimised over a suitable and finite time horizon. One of the most important features of the MPC scheme with respect to standard PID control relies on its ability to anticipate future behaviours, thus taking the required control actions accordingly. An example of the application of MPC to HVAC systems is shown, e.g., in [22] and compared with other control approaches. However, this work analyses viable control solutions with application to both the simulated and real system, in order to highlight the advantages and drawbacks of the suggested solutions.

In more detail, the MPC scheme generates a suitable control signal $u(t_k)$ by performing the minimisation of the cost function in the form of Equation (17) [46]:

$$J = \sum_{k=1}^{N_p} w_{y_k}\left(r(t_k) - y(t_k)\right) + \sum_{k=1}^{N_c} w_{u_k}\,\Delta u^2(t_k) \tag{17}$$

with w_{y_k} representing suitable weighting parameters indicating the relative importance of the sampled controlled output $y(t_k)$ with respect to the sampled reference $r(t_k)$. In the same way, the coefficients w_{u_k} represent weighting factors penalising possible variations of the actual control signal $u(t_k)$ at the instant $k\,T$ with respect to its previous value at the sampling time $t_{k-1} = (k-1)\,T$, i.e., $\Delta u(t_k) = u(t_k) - u(t_{k-1})$. Moreover, the cost function depends on appropriate values of the prediction horizon N_p and the control horizon N_c.

With this approach, by minimising the expression of Equation (17), the MPC strategy generates and applies only the first element $u(t_k)$ of the whole control sequence at the time sample t_k, whilst the future values of the sequence are dropped. On the other hand, at the next time instant t_{k+1}, the controlled output $y(t_{k+1})$ is measured, and the new control law is computed, thus generating a new control vector $u(t_{k+1})$ and its prediction sequence. This approach is recursively iterated in order to perform the complete simulation of the controlled system.

It is worth observing that the discrete-time MPC design is achieved in a straightforward way by exploiting the MPC toolbox in the Simulink® environment, which can require the knowledge of a state-space LTI model of the controlled process of Equation (4). A continuous-time LTI description of this dynamic process can be obtained by means of the linearisation of Equation (4), which leads to the state-space model in the form of Equation (18):

$$\begin{cases} \dot{x}(t) &= \mathbf{A}\,x(t) + \mathbf{B}\,u(t) \\ y(t) &= \mathbf{C}\,x(t) \end{cases} \tag{18}$$

where $x(t) \in \Re^4$ represents the state vector, whilst the state-space model matrices $\mathbf{A}$, $\mathbf{B}$ and $\mathbf{C}$ are defined by the linearisation at the operating point corresponding to the equilibrium state $x_e = \left[1.0093 \times 10^5, 444.3758, -87436, -0.6288\right]^T$ and inputs $u_e = [19.2351, 0.9093]^T$:

$$\mathbf{A} = \begin{bmatrix} -0.0550 & 0.0001 & 0 & 0 \\ 1.0000 & 0 & 0 & 0 \\ 0.0908 & 0.0007 & -0.1329 & -0.0007 \\ 0 & 0 & 1.0000 & 0 \end{bmatrix}, \mathbf{B} = \begin{bmatrix} 0.9093 & 19.2351 \\ 0 & 0 \\ 0 & -122.9626 \\ 0 & 0 \end{bmatrix}, \mathbf{C} = \begin{bmatrix} 0.0998 \\ 0.0008 \\ -0.0857 \\ -0.0008 \end{bmatrix}^T \tag{19}$$

With reference to the system in Figure 2, the air flow velocity $v(t)$ represents the control input; $y(t) = T_o(t)$ is the controlled output of the model of Equation (18), whilst $T_i(t)$ is the measurable disturbance $d(t)$. Therefore, the input vector in Equation (18) is $u(t) = [T_i(t), v(t)]^T$.

Note that the discrete-time MPC design is achieved by using the MPC toolbox in the Simulink® environment, which uses a state-space LTI description of the controlled process of Equation (4). A continuous-time model of this plant can be obtained by means of an identification procedure, for example based on the System Identification Toolbox in the MATLAB® environment. In this way, the subspace identification (N4SID) procedure has led to the state-space matrices in Equation (20) [41]:

$$\mathbf{A} = \begin{bmatrix} 0.005791 & -0.03346 & 0.06669 & 0.03293 \\ -0.03849 & -0.8226 & 3.345 & 1.513 \\ 0.03747 & 1.386 & -6.579 & -2.869 \\ -0.1521 & -1.89 & 7.195 & 2.452 \end{bmatrix}, \mathbf{B} = \begin{bmatrix} -0.2953 & 0.01477 \\ -13.28 & 0.9604 \\ 25.28 & -1.923 \\ -23.73 & 1.783 \end{bmatrix}, \mathbf{C} = \begin{bmatrix} 84.4 \\ -0.8255 \\ 0.1243 \\ 0.118 \end{bmatrix}^T \tag{20}$$

for a state-space model that is able to fit the identification data with an accuracy higher than 76% [41]. Similar procedures were proposed, e.g., in [47,48].

Note also that the MPC design can be performed also directly exploiting nonlinear formulations. In fact, this control package accepts also nonlinear models; in particular, using large-scale nonlinear programming solvers, such as the Advanced Process OPTimizer (APOPT) and Interior Point OPTimizer (IPOPT), which are available in the Optimization Toolbox in the MATLAB® and Simulink® environments. Therefore, this simulation code is able to implement the moving horizon estimation, dynamic optimisation and simulation, thus solving the nonlinear MPC problems [29]. The nonlinear input-output dynamic model used in simulation has been obtained again by exploiting the System Identification Toolbox in the MATLAB® environment. In particular, this estimation procedure performed via a prediction error method (PEM) has provided a nonlinear regression model with two inputs and one output, with standard regressors corresponding to the orders $n_a = n_b = 2$ for both the inputs and the output, without dead-times ($n_k = 1$) [41]. Moreover, the nonlinearity has been modelled via a sigmoidal network with 10 neurons. Therefore, this nonlinear regression model is able to fit the identification data with an accuracy higher than 90% [41]. Section 4 will show and compare the results achieved with the different models implementing both the linear and nonlinear MPC strategies.

Finally, also in this case, the discrete-time regulators obtained via the MPC approach are connected to the continuous-time TU system via D/A and A/D devices.

4. Simulation Results

The control strategies summarised in Section 3 were applied to the simulated TU process of Equation (4). The achieved results shown in this section have been obtained in the MATLAB® and Simulink® environments using the most appropriate development tools. These control solutions will be compared in terms of a performance index represented by the mean sum of squared error (*MSSE%*) computed via Equation (21):

$$MSSE\% = 100 \sqrt{\frac{\sum_{k=0}^{N} (r(t_k) - y(t_k))^2}{\sum_{k=0}^{N} r^2(t_k)}} \tag{21}$$

where N is the total number of samples.

As already remarked, many HVAC systems are controlled via standard PID regulators. Therefore, the first results are achieved by exploiting a regulator in the form of Equation (6) applied to the TU simulated model as represented in Figure 3. Note that in the scheme of Figure 3, the control input $u(t)$ is the inlet air speed $v(t)$, whilst the inlet air temperature $T_i(t)$ is considered as measurable disturbance $d(t)$, which is shown in Figure 4.

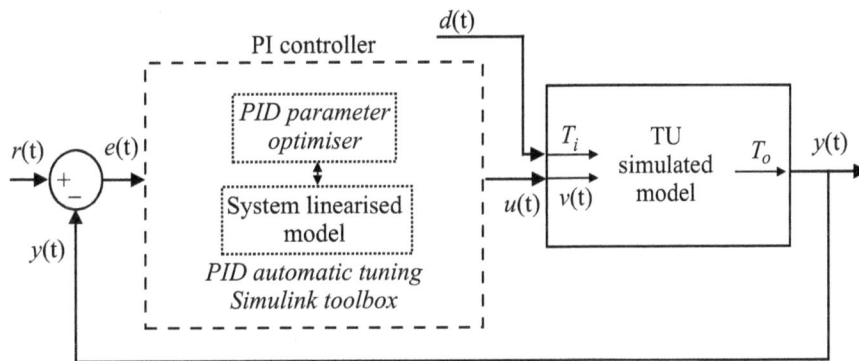

Figure 3. Block diagram of the TU simulator with the standard PID controller.

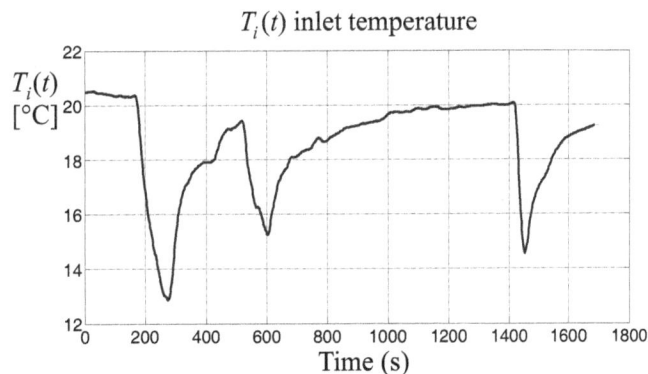

Figure 4. The inflow air temperature $T_i(t)$ considered as a disturbance $d(t)$ acting on the controlled system.

With reference to this control strategy, the optimal controller gains are computed using the automatic PID tuning procedure from the PID Simulink® block. The proportional, integral and derivative gains have been determined as $K_p = 1.4465$, $K_i = 0.0339$ and $K_d = 0.4228$, respectively. The derivative filter time–constant has been estimated as $T_f = 4.4034$.

Figure 5 represents the set-point $r(t)$ (blue continuous line) and the TU measured output $y(t) = T_o$ (red dashed line) regulated via the PID standard controller. With this methodology, the PID regulator is able to guarantee a response with settling time $T_s = 2.17$ s and maximum overshoot $S\% = 36.14\%$. These values are derived by applying a step change in the reference signal $r(t)$ from 39 °C to 40 °C. The tracking error evaluated via Equation (21) is $MSSE\% = 1.65\%$.

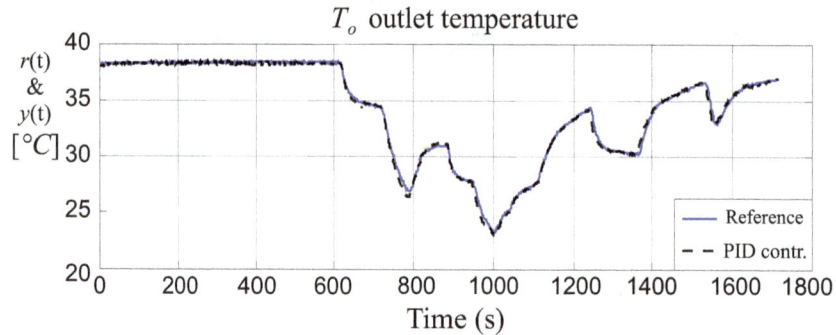

Figure 5. Controlled outlet temperature T_o with the PID regulator obtained via the auto-tuning procedure.

Note that the reference signal $r(t)$ considered for control purpose, i.e., the outlet air temperature $T_o(t)$ represented also one of the excitation signals exploited for the identification of the TU model described in [9]. Moreover, the authors have exploited this reference signal since it guarantees the correct working conditions and the validity of the identified model of Equation (4).

It is worth observing that PID standard controllers can provide sufficient robustness properties after a straightforward tuning phase, thus representing interesting and easy to use solutions with simple and viable implementation. However, despite these features, the achieved control laws might not be sufficiently efficient in terms, e.g., of energy consumption and maintenance costs, when applied to HVAC systems. Due to these possible limitations, the paper has investigated alternative control strategies for achieving improved performances. To this aim, the PID regulator obtained via the auto-tuning procedure is regarded as a reference controller for the computation of advanced and alternative control strategies.

First, a TS fuzzy model of the controller has been derived using the fuzzy identification method recalled in Section 3.2. The procedure exploited the so-called model reference control (MRC) approach described, e.g., in [49]. In this way, the TS fuzzy controller is derived via the ANFIS tool, with a sampling interval $T = 0.1$ s.

Figure 6 reports the diagram of this control solution, where this fuzzy regulator uses $K = 3$ Gaussian membership functions and a number of delayed input and output samples $n = 1$. Figure 6 highlights also that the antecedent vector of the ANFIS tool is $x(k) = [e(t_k), e(t_{k-1}), u(t_{k-1})]^T = [e_k, e_{k-1}, u_{k-1}]^T$.

On the other hand, Figure 7 reports the achieved performance of the regulator obtained with the ANFIS tool by comparing the reference $r(t)$ (continuous blue line) and controlled output $y(t)$ (red dashed line). In this case, the settling time is $T_s = 2.21s$, with a maximum overshoot $S\% = 38.22\%$ and $MSSE\% = 1.07\%$.

This work also proposes the derivation of a fuzzy controller in the form of Equation (11), whose structure and parameter estimation relies on the FMID tool. This tool is able also to provide the estimation of the fuzzy membership functions $\mu_{A_i}(.)$ in Equation (11).

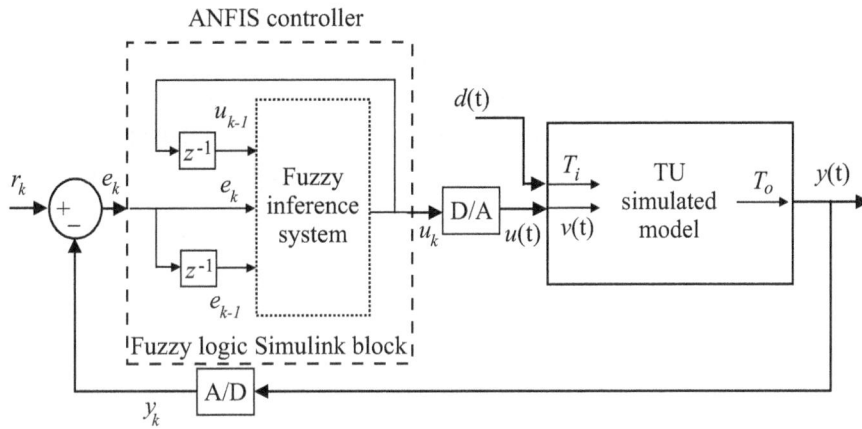

Figure 6. Diagram of the TU simulated model with the Adaptive Neuro-Fuzzy Inference System (ANFIS) fuzzy regulator.

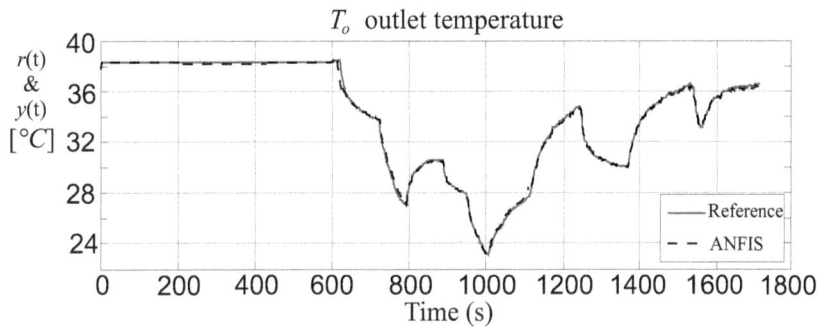

Figure 7. Outlet temperature regulated by the fuzzy controller achieved via the ANFIS tool.

Once the structure of this TS fuzzy regulator has been achieved, the obtained regulator is sketched in Figure 8, for an optimal number of clusters $K = 3$ and delays $n = 2$. For this fuzzy system, the antecedent vector is defined as $x(t_k) = [u(t_{k-1}), u(t_{k-2}), r(t_k), r(t_{k-1}), y(t_k), y(t_{k-1})]^T = [u_{k-1}, u_{k-2}, r_k, r_{k-1}, y_k, y_{k-1}]^T$.

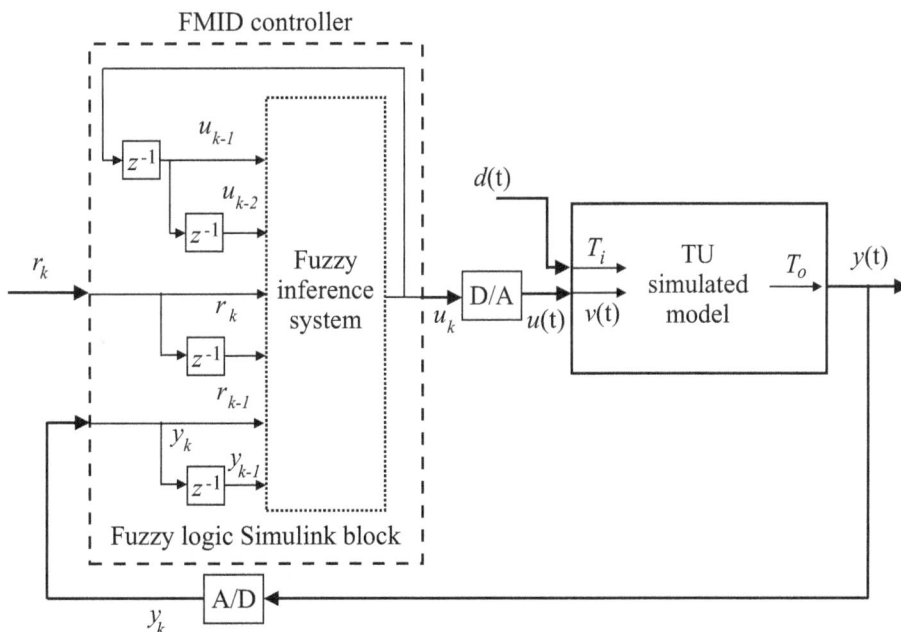

Figure 8. Diagram of the TU module with the TS fuzzy regulator identified from the FMID toolbox.

The results achieved by this TS fuzzy regulator obtained via the FMID library of the MATLAB® environment are summarised in Figure 9. In this situation, the set-point $r(t)$ (blue continuous line) is tracked with an $MSSE\% = 1.14\%$. On the other hand, the step transient response presents a settling time $T_s = 3.98$ s and a maximum overshoot $S\% = 41.65\%$. Finally, it is worth nothing that the high overshoot at the beginning of the simulation in Figure 9 is due to the initial conditions of the delay blocks of the fuzzy controller represented in Figure 9 that are zero.

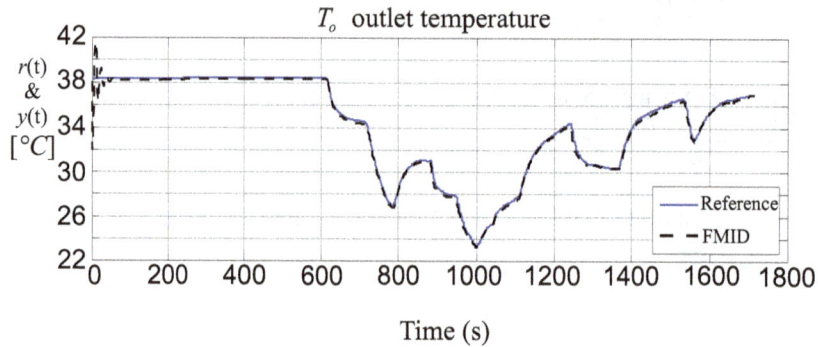

Figure 9. TU outflow air temperature with the Takagi–Sugeno (TS) fuzzy regulator derived via the FMID toolbox.

A further class of regulators has been considered in this work, and an adaptive controller has been developed according to the strategy recalled in Section 3.3.

The diagram of the adaptive compensator in the form of Equation (13) is shown in Figure 10, which is applied to the TU simulated model. The time-varying parameters of the difference model of Equation (12) have been recursively estimated by considering appropriate values of δ and ω_n in the polynomial $P(s)$ of Equation (16), which represent the damping factor and the natural resonance frequency of the closed-loop controlled system.

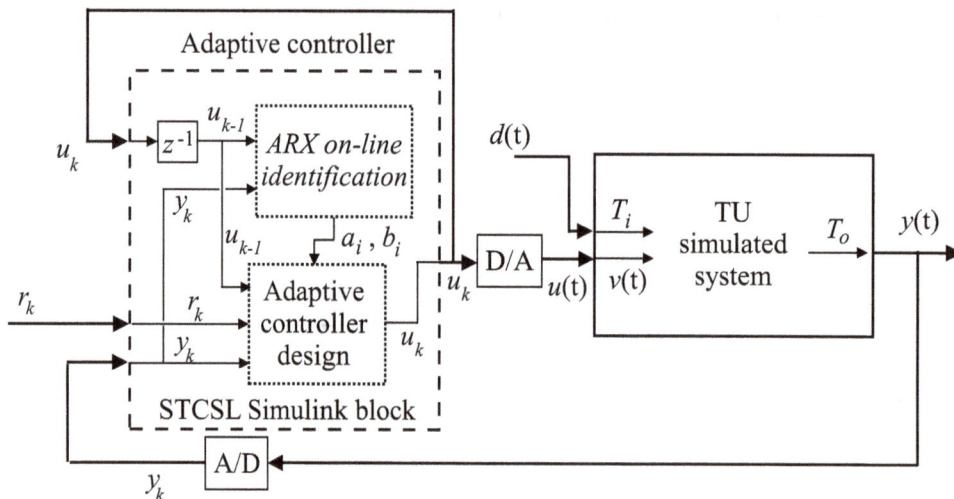

Figure 10. Diagram of the TU module controlled by the adaptive regulator.

The tracking performances of the designed adaptive controller are summarised Figure 11. In particular, the settling time of the step transient response is $T_s = 3.65$ s with a maximum overshoot $S\% = 40.18\%$, whilst the achieved tracking error corresponds to an $MSSE\% = 1.18\%$.

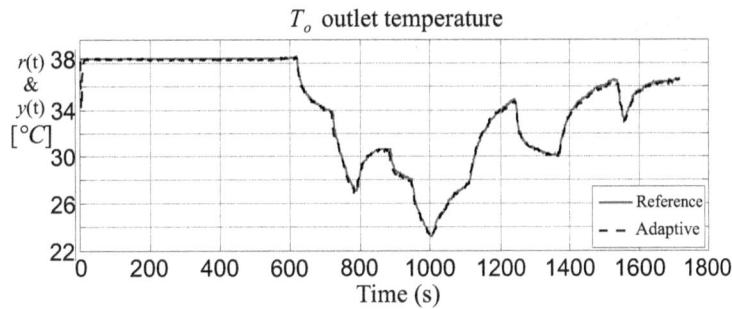

Figure 11. TU module outlet air temperature compensated by the adaptive regulator.

Finally, with reference to the MPC strategy summarised in Section 3.4, Figure 12 reports the block diagram of the MPC applied to the TU module via the D/A and A/D devices. The scheme of Figure 12 assumes that the disturbance signal $d(t) = T_i(t)$ can be measured and exploited by the MPC block.

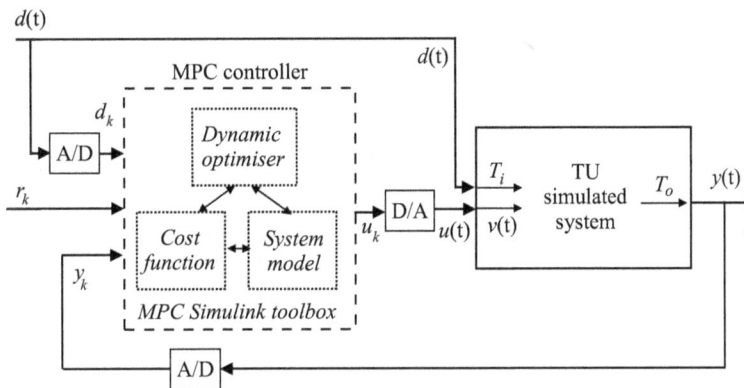

Figure 12. Diagram of the TU system controlled by the MPC scheme.

It is worth noting that in this case, the MPC design exploits a prediction horizon of $N_p = 10$ and a control horizon of $N_c = 2$ for the minimisation of the cost function J of Equation (17). Moreover, the weighting coefficients of this cost function J are settled to $w_{y_k} = 0.1$ and $w_{u_k} = 1$ in order to minimise possible abrupt changes of the control input $u(t_k)$ that would increase the energy consumption and the controlled system efficiency.

In this situation, as shown in Figure 13, the step transient response of the controlled TU module presents a settling time $T_s = 1.85$ s and a maximum overshoot $S\% = 35.51\%$, with a tracking error $MSSE\% = 0.41\%$. Figure 13 shows also the results obtained with the MPC control using the linearised state-space model of Equation (19), when the disturbance $d(t)$ does not feed the MPC block of Figure 12.

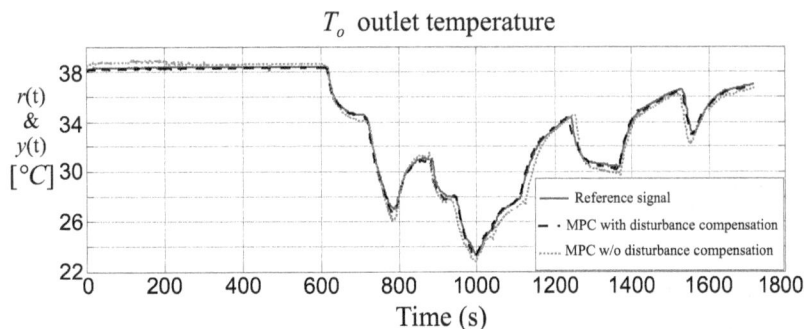

Figure 13. TU module outlet air temperature compensated by the linear MPC with and without disturbance $d(t)$ compensation.

Figure 13 highlights that the knowledge of the measured disturbance $d(t)$ that is exploited by the MPC block of Figure 12 improves the performance of the linear MPC strategy.

On the other hand, Figure 14 shows the comparison between the MPC design performed using the identified state-space model of Equation (20) and the nonlinear dynamic MPC scheme relying on a neural model of the controlled process sketched in Section 3.4.

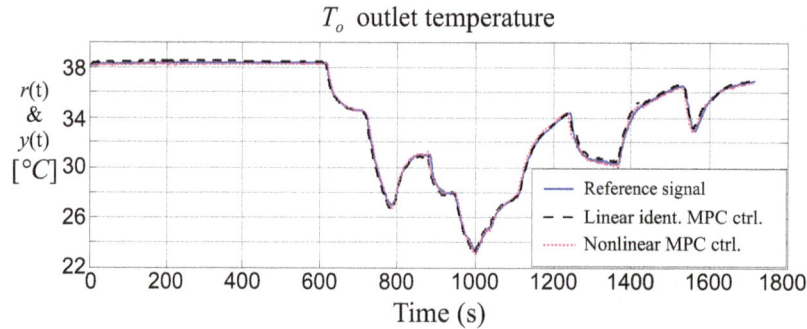

Figure 14. TU module outlet air temperature compensated by the identified linear and the nonlinear MPC solutions.

Figure 14 highlights that the nonlinear MPC leads to slightly better results with respect to the linear MPC with the identified state-space model of Equation (20).

In order to analyse the obtained performance and compare the results achieved with the application of the control strategies proposed in this work, Table 2 summarises the features of these regulators in terms of step response settling time T_s, maximum overshoot $S\%$ and tracking error $MSSE\%$.

Table 2. Performances with the proposed controllers.

Controller Type	Settling Time T_s	Overshoot $S\%$	Tracking Error $MSSE\%$
Auto-tuning PID	2.17 s	36.14%	1.65%
ANFIS Fuzzy	2.21 s	38.22%	1.07%
FMID Fuzzy	3.98 s	41.65%	1.14%
Adaptive	3.65 s	40.18%	1.18%
Linear MPC with disturbance compensation	1.85 s	35.51%	0.41%
Linear MPC w/o disturbance compensation	1.97 s	37.64%	1.27%
Linear MPC with identified model	1.82 s	35.02%	0.33%
Nonlinear MPC with identified neural model	1.79 s	29.73%	0.14%

With reference to Table 2, the MPC schemes allow one to achieve the best performances with respect to the proposed indices. The superior results obtained by this control scheme seem to derive from the ability of the MPC methodology to anticipate future events, thus being able to take control actions accordingly. Moreover, the benefits of exploiting both an identified state-space model and a nonlinear prototype of the controlled process in connection with the knowledge of the measured disturbance $d(t)$ seem quite clear from the results reported in Table 2.

On the other hand, also the TS fuzzy controller derived via the ANFIS toolbox in connection with the MRC principle seems to present interesting features in terms of step response settling time, maximum overshoot and tracking error when compared to the other methodologies. This property can be due to the ANFIS strategy that relies on a fuzzy inference system. In fact, the ANFIS scheme includes the capabilities of both the neural network and the fuzzy logic tools, with the advantage of integrating their benefits in one whole structure. Moreover, as already observed, the proposed fuzzy approach that integrates learning capability is able to approximate the process nonlinear behaviour with an error depending on the required accuracy level. The adaptation strategy implemented in ANFIS presents interesting computationally-efficient features since it relies on genetic algorithms used to estimate the best model structure and its parameters [49].

Note, however, that the standard PID regulator leads to achieving the second best step response settling time since it exploits the auto-tuning scheme implemented in the Simulink® PID block in order to optimise its parameters. This feature can represent an important aspect when a good trade-off between control performance and implementation simplicity is required. In general, the standard PID control law is usually based on the process controlled variable and not on the knowledge of the underlying process behaviour. However, on the one hand, an automatic tuning of its parameters allows the PID controller to manage general control requirements, also in terms of step response rise time, closed-loop bandwidth, maximum overshoot and system oscillation amplitude. On the other hand, PID controllers cannot guarantee any control optimality and, in some situations, the overall system stability, but provides a viable and easy-to-use tool for providing a simple control law with acceptable performance.

4.1. Control Solution Sensitivity Evaluation

This study has considered further simulations that are useful for analysing the reliability and robustness characteristics of the considered control solutions with respect to possible parameter variations. This approach represents a way for analysing the well-known model-reality mismatch issue that can represent a limitation of the achievable performance of the proposed controller designs.

To this aim, the Monte Carlo (MC) tool is the key point since the controller behaviour and the design strategy depend on this model-reality mismatch, which can derive, e.g., from the model nonlinearity and its approximation, the uncertainty and disturbance terms, as well as input and output measurement errors. Therefore, the MC analysis simulates the behaviour of the controlled TU model when its parameters are described as Gaussian variables with mean values equal to the nominal ones and standard deviations of $\pm 20\%$ of the corresponding parameter values.

Under these assumptions, the analysis of the closed-loop control schemes has been performed by computing the best, average and the worst values of the $MSSE\%$ index of Equation (21) evaluated over 500 MC runs. These values are summarised in Table 3.

Table 3. $MSSE\%$ values obtained via the MC analysis for controller performance evaluation.

Control Scheme	$MSSE\%$ Best Case	$MSSE\%$ Worst Case	$MSSE\%$ Average Value
Auto-tuning PID	1.44%	3.14%	1.65%
ANFIS Fuzzy	1.03%	2.33%	1.07%
FMID Fuzzy	1.10%	2.47%	1.14%
Adaptive	1.06%	1.78%	1.18%
Linear MPC with disturbance compensation	0.38%	0.81%	0.41%
Linear MPC w/o disturbance compensation	0.97%	1.98%	1.27%
Linear MPC with identified model	0.28%	0.67%	0.33%
Nonlinear MPC with identified neural model	0.09%	0.31%	0.14%

On the basis of the results of Table 3, it seems clear that the MPC designs lead to the best performance when the modelling of the controlled system and the measured disturbance $d(t) = T_i(t)$ is taken into account. On the one hand, the MPC design can rely on the knowledge of a state-space model in the form of Equation (18), derived from both a linearisation procedure or an identification experiment. On the other hand, the MPC design can use an identified nonlinear model of the process, as remarked at the end of Section 3.4. The overall methodology is based on the optimisation of the cost function of Equation (17). However, once the description of the controlled process has been available as a linear or nonlinear dynamic model, the MPC design is quite simple and straightforward.

The control schemes relying on the ANFIS and the FMID tools can lead to interesting control performance, but with a learning phase that can be computationally heavy and time consuming, especially when the numbers of rules K, the antecedents and the model delays n are high.

Similar considerations hold for the adaptive controllers, which can track possible variations of the controlled model parameters, but with possibly complex design procedures required for computing the controller coefficients.

Standard PID regulators require simple design and simple implementation, but in general, they lead to limited performance, which can imply lower efficiency when applied, e.g., to HVAC systems. The same remarks are valid for the fuzzy controllers relying on AI tools, whose parameters can be easily estimated from the input-output data acquired from the controlled process. However, a further optimisation stage can be required, which sometimes is time consuming, but these fuzzy solutions can enhance the achievement of advanced performance indices.

It is worth observing that the MC methodology proposed in this work seems to represent the key point for the validation and the verification of the proposed control solutions when applied to the TU module in the presence of modelling and measurement errors, uncertainty and disturbance terms.

Note finally that the control methodologies followed by the analysis procedures shown in Sections 3 and 4 are developed using the MATLAB® and Simulink® software tools, in order to automate the overall design and simulation phases. These feasibility and reliability studies are of paramount importance for real application of control strategies once implemented for future air conditioning system installations.

5. Conclusions

In this work, several data-driven and model-based control strategies were recalled, designed and applied to a nonlinear thermal unit model, which can be considered as a fundamental module of phase change material larger systems exploited in passive air conditioning devices. The feasibility of the obtained solutions and the reliability features of the proposed methodologies were analysed in simulation using the measurements acquired from a realistic test-rig of a passive air conditioning system. The Monte Carlo tool represented the practical method for validating the features of these control schemes in the presence of modelling, disturbance and measurement errors. The achieved results highlighted that the controllers designed for example with artificial intelligence schemes were able to provide interesting behaviour in terms of settling time, maximum overshoot and tracking error, even if the adaptation phase can be time consuming. Optimal results were obtained using the model predictive control methodology, even if the derivation of appropriate descriptions of the controlled process and the minimisation of a cost function are required. Finally, future works will investigate the control design and its application for the regulation of different comfort and health parameters of real air conditioning systems, such as the relative humidity, since they directly affect the air conditioning system operating costs in terms of energy.

Acknowledgments: The research works have been supported by the FAR2014 local fund from the University of Ferrara. On the other hand, the costs to publish in open access have been covered by the FIR2016 local fund from the University of Ferrara.

Author Contributions: Cihan Turhan and Silvio Simani conceived of and designed the simulations. Ivan Zajic and Gulden Gokcen Akkurt analysed the methodologies and the achieved results. Silvio Simani wrote the paper.

Conflicts of Interest: The authors declare no conflict of interest.

References

1. Payam, N.; Fatemeh, J.; Mohammad, M.T.; Mohammad, G.; Muhd Zaimi, M.A. A global review of energy consumption, CO_2 emissions and policy in the residential sector (with an overview of the top ten CO_2 emitting countries). *Renew. Sustain. Energy Rev.* **2015**, *43*, 843–862.

2. Lindelof, D.; Afshari, H.; Alisafaee, M.; Biswas, J.; Caban, M.; Mocellin, X.; Viaene, J. Field tests of an adaptive, model-predictive heating controller for residential buildings. *Energy Build.* **2015**, *99*, 292–302.

3. Ferreira, P.M.; Ruano, A.E.; Silva, S.; Conceicao, E.Z.E. Neural networks based predictive control for thermal comfort and energy savings in buildings. *Energy Build.* **2012**, *55*, 238–251.

4. Fanger, P.O. *Thermal Comfort Analysis and Application in Environmental Engineering*, 1st ed.; McGraw Hill: New York, NY, USA, 1992.

5. Mirinejad, H.; Sadati, S.H.; Ghasemian, M.; Torab, H. Control techniques in heating, ventilating and air conditioning (HVAC) systems. *J. Comput. Sci.* **2008**, *4*, 777–783.

6. Wang, Q.G.; Lee, T.H.; Fung, H.W.; Bi, Q.; Zhang, Y. PID tuning for improved performance. *IEEE Trans. Control Syst. Technol.* **1999**, *7*, 457–465.

7. Rahmati, A.; Rashidi, F.; Rashidi, M. A hybrid fuzzy logic and PID controller for control of nonlinear HVAC systems. In Proceedings of the IEEE International Conference on Systems, Man and Cybernetics, Washington, DC, USA, 5–8 October 2003; Volume 3, pp. 2249–2254.

8. Guo, W.Q.; Zhou, M.C. Technologies toward thermal comfort-based and energy-efficient HVAC systems: A review. In Proceedings of the IEEE International Conference on Systems, Man and Cybernetics, San Antonio, TX, USA, 11–14 October 2009; pp. 3883–3888.

9. Zajic, I.; Iten, M.; Burnham, K.J. Modelling and data-based identification of heating element in continuous-time domain. *J. Phys. Conf. Ser.* **2014**, *570*, doi:10.1088/1742-6596/570/1/012003.

10. Krarti, M. An overview of artificial intelligence-based methods for building energy systems. *J. Sol. Energy Eng.* **2003**, *125*, 331–342.

11. He, M.; Cai, W.J.; Li, S.Y. Multiple fuzzy model-based temperature predictive control for HVAC systems. *Inf. Sci.* **2005**, *169*, 155–174.

12. Kumar, P.; Singh, K.P. Comparative analysis of air conditional system using PID and neural network controller. *Int. J. Sci. Res. Publ.* **2013**, *3*, 1–6.

13. Dounis, A.I.; Caraiscos, C. Advanced control systems engineering for energy and comfort management in a building environment—A review. *Renew. Sustain. Energy Rev.* **2009**, *13*, 1246–1261.

14. Etik, N.; Allahverdi, N.; Sert, I.U.; Saritas, I. Fuzzy expert system design for operating room air-condition control systems. *Expert Syst. Appl.* **2009**, *36*, 9753–9758.

15. Soyguder, S.; Alli, H. An expert system for the humidity and temperature control in HVAC systems using ANFIS and optimization with fuzzy modelling approach. *Energy Build.* **2009**, *41*, 814–822.

16. Moon, J.; Jung, S.K.; Kim, Y.; Han, S.H. Comparative study of artificial intelligence-based building thermal control—Application of fuzzy, adaptive neuro-fuzzy inference system and artificial neural network. *Appl. Therm. Eng.* **2011**, *31*, 2422–2429.

17. Jassar, S.; Liao, Z.; Zhao, L. Adaptive neuro-fuzzy based inferential sensor model for estimating the average air temperature in space heating systems. *Build. Environ.* **2009**, *44*, 1609–1616.

18. Ku, K.L.; Liaw, J.S.; Tsai, M.Y.; Liu, T. Automatic control system for thermal comfort based on predicted mean vote and energy saving. *IEEE Trans. Autom. Sci. Eng.* **2015**, *12*, 378–383.

19. Privera, S.; Siroky, J.; Ferkl, L.; Cigler, J. Model predictive control of a building heating system: The first experience. *Energy Build.* **2011**, *43*, 564–572.

20. Ma, J.; Qin, J.; Salsbury, T.; Xu, P. Demand reduction in building energy systems based on economic model predictive control. *Chem. Eng. Sci.* **2011**, *67*, 92–100.

21. Lefort, A.; Bourdais, R.; Ansanay-Alex, G.; Gueguen, H. Hierarchical control method applied to energy management of a residential house. *Energy Build.* **2013**, *64*, 53–61.

22. Afram, A.; Janabi-Sharifi, F. Theory and applications of HVAC control systems—A review of model predictive control (MPC). *Build. Environ.* **2014**, *72*, 343–355.

23. Turhan, C.; Simani, S.; Zajic, I.; Gokcen, G. Application and comparison of temperature control strategies to a heating element model. In Proceedings of the International Conference on Systems Engineering (ICSE 2015), Coventry, UK, 8–10 September 2015.

24. Turhan, C.; Simani, S.; Zajic, I.; Gokcen Akkurt, G. Comparative analysis of thermal unit control methods for sustainable housing applications. In Proceedings of the 12th Federation of European Heating and Air Conditioning Associations (REHVA) World Congress CLIMA 2016, Aalborg, Denmark, 22–25 May 2016; Volume 8, pp. 1–10.

25. Turhan, C.; Simani, S.; Zajic, I.; Gokcen Akkurt, G. Analysis and application of advanced control strategies to a heating element nonlinear model. In Proceedings of the 13th European Workshop on Advanced Control and Diagnosis—ACD2016, Lille, France, 17–18 November 2016; pp. 1–12.

26. Simani, S.; Castaldi, P. Data-driven and adaptive control applications to a wind turbine benchmark model. *Control Eng. Pract.* **2013**, *21*, 1678–1693.

27. Simani, S.; Alvisi, S.; Venturini, M. Study of the time response of a simulated hydroelectric system. *J. Phys. Conf. Ser.* **2014**, *570*, doi:10.1088/1742-6596/570/5/052003.

28. Simani, S.; Alvisi, S.; Venturini, M. Fault tolerant control of a simulated hydroelectric system. *Control Eng. Pract.* **2016**, *51*, 13–25.

29. Huang, G. Model predictive control of VAV zone thermal systems concerning bi-linearity and gain nonlinearity. *Control Eng. Pract.* **2011**, *19*, 700–710.

30. Huang, S.; Nelson, R.M. A PID-law-combining fuzzy controller for HVAC application. *ASHRAE Trans.* **1991**, *97*, 768–774.

31. Åström, K.J.; Wittenmark, B. *Computer Controlled Systems: Theory and Design*, 3rd ed.; Prentice-Hall: Englewood Cliffs, NJ, USA, 1990.

32. Åström, K.J.; Hägglund, T. *Advanced PID Control*; ISA—The International Society of Automation: Research Triangle Park, NC, USA, 2006.

33. Zadeh, L. The Concept of a linguistic variable and its application to approximate reasoning, Part 1 and 2. *Inf. Sci.* **1975**, *8*, 301–357.

34. Zadeh, L.A. Is there a need for fuzzy logic? *Inf. Sci.* **2008**, *178*, 2751–2779.

35. Gouda, M.M.; Danaher, S.; Underwood, C.P. Thermal comfort based fuzzy logic controller. *Build. Serv. Eng. Res. Technol.* **2001**, *22*, 237–253.

36. Homod, R.Z.; Sahari, K.S.M.; Almurib, H.A.F.; Nagi, F.H. Gradient auto-tuned Takagi-Sugeno Fuzzy Forward control of a HVAC system using predicted mean vote index. *Energy Build.* **2012**, *49*, 254–267.

37. Takagi, T.; Sugeno, M. Fuzzy identification of systems and its application to modeling and control. *IEEE Trans. Syst. Man Cybern.* **1985**, *SMC-15*, 116–132.

38. Jang, J.S.R. ANFIS: Adaptive-network-based fuzzy inference system. *IEEE Trans. Syst. Man Cybern.* **1993**, *23*, 665–684.

39. Zadeh, L. Fuzzy sets. *Inf. Control* **1965**, *8*, 338–353.

40. Simani, S.; Fantuzzi, C.; Rovatti, R.; Beghelli, S. Parameter identification for piecewise linear fuzzy models in noisy environment. *Int. J. Approx. Reason.* **1999**, *22*, 149–167.

41. Ljung, L. *System Identification: Theory for the User*, 2nd ed.; Prentice Hall: Englewood Cliffs, NJ, USA, 1999.

42. Fantuzzi, C.; Rovatti, R. On the approximation capabilities of the homogeneous Takagi-Sugeno model. In Proceedings of the Fifth IEEE International Conference on Fuzzy Systems, New Orleans, LA, USA, 11 September 1996; pp. 1067–1072.

43. Babuška, R. *Fuzzy Modeling for Control*; Kluwer Academic Publishers: Boston, MA, USA, 1998.

44. Bobál, V.; Böhm, J.; Fessl, J.; Machácek, J. *Digital Self-Tuning Controllers: Algorithms, Implementation and Applications*, 1st ed.; Advanced Textbooks in Control and Signal Processing; Springer: Berlin/Heidelberg, Germany, 2005.

45. Bobál, V.; Chalupa, P. *Self-Tuning Controllers Simulink Library*; Tomas Bata University in Zlín, Faculty of Technology: Zlín, Czech Republic, 2002.

46. Camacho, E.; Bordons, C. *Model Predictive Control*, 2nd ed.; Advanced Textbooks in Control and Signal Processing; Springer: Berlin/Heidelberg, Germany, 2007.

47. Wallace, M.; Das, B.; Mhaskar, P.; House, J.; Salsbury, T. Offset–free model predictive controller for Vapor Compression Cycle. In Proceedings of the 2012 American Control Conference (ACC), Montreal, QC, Canada, 27–29 June 2012; pp. 398–403.

48. Wallace, M.; House, J.; Salsbury, T.; Mhaskar, P. Offset–free model predictive control of a heat pump. *Ind. Eng. Chem. Res.* **2015**, *54*, 994–1005.

49. Brown, M.; Harris, C. *Neurofuzzy Adaptive Modelling and Control*; Prentice Hall: Englewood Cliffs, NJ, USA, 1994.

9

Real-Time Velocity Optimization to Minimize Energy Use in Passenger Vehicles

Thomas Levermore [1,*], M. Necip Sahinkaya [1], Yahya Zweiri [1] and Ben Neaves [2]

[1] Faculty of Science, Engineering and Computing, Kingston University London, London SW15 3DW, UK; M.Sahinkaya@kingston.ac.uk (M.N.S.); Y.Zweiri@kingston.ac.uk (Y.Z.)
[2] Jaguar Land Rover Limited, Gaydon CV35 0RR, UK; bneaves@jaguarlandrover.com
*   Correspondence: t.levermore@gmail.com

Academic Editor: K.T. Chau

Abstract: Energy use in internal combustion engine passenger vehicles contributes directly to CO_2 emissions and fuel consumption, as well as producing a number of air pollutants. Optimizing the vehicle velocity by utilising upcoming road information is an opportunity to minimize vehicle energy use without requiring mechanical design changes. Dynamic programming is capable of such an optimization task and is shown in simulation to produce fuel savings, on average 12%, compared to real driving data; however, in this paper it is also applied in real time on a Raspberry Pi, a low cost miniature computer, in situ in a vehicle. A test drive was undertaken with driver feedback being provided by a dynamic programming algorithm, and the results are compared to a simulated intelligent cruise control system that can follow the algorithm results precisely. An 8% reduction in fuel with no loss in time is reported compared to the test driver.

Keywords: dynamic programming; optimization; fuel; fuel consumption

---

1. Introduction

The increasing level of CO_2 in the Earth's atmosphere is known to be one of the leading causes of global warming [1]. The contribution of the transport sector globally is estimated to be 23% of which road transport was responsible for three quarters in 2013 [2]. Combustion of both gasoline and diesel fuel produces CO_2 [3], along with a number of undesirable emissions. In order to reduce the impact of passenger vehicle CO_2 emissions on global warming, regulations have been put in place by governments across the globe [4] and are increasingly being tightened. In addition to the regulation of emissions, there is an economic incentive to reduce energy use and fuel consumption in passenger vehicles.

The approaches to fuel consumption reduction can be grouped into two types: the first involves changes to the mechanical design of the vehicle, such as weight reduction, aerodynamic improvements, as well as more complex changes, such as the introduction of hybrid electric powertrains; the second type of approach involves modifying driver behaviour by such means as navigation systems that select the most economical route [5], car sharing [6] and training or guidance to improve the style of driving [7]. Such approaches can be applied to the existing vehicle population, thus having an impact in a fast and cost-effective manner.

It is the modification of driving style that will be the focus of this paper, specifically the longitudinal velocity and gear selection. Training drivers in economical or eco-driving has been shown to improve fuel consumption [7–9]; however, the increased mental workload in complex traffic environments means eco-driving techniques cannot always be applied. The following general rules were presented in an EcoWILLpublication [10] for providers of driver training:

1.   Anticipate traffic flow
2.   Maintain a steady speed at low engine speed
3.   Shift up early

The ability to anticipate traffic flow depends on the road layout and visibility amongst other factors and, therefore, is unreasonable to be expected consistently of a driver; however, the use of map data is becoming more prevalent in modern vehicles allowing a real-time eco-driving guide to be implemented in a vehicle. From the rules noted above, the velocity and gear selection are clearly the most important factors for the eco-driving guide to focus on.

Optimization of vehicle velocity has been frequently investigated in the literature for Heavy Goods Vehicles (HGVs) due to the cost of fuel consumption to haulage companies [11]. Publication [12] proposes a Model Predictive Control (MPC) algorithm to minimize fuel consumption, as well as deviation from a desired velocity. This approach is simulated with an HGV with artificial road profiles. An MPC algorithm is also developed in [13] to produce a time and fuel optimal velocity profile, with an average computation time of 1.26 s in simulation, but real-time implementation is considered only for further work.

Dynamic Programming (DP) [14] is applied to the problem of HGV velocity optimization in [15] at free-way speeds with a reduction in fuel consumption of 3.5% shown compared to a standard cruise control system. As HGVs are limited to a maximum speed that is lower than a passenger vehicle and drive predominantly on free-ways, the optimization problem is less complex than that of a passenger vehicle. DP is applied to a hybrid electric passenger vehicle in [16] considering road sections with varying speed limits on both highway and free-way routes. Traffic conditions are included in the optimization system presented in [17] with different levels of traffic information provided to the optimization system, and the effects are simulated. Breaking a journey into 2.5 km sections and optimizing the velocity for consecutive sections results in a fuel savings of 4% over a fixed velocity profile with no loss of time. Hybrid electric buses are the application of DP for energy management in [18] and in [19], where gear shift scheduling is the focus with fuel savings shown over the standard shift schedule.

It is seen then that DP shows promise as a method both for vehicle velocity and gear optimization; however, due to the dynamic nature of the road environment, there is a requirement to regularly update the control policy based on the current situation the vehicle faces. The application of DP in optimizing both vehicle velocity and gear selection is detailed here with emphasis on the real-time implementation of such an algorithm. As DP reduces a complex optimization problem into a multi-stage discrete optimization problem, consideration must be given to the number of stages and the discretisation of the system. The exponential increase in algorithm complexity with an increased decision variable range is investigated to ensure a compromise between the quality of results and algorithm computation time and, thus, the feasibility of real-time implementation.

2. Methodology

In [20], a DP problem is presented as having two main features, a cost function that increases cumulatively and a system that can be described by a discrete time model. Such a model can be described as follows:

$$x_{k+1} = f_k(x_k, u_k), \quad k = 0, 1, ..., N-1. \tag{1}$$

where x_{k+1} represents the state of the system currently and f_k is the transition relationship, which is a function of the current state x_k and u_k the control decision. The state and control vectors are:

$$x_k = \begin{bmatrix} v & g \end{bmatrix}^T \tag{2}$$

$$u_k = \begin{bmatrix} u_v & u_g \end{bmatrix}^T \tag{3}$$

where v is the vehicle velocity (m·s^{-1}), g is the vehicle current gear, u_v is the demanded velocity (m·s^{-1}) and u_g is the demand gear selection. The other feature noted above, the cost function, is defined by the following:

$$J_k = \frac{\lambda}{\mu_t} J_t + \frac{1 - \lambda}{\mu_f} J_f \tag{4}$$

where J_k is the total cost for a discrete step, k, J_t is the cost due to the time taken, J_f is the cost due to the fuel consumed, λ is a weighting factor to prioritize time or fuel and μ_t and μ_f are normalisation factors for time and fuel, respectively. As the units of fuel and time are not comparable, normalisation is required to ensure that the maximum values for each correspond to a cost value of one in the overall cost function (4). The cost due to the time taken is calculated according to:

$$J_t = \frac{\Delta s}{v_{avg}} \tag{5}$$

where Δs (m) is the distance covered during the step and v_{avg} (m·s^{-1}) is the average speed for the step.

$$J_f = \dot{m}_f(\tau_e, \omega_e) \left(\frac{\Delta s}{v_{avg}} \right) \tag{6}$$

where $\dot{m}_f$ is the fuel consumption rate (g·s^{-1}) multiplied by the duration, in seconds, as calculated in Equation (5). The fuel consumption rate assumes a steady state engine model based on torque, τ_e, and speed, ω_e. The total cost for a velocity profile is described by the summation of each step cost and a terminal cost, $g(x_N)$:

$$J = g(x_N) + \sum_{k=0}^{N-1} J_k(J_t, J_f) \tag{7}$$

The possible decisions at each step can be presented as a multi-dimensional grid, known as a search space. In this case, the distance along the road, the vehicle velocity and the gear selection make up a three-dimensional search space, as shown in Figure 1. Three transitions are shown illustrating velocity and gear changes, where transition (a) is to a higher velocity, while shifting to a higher gear (b) is to a lower velocity while remaining in the same gear and (c) is maintaining velocity while shifting to a lower gear.

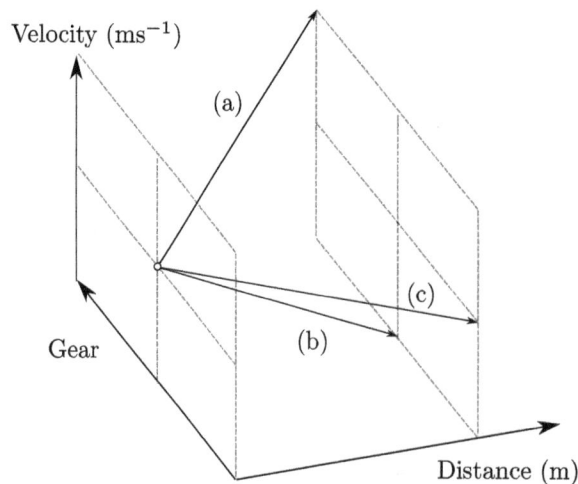

Figure 1. Dynamic programming search space with transitions (**a**) increasing velocity and selecting a higher gear; (**b**) decreasing velocity and (**c**) maintaining velocity and selecting a lower gear.

The complexity of the algorithm (O) can be calculated from the dimensions of the search space:

$$O(N_s \cdot N_v^2 \cdot N_g^2) \tag{8}$$

where N_s is the number of distance intervals between s_0 and s_{max}, N_v is the number of velocity intervals and N_g is the number of gears. As can be seen, the number of distance intervals has a linear relationship with complexity as opposed to the velocity and gear intervals, which both have an exponential relationship with complexity. The gear interval spacing is fixed by the discrete nature of the automatic transmission; however, the relationship described in Equation (8) assumes that all gears N_g are available to be reached from each of the gears, which in reality is not feasible. With the structure of the algorithm and search space specified, the calculation of each transition is detailed.

2.1. Vehicle Model

An essential part of the optimization algorithm is the vehicle model that is used in each transition calculation. It is important that the model is accurate, but can also produce results in a sufficient time so as to not impact the real-time deployment of the algorithm. For a given transition, the start and end velocity are known, and so, a backward or quasi-static vehicle model [21] is utilised. The vehicle velocity is dictated by the forces acting to accelerate or decelerate the vehicle. These forces are presented according to Newton's second law and developed from [22] in the time domain. As the road state changes with position and this can be measured directly rather than estimated based on velocity and time, there are advantages to presenting the model with position as an input variable. The drawback with this is that standard equations with time as a variable need to be converted to consider position instead. The relationship noted in [23] between acceleration in the time domain and the spatial domain equivalent is an application of the chain rule:

$$\frac{dv}{dt} = \frac{ds}{dt}\frac{dv}{ds} = v\frac{dv}{ds} \tag{9}$$

where s is the distance along the road (m), v is the vehicle velocity (m·s^{-1}) and t is the time (s). This results in an equivalent to the equation from [22] as follows:

$$m_v v(s)\frac{dv}{ds} = F_t - (F_a(v, v_w) + F_r(\alpha(s)) + F_g(\alpha(s))) \tag{10}$$

where m_v is the vehicle mass including a nominal equivalent inertial mass (kg) and $v(s)$ is the vehicle velocity (m·s^{-1}), as a function of position. F_t is the tractive force applied by the tyres of the driven wheels (N); F_a is the aerodynamic drag force of the vehicle (N), a function of vehicle and wind velocity, v_w. F_r is the rolling resistance (N); F_g is the gravitational force acting on the vehicle (N); and both are functions of the road gradient, $\alpha(s)$, which itself is a function of road distance, s.

The sensitivity of the vehicle model to changes in parameters is highlighted in Figure 2 with road gradient changes having by far the most impact on the total resistive forces under normal weather conditions, followed by wind speed, which becomes significant when above average wind speeds are considered. Both the gradient and wind speed can change during the course of a given journey. The road gradient is known from enhanced map data, and the wind speed can be estimated using historical local data or weather forecast data. The vehicle mass on the other hand has less effect, and while it can change between journeys with additional passengers or luggage, it is unlikely to change significantly during the course of a single journey being optimized.

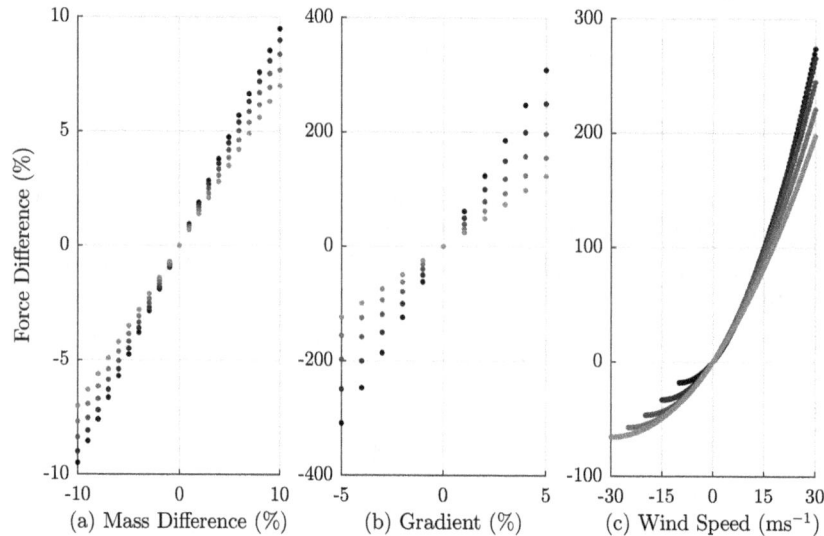

Figure 2. Vehicle model resistive force sensitivity to change in (**a**) mass; (**b**) gradient and (**c**) wind speed. The difference in total force resulting from a change from the default to the maximum and minimum of each variable is plotted for vehicle speeds from $10\,\mathrm{m\cdot s^{-1}}$ to $30\,\mathrm{m\cdot s^{-1}}$. The range of values are $+10$---10% for mass, $+5\%$---5% for gradient and maximum headwind to maximum tailwind of $30\,\mathrm{m\cdot s^{-1}}$.

In order to find the fuel consumption for a given transition, the engine speed and torque are required to be calculated. For a transition between v_0 and v_1, the acceleration, which is assumed constant, can be calculated and used in Equation (10) to find the required traction force at the wheels, F_t. Using this force and the wheel radius, r_w (m), the torque equivalent at the wheels (τ_w) in (N·m) is found. This traction torque is related to the torque available from the engine via the final drive and transmission ratios. Similarly, the engine speed, ω_e, can be calculated from the drivetrain ratios. The engine speed and torque are then used to find the fuel consumed from a Brake-Specific Fuel Consumption (BSFC) map produced by the vehicle manufacturer to identify the efficiency of the engine over a wide range of operating points. Where the engine operating point does not correspond exactly with a data point in the map, interpolation is required.

With the optimization problem and vehicle model defined, the algorithm was tested in the offline simulation on a number of real road profiles that were constructed based on measured data from real journeys undertaken unassisted by any eco-feedback system and with data logging equipment onboard. Using a commercial datalogger [24] to record GPS position, as well as all messages on the Controller Area Network (CAN) bus, journeys could be analysed retrospectively by reconstructing the road profile from the GPS longitude, latitude and elevation measurements. The position coordinate pairs were used to incorporate speed limit data into the road model by requesting such information from the Here routing service [25] at each position in the road. Pertinent data recorded from the CAN bus included vehicle speed, engine speed, calculated engine torque and a rolling fuel consumption measurement. The recorded fuel consumption is measured from the requested fuel injection quantities and updated at a rate of 5 Hz.

The velocity and gear profiles generated by the unassisted driver were used by the vehicle model to simulate fuel consumption and to compare to the algorithm results, and on average with $\lambda = 0.5$, the fuel savings were 12% and the time savings 3%. In order to assess if these improvements could be transferred from simulation to an in-vehicle eco-feedback system, such a system was developed.

2.2. Hardware

While the approach of DP aims to represent problems in a format that is suited to evaluation by a computer program, the process of implementing such a program to solve the DP problem still involves several obstacles. Two scenarios were considered where the DP algorithm would have to be implemented, the first being a simulation environment to allow testing of various scenarios and the second being implementation on a hardware platform capable of running the algorithm in real time in a test vehicle. MATLAB and Simulink are utilised across many academic and industrial fields, particularly the automotive industry [26] for modelling and simulation work, and were used to simulate the DP algorithm for testing.

(a)

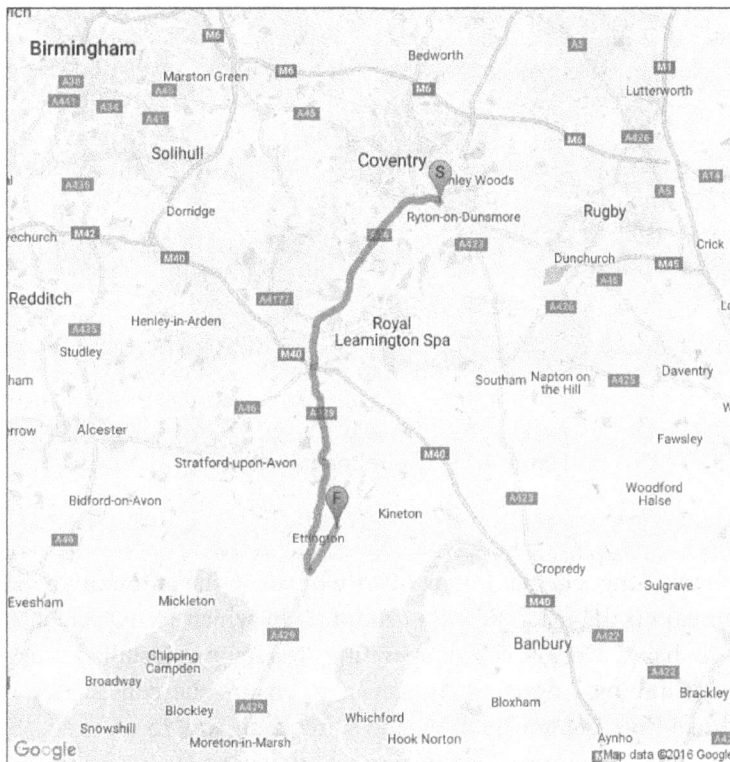

(b)

Figure 3. In-vehicle deployment of the eco-feedback system (**a**) Test route and (**b**) on the extra urban roads, including multi-lane free-way and single-lane country roads.

To run the DP algorithm in a vehicle, a hardware platform was required that was compact, portable, had a flexible operating system and enough processing power to run the necessary software and hardware components that would form the complete system. The Raspberry Pi 2B single-board computer with a quad-core 900-MHz processor and 1 Gb of RAM was used for this application. The in-vehicle implementation is shown in Figure 3 along with the route used for testing. Details of the hardware are given in Table 1 and Figure 4. As the software implementation of the algorithm had to be flexible enough to perform adequately in both simulation and real-time implementation, the choice of programming language used to execute the algorithm is a crucial factor decision.

Table 1. Hardware components' details.

	Item	Model
1	Mini PC	Raspberry Pi 2 Model B
2	Display	Raspberry Pi Touch screen [27]
3	USB GPS Receiver	GlobalSat BU-353-S4 [28]
4	CAN Bus Interface	SK Pang PiCAN2 Board [29]
5	USB 4G Modem	ZTE MF823
6	CAN Bus OBD Connector	SK Pang OBDII to DB9F

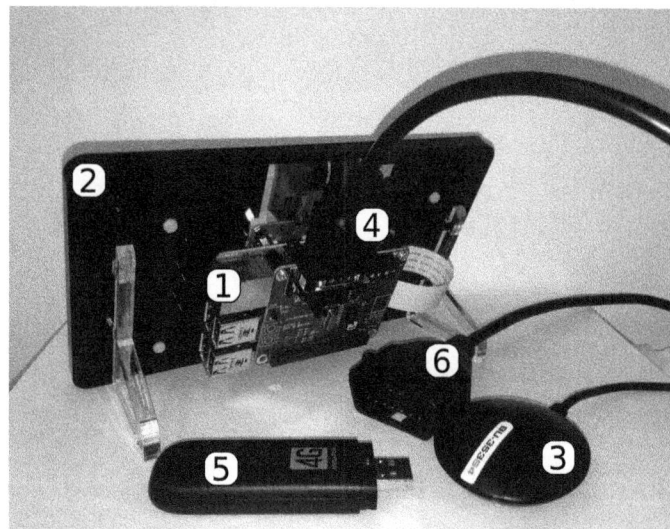

Figure 4. Hardware component setup.

2.3. Software

The C programming language has a long history of use in the automotive industry [30], and its speed of execution made it the most suitable candidate in which to implement the DP algorithm. The Raspberry Pi 2B has a Debian-based operating system pre-installed, capable of running a wide variety of programming languages; however, to ensure the consistency of results between the simulation and real-time implementation, the same code was to be used. Simulink is able to run code that has been produced in C as an S-function compiled using the MATLAB Executable (MEX) compiler, and so, the same DP algorithm code could be used in both simulation and real-time implementation, giving confidence that the simulated results would be transferable. While the core of the DP algorithm was to be written in C, an interface with external inputs, such as GPS position, was required. This interface was written as a Python script due to the ease of development and existing libraries for interacting with peripherals, such as GPS receivers and CAN bus interfaces. The DP algorithm implemented in C code can be run from within the Python script by compiling the

C code and linking it so that it can be accessed in the same way as a standard Python library function. The C code is compiled using the GNU Compiler Collection (GCC) [31].

The overall software package considers the acquisition of data, the integration of the DP algorithm, a road information database, data logging and the displaying of relevant data on a Graphical User Interface (GUI) along with audible feedback. In order to provide each of these tasks with sufficient computational power, the multiple cores of the Raspberry Pi 2 were utilised with the multiprocess library in Python [32].

Data Acquisition

The system relies on a number of data sources that need to be considered for data logging and extraction of the relevant data.

To ensure that the DP algorithm provides usable and relevant results, it is vital that the current status of the vehicle be known at the algorithm start. These data are available for transmission on the internal communication network of the vehicle, the CAN bus. In order to communicate on this network, a hardware interface is required between the Raspberry Pi and the vehicle On-Board Diagnostics (OBD) port, which has a connection to the CAN bus. Such an interface board is commercially available as the SK Pang PiCAN2 CAN Bus Interface Board [29], and an existing Python library [33] can be utilised to access the data on the network. Once the Raspberry Pi has been configured to communicate with the PiCAN board using the General Purpose IO connector (GPIO) and the device is set up as a network interface, the Python Library is able to access the CAN bus.

All control modules in the vehicle communicate on the CAN bus, resulting in a large amount of data passing through the bus network. The CAN specification [34] details a standard format for messages that includes an 11-bit identifier to allow filtering so that only messages required by the relevant system are read from the CAN bus. Using confidential manufacturer-specific message identifiers, provided by Jaguar Land Rover Limited, the required data can be extracted in this way.

Once retrieved, the required messages have to be converted so that the relevant data are in a readable format in the Python script and are subsequently logged, as well as used to update the current vehicle status. This process was developed using a Microsoft Windows-based CAN bus logging and replaying software package, BUSMASTER [35], which is able to replay previously-recorded CAN bus messages to simulate the network traffic of a particular vehicle from which CAN bus data have been recorded.

In order to locate the vehicle within the road section database, a Global Position System (GPS) receiver is used to measure the vehicle's longitude and latitude. A USB GPS receiver [28], a GlobalSat WorldCom Corporation Model BU-353-S4 is used with an existing Python library [36] that provides a conversion from a variety of GPS communication protocols to a standard, readable JavaScript Object Notation (JSON) format. With the vehicle longitude and latitude known, the road section data are interrogated to identify the closest road section to the current vehicle position using a brute-force search approach.

As the traffic conditions play an important role in the ability to follow an optimal velocity profile, a method for receiving real-time traffic data in the vehicle is required. A ZTE MF823 4G USB Modem is used to establish a mobile Internet connection, enabling traffic data to be acquired from the Here routing service [25]. A request is sent to the service for each pair of coordinates on the current road section, and the response is parsed to extract the "SpeedLimit" and "TrafficSpeed" data, which both contain the speed in $m \cdot s^{-1}$.

2.4. Algorithm Implementation

When all of the current data required by the DP algorithm have been made available in the Python script, they are passed as a number of arguments to the DP C program for execution. The current vehicle speed, gear and engine torque are provided along with the weighting factors used in Equation (4). The current position in the road data is also provided along with a road identifier to allow retrieval of the relevant road section from a database of road data. This database is developed to replicate the data fields from the Advanced Driver Assistance Interface Standard (ADASIS) that are relevant for the DP algorithm and would be provided by a commercial eHorizon system, but without the cost of such a system. Using a combination of GPS data recorded from previously-driven routes with speed limit data from the Here routing service mentioned above, the road database is constructed. Using the current vehicle longitude and latitude coordinate pair, the distance to each data point of the road can be calculated using the haversine formula [37] used in the navigation. The nearest point of road data to the current position is used as the starting point of the horizon data.

With the horizon data and current vehicle status provided to the DP algorithm, the optimisation can begin. A forward DP algorithm is implemented that calculates the transition cost for each step in the horizon from the current position to the end of the horizon. In order to reduce the computational load, an assessment is made at each step of the physically-realisable states possible in the next step, and only feasible transition costs are calculated. The physically-realisable states are assessed based on the maximum traction force available at the wheels that can be generated at the current engine speed and the maximum braking torque, each of which are used in Equation (10) to find a maximum and minimum velocity, respectively. During a gear shift, an idle fuel consumption is used during the disengagement, and an intermediate velocity is calculated with no propulsion force. This is followed by the engaged portion of the shift where the velocity of the next step is achieved and fuel consumption calculated as a transition from the intermediate velocity to the next step velocity. As the algorithm is to be used repeatedly with a receding horizon, the cost function includes a terminal cost to penalise velocity trajectories that benefit the current horizon at the cost of future horizons, for instance by reducing the velocity drastically at the end of the horizon and, thus, requiring a high acceleration at the beginning of the following horizon. An overview of the DP algorithm software implementation is shown in Figure 5.

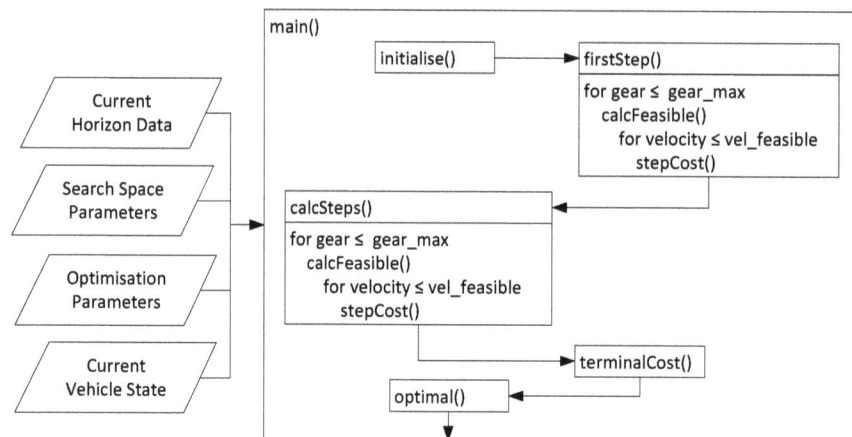

Figure 5. Dynamic programming software overview.

2.5. User Interface

To test the algorithm in a vehicle, it was required to provide a User Interface (UI) that would present information to both test personnel and the test driver. The operating status of the system components was required to be presented for the test personnel without interfering with the driver feedback system. Driving feedback had to be presented to the driver in the most straightforward way to ensure that the feedback could be followed without providing too much distraction. A combination of audible and visual feedback was implemented allowing driver preference to enable either or both of the elements. The design of UI is a discipline in its own right, and the design of automotive UI in particular presents a variety of challenges [38]; so the merits of the UI design in this case are out of the scope of this work. On completion of the DP algorithm, the initial portion of the optimal velocity profile is compared to the current velocity, and the driver feedback is developed from this to advise either maintaining, increasing or decreasing speed to follow the optimal profile. The driver feedback is provided by means of a GUI deployed on a 7-inch touch screen display [27]. A tick icon or coloured arrow pointing up or down represents each piece of advice and is displayed on the GUI, as shown in Figure 6, along with a voice command generated from a text to speech library [39]. Current and aim velocity and gear selection are also displayed along with a number of system status indicators. The optimisation results are converted to whole miles per hour as the standard units on U.K. roads. A plot of the current optimal velocity profile and legal speed limit is also displayed for testing purposes.

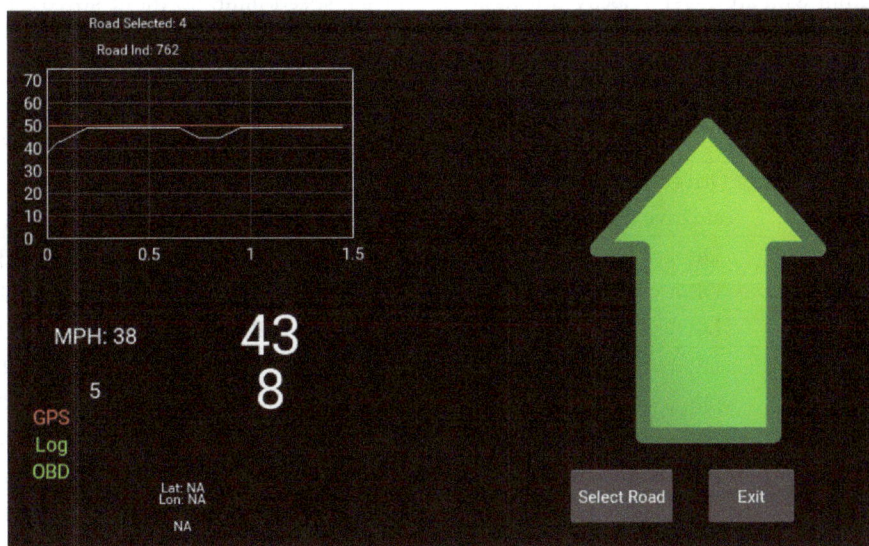

Figure 6. Driver feedback GUI with guidance to increase speed based on current speed (38) and aim speed (43) with current gear (5) and guidance gear (8). Status of GPS, datalogger (Log) and vehicle data connection (OBD) shown for testing purposes. Optimal Velocity profile shown on plot with speed limit for current road selection and position along that road.

3. Results and Discussion

Following deployment of the algorithm and associated interface on the Raspberry Pi hardware, the driver eco-guidance system was tested in a mid-sized Sports Utility Vehicle (SUV). Data were recorded from the vehicle during a journey on extra urban roads with driver guidance provided with a weighting of $\lambda = 0.5$, as a balance between fuel and time, and a receding horizon of 1.5 km. The results of this test drive were compared with simulated results to assess the ability of the driver to follow the eco-guidance, as well as the ability of the system to present guidance that is applicable to the current road conditions.

The velocity and gear profiles recorded during the test were fed into the vehicle model and used to verify that the fuel consumption produced by the vehicle model was accurate. As shown in Table 2, the model calculates a fuel consumption of 4.90 l/100 km, which is 2.9% higher than the recorded fuel consumption of 4.76 l/100 km. Similarly accurate fuel consumption results were produced across a range of recorded journeys, where historical wind speed data were available from the U.K. Meteorological Office [40]. To validate the vehicle model results, firstly, the recorded engine speed and torque were used with the BSFC map to test its accuracy, then the recorded vehicle and engine speeds and gear were used to test the transmission model. Finally, the entire model was tested with recorded vehicle speed and gear and the road profile as inputs and fuel consumption as the output. This simulated test drive fuel consumption was compared to simulated data generated by the vehicle model following the

DP profile precisely, as seen in Figure 7, and the fuel consumption and road section times are shown in Table 2. The DP algorithm was applied with fuel and time weightings of $\lambda = 0.3$, 0.5 and 0.7. The wind speed for the area of the test was on average 6.5 m·s^{-1} [40] in a north-westerly direction producing a tailwind for the duration of the journey. The use of simulated fuel consumption for the comparison as opposed to that recorded was to ensure that any difference in fuel consumption was due to the velocity and gear profile used rather than discrepancies between the model and the real vehicle.

Table 2. Fuel consumption and time from a 5 km section of the test drive. The recorded test drive velocity and gear profile are used to simulate fuel consumption and compared to results from the DP algorithm with an upper velocity constraint of (1) legal speed limit, (2) traffic speed +10% and (3) driver speed +10%.

Constraint		Fuel Consumption (l/100 km)	Difference (%)	Time (min)	Difference (%)
	Test Drive (Recorded)	4.76		3.88	
	Test Drive (Sim)	4.90	0	3.88	0
Legal Limit	DP, $\lambda = 0.3$	4.61	−5.95	3.36	−13.46
	DP, $\lambda = 0.5$	5.52	12.64	3.03	−21.98
	DP, $\lambda = 0.7$	5.91	20.46	2.95	−23.94
Traffic Speed	DP, $\lambda = 0.3$	4.44	−9.36	4.15	6.87
	DP, $\lambda = 0.5$	4.47	−8.80	4.13	6.50
	DP, $\lambda = 0.7$	4.85	−1.14	4.06	4.70
Driver Speed	DP, $\lambda = 0.3$	4.50	−8.28	3.88	0.01
	DP, $\lambda = 0.5$	4.99	1.71	3.74	−3.60
	DP, $\lambda = 0.7$	5.28	7.76	3.71	−4.49

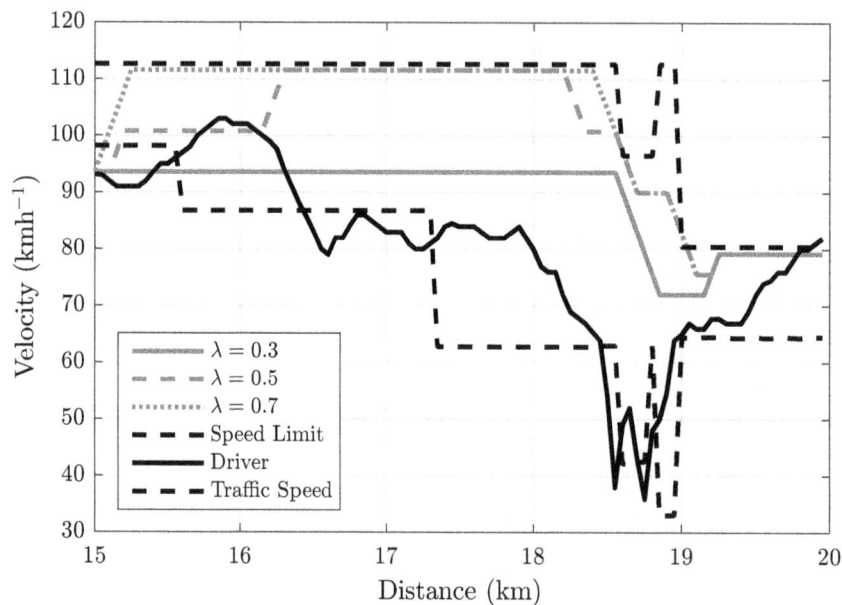

Figure 7. Speed limit-constrained velocity profiles from driver data (solid black) and DP optimal velocity (solid grey $\lambda = 0.3$, dashed grey, $\lambda = 0.5$ and dotted grey, $\lambda = 0.7$) with the speed limit upper velocity constraint. Legal speed limit and real-time traffic shown as the dashed black line, upper and lower, respectively.

The velocity profile produced by the driver is shown in Figure 7 with the legal speed limit and real-time traffic information, while the three DP velocity profiles are also shown. The driver velocity profile includes a reduction in speed prior to 19 km, which does not correspond with the speed limit data, but follows more closely the traffic velocity profile due to a roundabout. In this example, the DP algorithm only considers the legal speed limit and not the traffic speed, hence the deviation.

Using the real-time traffic information recorded during the journey, the effect of traffic on the DP algorithm can be investigated retrospectively. As seen in Figure 8, the driver velocity follows the traffic velocity rather than the speed limit; however, there is still deviation from the real-time traffic information around 16 km and 17.5 km. As such, the DP algorithm is unfairly disadvantaged by the implementation of the traffic speed as an upper limit; for this reason, a +10% upper bound is allowed above the traffic speed. The results are shown in the middle section of Table 2 with improvements in fuel consumption for $\lambda = 0.3$ and $\lambda = 0.5$ coupled with journey times that decrease, but do not meet the test drive time. Even for $\lambda = 0.7$, the results are still worse for time, which can be attributed to the upper speed limitation. For an entirely unbiased comparison, it would be necessary for the DP algorithm to exceed the traffic speed by more than the allocated 10% as the driver does. Increasing the reliability and timeliness of the traffic information would minimise this issue. In order to see the impact of a fairer comparison, a window of +10% was applied to the driver velocity and used as an upper limit for the DP algorithm, as shown in the lower section of Table 2 with the velocity profiles shown in Figure 9. This allows both lower journey times and fuel consumption results to be achieved with a 8.28% savings in fuel with no loss in time for $\lambda = 0.3$. A 3.6% reduction in time with a 1.71% increase in fuel consumption is achieved with $\lambda = 0.5$, with these results highlighting the potential for the DP algorithm when followed precisely by an intelligent cruise control system rather than driver feedback.

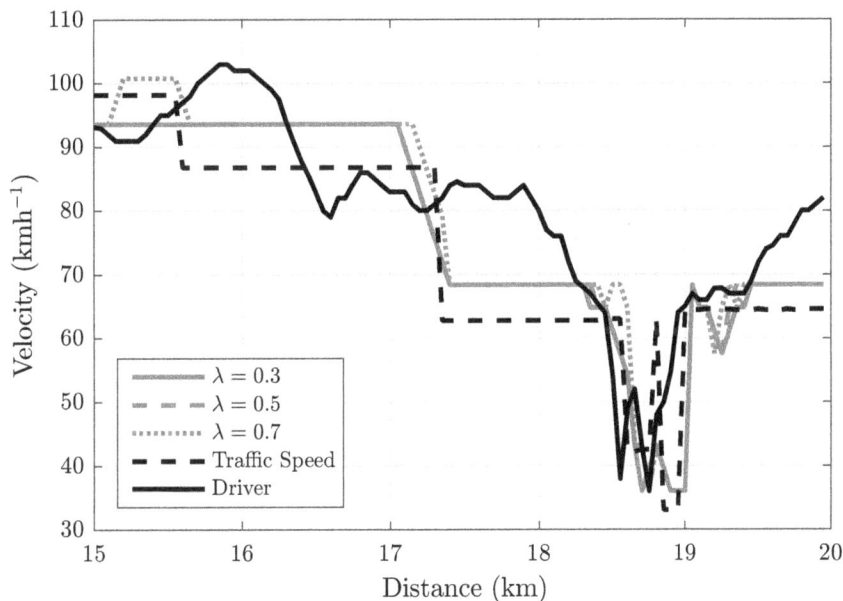

Figure 8. Traffic speed-constrained velocity profiles from DP and driver data with a 10% limit above the traffic speed-constraining DP algorithm. Driver velocity (solid black), traffic speed (dashed black), algorithm profiles with $\lambda = 0.7$ (dotted), $\lambda = 0.5$ (dashed) and $\lambda = 0.3$ (solid grey).

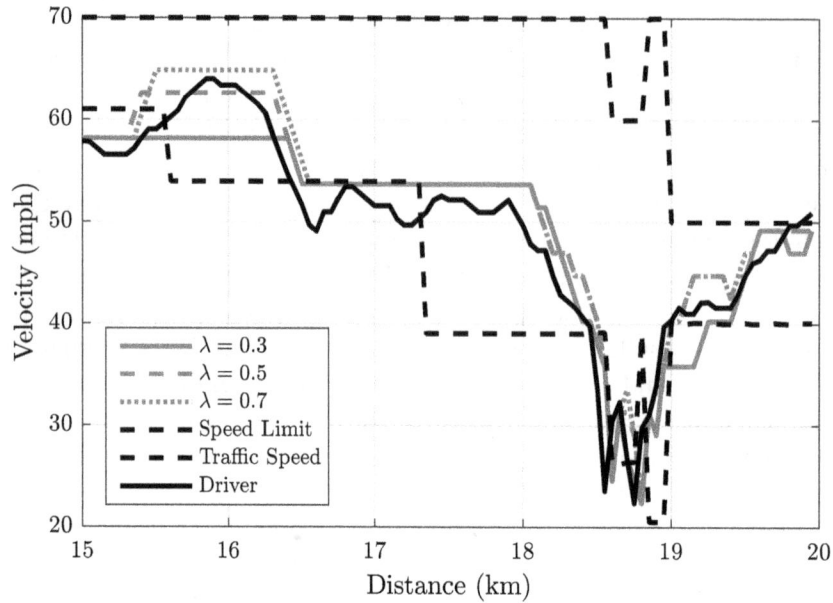

Figure 9. Driver speed-constrained velocity profiles from DP and driver data with a 10% limit above the driver velocity-constraining DP algorithm. Driver velocity (solid black), traffic speed (dashed black), algorithm profiles with $\lambda = 0.7$ (dotted), $\lambda = 0.5$ (dashed) and $\lambda = 0.3$ (solid grey).

The gear selection of the optimization algorithm and the driver are shown in Figure 10, highlighting the foresight of the algorithm; as the driver and traffic speed reduces at 18.5 km, the algorithm initiates coasting in neutral gear. It is also observed around 17 km that the driver uses Gear 8, while the algorithm remains in Gear 9; this behaviour was observed occasionally during the test run due to the automatic gearbox overriding the driver selection of Gear 9. The test road section elevation and gradient are shown in Figure 11 with a mostly downhill road slope, which along with the speed limit restrictions has the most influence on the DP algorithm velocity and gear profile.

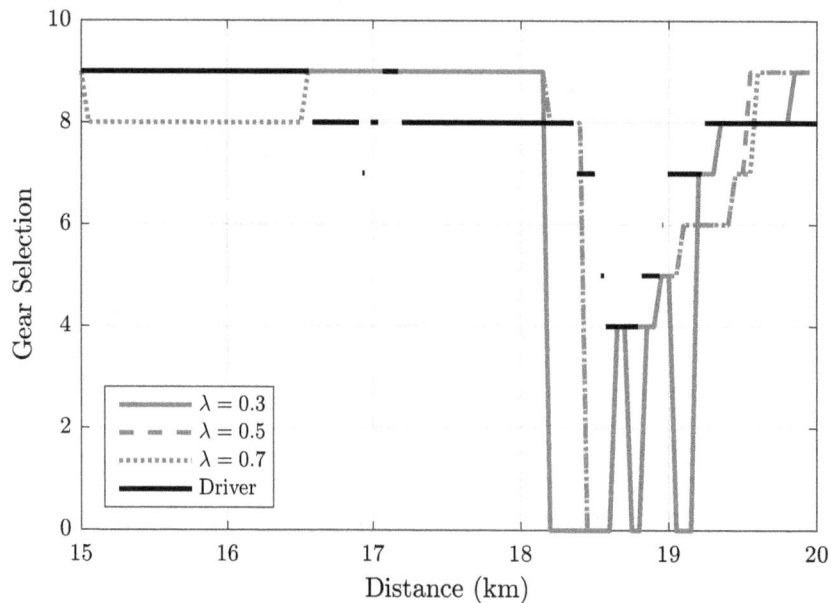

Figure 10. Gear profiles of the optimization algorithm and driver selection where zero corresponds to neutral, and the discontinuity in the driver gear selection is due to gear shift operations where no gear is selected.

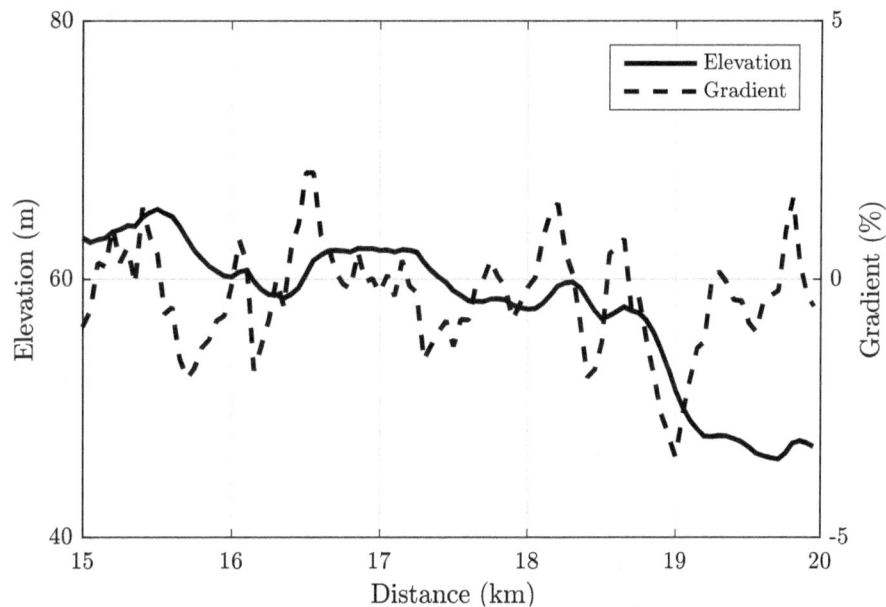

Figure 11. Test road section profile with elevation (left axis) and gradient (right axis).

Algorithm Sensitivity

The search space considered for the DP algorithm implemented in the test was sized to balance the quality of results with computation time, with a distance step interval of 50 m and 30 steps in the horizon, giving a total of 1.5 km. The velocity interval was $1 \, \text{m·s}^{-1}$ as a compromise, as noted above, but also with consideration given to the ability of either a human driver or cruise control system to track precisely a target velocity. As a driver aid, the feedback system shows potential to save fuel; however, even greater benefits could be realised with automation of the vehicle velocity regulation in an intelligent cruise control system. In this scenario, the algorithm output is not limited by the driver, and a more detailed velocity profile can be accurately followed. Further investigation is presented here into the sensitivity of the algorithm to vary the search space parameters of the velocity discretisation interval and horizon length.

As noted in Equation (8), the complexity of the algorithm is exponentially related to the number of velocity intervals. This behaviour is confirmed by the simulation in Figure 12, which shows the number of calculations and computation time for a fixed horizon length of 1.5 km with a varying velocity discretisation interval. The total number of calculations of the vehicle model reduces from 3×10^8 down to 3×10^4 as the velocity interval increases from 0.1–2. Similarly, the calculation time reduces from 19 s down to 0.05 s. This relationship between complexity and velocity discretisation is consistent across a variety of roads with the caveat that the absolute values of calculation time and count decrease when the search space is restricted by speed limit, as shown by the two curves in Figure 12. The influence of this interval on the quality of the optimization results is shown in Figure 13, where the total cost (lowest plot) of the optimal profile is shown to increase as the discretisation interval increases. The total cost is a product of the journey time and fuel consumption, as in Equation (4); however, neither increase linearly with the interval size in Figure 13, which can be attributed to different velocity profiles being possible at a given interval, which presents opportunities to reduce the fuel consumption at the expense of journey time, and vice versa.

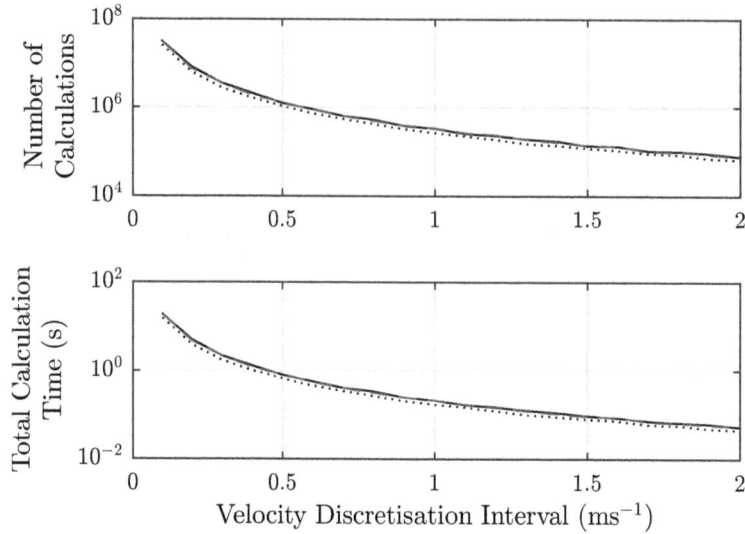

Figure 12. Optimization computational load and calculation time as a function of the velocity discretisation interval. Results shown for three different roads.

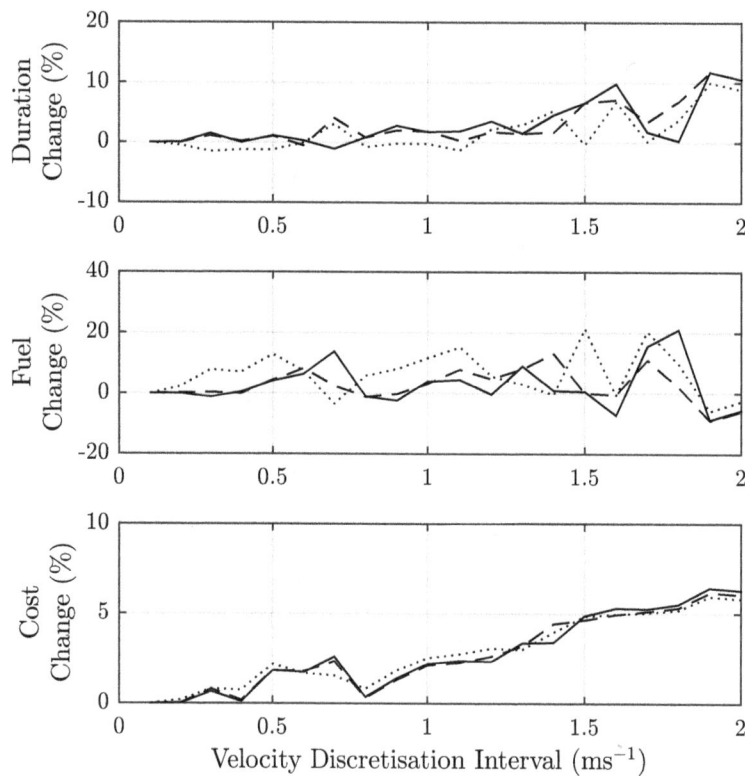

Figure 13. Optimization results as a function of the velocity discretisation interval, with journey time change (top), fuel consumption change (middle) and total cost change (bottom), using $\lambda = 0.5$ and three separate road sections.

4. Conclusions

This work has highlighted the potential of real-time vehicle speed optimization as a mechanism for energy use reduction. By balancing computation time with the accuracy of the results, it is shown that such optimization can be applied in real time and deployed in a vehicle, such that changes to route or driving conditions can be updated and immediately impact the algorithm results. A driver feedback system is demonstrated in a test vehicle that uses the optimal velocity profile for the upcoming road to guide the driver to more economical driving. Even greater benefits are demonstrated in simulations of following the optimal velocity profile exactly, with an 8% reduction in fuel consumption produced. The development of an intelligent cruise control system to implement the optimal velocity profile could achieve similar results. The impact of traffic and wind speed on these results is highlighted as an area where improvements in live data sources for both would be beneficial.

Further tests are required to identify the energy saving potential of the system across a wide range of drivers and road profiles. The DP algorithm interface can be extended to communicate with a commercial map database as part of an eHorizon system using the Advanced Driver Assistance System Interface Specification (ADASIS) to allow testing on a large number of roads without additional road measurement required. Improvement in the feedback system user experience is not considered in this work; however, the integration of the algorithm with an eco-cruise control system would remove the need for driver feedback, thus improving the results and driver experience.

Acknowledgments: This work was part of a Ph.D. project funded by Jaguar Land Rover Limited. The provision of a test vehicle and driver is gratefully acknowledged.

Author Contributions: M. Necip Sahinkaya and Yahya Zweiri supervised the design of experiments and algorithm development; Thomas Levermore developed the algorithm, implemented the in-vehicle system and undertook the experiments; Ben Neaves organised the test vehicle and driver and provided vehicle data; Thomas Levermore wrote the paper.

Conflicts of Interest: Jaguar Land Rover Limited, the funding sponsor, collected vehicle data and made available a vehicle and driver for undertaking the testing described in this publication. The funding sponsor were involved in the decision to publish the results.

Abbreviations

BSFC	Brake-Specific Fuel Consumption
CAN	Controller Area Network
DP	Dynamic Programming
HGV	Heavy Goods Vehicle
MDPI	Multidisciplinary Digital Publishing Institute

References

1. Stocker, T.; Qin, D.; Plattner, G.; Tignor, M.; Allen, S.; Boschung, J.; Nauels, A.; Xia, Y.; Bex, V.; Midgley, P. *Climate Change 2013: The Physical Science Basis;* Fifth Assessment Report; Intergovernmental Panel on Climate Change: Geneva, Switzerland, 2013.

2. International Energy Agency. *CO_2 Emissions From Fuel Combustion Highlights 2015;* Technical Report; International Energy Agency: Paris, France, 2015.

3. Heywood, J.B. *Internal Combustion Engine Fundamentals;* McGraw-Hill: New York, NY, USA, 1988.

4. An, F.; Sauer, A. *Comparison of Passenger Vehicle Fuel Economy and Greenhouse Gas Emission Standards Around the World;* Center for Climate and Energy Solutions: Arlington, VA, USA, 2004.

5. Boriboonsomsin, K.; Barth, M.J.; Zhu, W.; Vu, A. Eco-routing navigation system based on multisource historical and real-time traffic information. *IEEE Trans. Intell. Transp. Syst.* **2012**, *13*, 1694–1704.

6. Minett, P.; Pearce, J. Estimating the energy consumption impact of casual carpooling. *Energies* **2011**, *4*, 126.

7. Barkenbus, J.N. Eco-driving: An overlooked climate change initiative. *Energy Policy* **2010**, *38*, 762–769.

8. Hiraoka, T.; Terakado, Y.; Matsumoto, S.; Yamabe, S. Quantitative evaluation of eco-driving on fuel consumption based on driving simulator experiments. In Proceedings of the 16th World Congress on Intelligent Transport Systems, Stockholm, Sweden, 21–25 September 2009; pp. 21–25.

9. Beusen, B.; Broekx, S.; Denys, T.; Beckx, C.; Degraeuwe, B.; Gijsbers, M.; Scheepers, K.; Govaerts, L.; Torfs, R.; Panis, L.I. Using on-board logging devices to study the longer-term impact of an eco-driving course. *Transp. Res. Part D Transp. Environ.* **2009**, *14*, 514–520.

10. ECOWILL Project. Available online: http://www.ecodrive.org/download/downloads/ecowill_brochure.pdf (accessed on 8 January 2016).

11. Kock, P.; Ordys, A.; Collier, G.; Weller, R. Intelligent predictive cruise control application analysis for commercial vehicles based on a commercial vehicles usage study. *SAE Int. J. Commer. Veh.* **2013**, *6*, 598–603.

12. Terwen, S.; Back, M.; Krebs, V. Predictive powertrain control for heavy duty trucks. In Proceedings of the International Federation for Automatic Control (IFAC) Symposium on Advances in Automotive Control, Salerno, Italy, 19–23 April 2004; pp. 451–457.

13. Passenberg, B.; Kock, P.; Stursberg, O. Combined time and fuel optimal driving of trucks based on a hybrid model. In Proceedings of the 2009 European Control Conference (ECC), Budapest, Hungary, 23–26 August 2009; pp. 4955–4960.

14. Bellman, R. *Dynamic Programming*; Princeton University Press: Princeton, NJ, USA, 1957.

15. Hellström, E.; Ivarsson, M.; Aslund, J.; Nielsen, L. Look-ahead control for heavy trucks to minimize trip time and fuel consumption. *Control Eng. Pract.* **2009**, *17*, 245–254.

16. Wahl, H.G.; Bauer, K.L.; Gauterin, F.; Holzäpfel, M. A real-time capable enhanced dynamic programming approach for predictive optimal cruise control in hybrid electric vehicles. In Proceedings of the 16th International IEEE Conference on Intelligent Transportation Systems (ITSC 2013), Hague, The Netherlands, 6–9 October 2013; pp. 1662–1667.

17. Jiménez, F.; Cabrera-Montiel, W.; Tapia-Fernandez, S. System for road vehicle energy optimization using real time road and traffic information. *Energies* **2014**, *7*, 3576.

18. Wang, X.; He, H.; Sun, F.; Zhang, J. Application study on the dynamic programming algorithm for energy management of plug-in hybrid electric vehicles. *Energies* **2015**, *8*, 3225–3244.

19. Shen, W.; Yu, H.; Hu, Y.; Xi, J. Optimization of shift schedule for hybrid electric vehicle with automated manual transmission. *Energies* **2016**, *9*, 220.

20. Bertsekas, D. *Dynamic Programming and Optimal Control*; Athena Scientific: Belmont, MA, USA, 2005.

21. Fröberg, A.; Nielsen, L. *Dynamic Vehicle Simulation-Forward, Inverse and New Mixed Possibilities for Optimized Design and Control*; SAE Technical Paper; SAE International: Warrendale, PA, USA, 2004.

22. Rajamani, R. *Vehicle Dynamics and Control*; Springer: New York, NY, USA, 2011.

23. Monastyrsky, V.; Golownykh, I. Rapid computation of optimal control for vehicles. *Transp. Res. Part B Methodol.* **1993**, *27*, 219–227.

24. Influx Technology. Available online: http://www.influxtechnology.com/SharedFiles/Documentation/Rebel/Rebel%20XT%20Product%20Flyer.pdf (accessed on 8 January 2016).

25. HERE Routing API Developer's Guide. Available online: https://developer.here.com/rest-apis/documentation/routing (accessed on 8 July 2016).

26. Pretschner, A.; Broy, M.; Kruger, I.H.; Stauner, T. Software engineering for automotive systems: A roadmap. In Proceedings of the Future of Software Engineering, Washington, DC, USA, 23–25 May 2007; pp. 55–71.

27. Raspberry Pi Touch Display. Available online: https://www.raspberrypi.org/products/raspberry-pi-touch-display/ (accessed on 6 September 2016).

28. US GlobalSat Inc. Available online: http://usglobalsat.com/p-688-bu-353-s4.aspx (accessed on 8 July 2016).

29. SK Pang Electronics. *PiCAN 2 User Guide*, version 1.1; SK Pang Electronics: Essex, UK, 2016.

30. Broy, M.; Kruger, I.H.; Pretschner, A.; Salzmann, C. Engineering automotive software. *Proc. IEEE* **2007**, *95*, 356–373.

31. GNU Project. Availiable online: https://gcc.gnu.org/ (accessed on 8 July 2016).

32. Python Software Foundation. Available online: https://docs.python.org/2/library/multiprocessing.html (accessed on 28 June 2016).

33. Python-CAN SocketCAN. Available online: https://python-can.readthedocs.io/en/latest/interfaces/socketcan.html (accessed on 21 September 2016).

34. *CAN Specification, v2.0*; Robert Bosch GmbH: Gerlingen, Germany, 1991.

35. BUSMASTER. Availiable online: http://www.etas.com/en/products/applications_busmaster.php (accessed on 18 July 2016).

36. GPSD Service Daemon. Available online: http://www.catb.org/gpsd/ (accessed on 8 July 2016).

37. Robusto, C.C. The cosine-haversine formula. *Am. Math. Mon.* **1957**, *64*, 38–40.

38. Kern, D.; Schmidt, A. Design space for driver-based automotive user interfaces. In Proceedings of the 1st International Conference on Automotive User Interfaces and Interactive Vehicular Applications (AutomotiveUI 2009), Essen, Germany, 21–22 September 2009; pp. 3–10.

39. eSpeak Text to Speech. Available online: http://espeak.sourceforge.net/ (accessed on 8 July 2016).

40. Met Office. Available online: https://data.gov.uk/metoffice-data-archive (accessed on 8 July 2016).

10

Modeling of a Photovoltaic-Powered Electric Vehicle Charging Station with Vehicle-to-Grid Implementation

Azhar Ul-Haq [1,2,*], Carlo Cecati [1] and Essam A. Al-Ammar [3]

[1]   DSIM, University of L'Aquila, 67100 L'Aquila, Italy; carlo.cecati@univaq.it
[2]   College of E&ME, National University of Science and Technology (NUST), H-12 Islamabad, Pakistan
[3]   Department of Electrical Engineering, King Saud University, Riyadh 12372, Saudi Arabia; essam@ksu.edu.sa
*   Correspondence: aulhaq@uwaterloo.ca

Academic Editors: Michael Gerard Pecht and Kuohsiu Huang

Abstract: This paper is aimed at modelling of a distinct smart charging station for electric vehicles (EVs) that is suitable for DC quick EV charging while ensuring minimum stress on the power grid. Operation of the charging station is managed in such a way that it is either supplied by photovoltaic (PV) power or the power grid, and the vehicle-to-grid (V2G) is also implemented for improving the stability of the grid during peak load hours. The PV interfaced DC/DC converter and grid interfaced DC/AC bidirectional converter share a DC bus. A smooth transition of one operating mode to another demonstrates the effectiveness of the employed control strategy. Modelling and control of the different components are explained and are implemented in Simulink. Simulations illustrate the feasible behaviour of the charging station under all operating modes in terms of the four-way interaction among PV, EVs and the grid along with V2G operation. Additionally, a business model is discussed with comprehensive analysis of cost estimation for the deployment of charging facilities in a residential area. It has been recognized that EVs bring new opportunities in terms of providing regulation services and consumption flexibility by varying the recharging power at a certain time instant. The paper also discusses the potential financial incentives required to inspire EV owners for active participation in the demand response mechanism.

Keywords: electric vehicles; PV power; vehicle-to-grid; power management

1. Introduction

Envisioned large-scale penetration of electric vehicles (EVs) in the system would trigger a need for readily-available charging facilities. A huge amount of EVs' charging load on the power system could arouse various technical issues, including voltage regulation, harmonic contamination, frequency variations, etc. A smart grid technology may come up with a technique to manage EVs' charging load and its scheduling in an efficient way with seamless integration and operation of renewable energies in the power system, which can help keep stress on the power grid to a minimum. In the wake of the Smart Grid Initiative, there has been much attention on the impact of large-scale EVs carrying out vehicle-to-grid (V2G) operation through smart control approaches [1,2]. In order to deal with grid limitations during peak load hours, the implementation of the V2G concept enables the use of EVs as an auxiliary power source that can be discharged to feed the grid, and thus, it contributes towards the latter's stability.

The PV-based power generation can be located on the roof-tops of parking lots for recharging EVs with an effective strategy of power flow management among PVs, EVs and the grid. Different grid-connected EV charging levels (Level 1, Level 2 and Level 3) are defined by the SAEJ1772, IEC62196-2 and CHAdeMO, for single-phase and three-phase slow, medium and DC charging, respectively [3–5]. Keeping in view the driving needs of EVs, there is a need to develop fast charging stations with the feature of lesser charging time, and this has to be supported through the deployment of appropriate infrastructure in the power grid [6], including smart meters, power electronic converters and smart load management strategies [7–9].

Smart DC charging is based on direct DC charging in which EVs can be supplied either from PV power or the utility grid, and EVs could be discharged for supplying the grid. A DC bus-based system has many advantages over an AC system, such as involving a few stages of power conversion with reduced losses and hardware costs [10–12]. This work presents a solar carport-based charging point that can be supplied by DC power for efficient and quick recharging of electric cars. A charging station supplied by a PV source may provide an economical recharging facility. The charging station involves vehicle driver's participation in the electricity market by selling extra stored energy in the vehicle, when not needed, back to the charging station during peak hours for better economy. The idea of utilizing the electric vehicle as a power source may serve to a great extent to satisfy the mean power demand at economical cost. It is considered a potential concept in dealing with the intermittency and fluctuating nature of renewable energy sources. A block diagram of the system is shown in Figure 1.

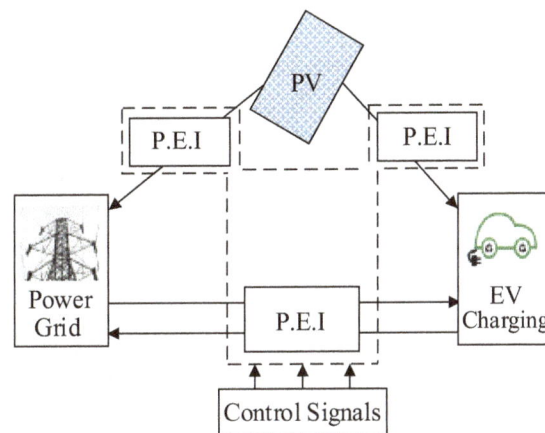

Figure 1. Block diagram of the system.

There have been many attempts by researchers to present DC charging station [13,14] with a limited discussion on the potential solution through efficient power management. For instance, the authors in [15] proposed an intelligent energy management system to allocate power for recharging EVs with the optimal use of supplied and recharging time, but the power electronic converter-based implementation of the management system is not significantly reported. The authors in [16,17] have presented numerous EV control and modelling strategies, including reactive power compensation given in [18,19]. Large-scale placement of photovoltaic-powered charging facilities in a public parking lot is analysed in [20]. A practical control strategy for an electric vehicle charging station consists of a photovoltaic array, the power grid and a lithium ion battery emulator, as illustrated in [21,22]. The economics of a PV-powered charging station are investigated in [23,24]. However, none of those ensure an effective and efficient strategy for the management of EVs' charging load though multiple-mode operation. The presented charging station is made to operate in three different modes, including V2G operation, i.e., EV discharging to the grid with seamless transition from charging to discharging mode. The presented strategy is validated through simulations carried out in MATLAB/Simulink. Additionally, the positive impact of EVs is given in [25–27]. In this paper, a business model is envisioned with a cost-benefit analysis under EVs' penetration for the assessment of the cost for charging infrastructure and external factors, such as health and climate benefits.

In principle, a distribution system must be designed to withstand the power demand during peak load hours. An additional load of EVs can be supplied during the off-peak period without additional generation capacity. In general, an optimal time to recharge EVs is between the hours of the evening and early morning, and thus, it needs to be managed through the enforcement of some incentive-based techniques. In other words, EVs can actively participate in the demand response by offering financial benefits to EV owners. The implementation of vehicle-to-grid (V2G) technology depends on whether an EV owner is willing to trade his/her surplus energy for some money versus his/her convenience. In this regard, the efficiency of energy delivery and its effect on battery life are also of great concern. A traditional means lies in offering regulation by modulating the required power level during EV recharging and discharging. A principle adopted in [6] is meant to maximize the aggregator revenue from EV users and the grid. The authors in [28] presented an algorithm to manage an EV fleet assisting wind power generation to preserve a stable output. A technique based on variation in real time for recharging an EV may respond to regulation signals sent from the system operator with the purpose of power exchange for remuneration. For instance, a regulation scheme given in [6] entitles EV owners to freely decide whether to go for the regulating-while-recharging option. With the advent of recent technological changes in distribution engineering, it can be seen that the pricing factor has gained significance in smart grid operations [29]. A discussion on dynamic EV charging pricing was started in [30], and the relevant concepts are broadly debated for demand response options in [31,32]. Incentives are considered important to actuate demand response in both the residential, as well as the industrial sector. Smart grids and EVs' penetration in the market move the discussion of dynamic pricing to the level of distribution networks [33]. Nonetheless, the response will be assisted by innovative automation technology.

This paper is organized as follows: The charging station architecture is illustrated in Section 2. The description of the EV battery is given in Section 3. Section 4 contains an explanation of the various components of the smart charging station, and simulations are detailed in Section 6. A business model of the EV charging infrastructure along with incentives for EVs' participation in demand response and regulation are explained in Section 5. The paper is concluded in Section 7.

2. Smart Charging Station Architecture

The proposed charging station is primarily supplied by PV power generation, and if necessary, power could be drawn from the utility grid via the corresponding DC/DC boost converter and bidirectional power converter, respectively. EVs can be operated in V2G mode, i.e., EVs are discharged to the grid during peak load hours provided that extra energy is available in them. The charging station architecture is shown in Figure 2. EV charging is supplied by a common DC bus that is energized by either PV or the utility grid depending on the availability and operating conditions, respectively. Major components of the whole system include the bidirectional grid-interfaced power converter, the DC-DC boost converter with the function of maximum power point tracking (MPPT) that delivers PV power generation to the variable DC bus and an EV battery charger with a bidirectional DC-DC buck boost converter for recharging and discharging of EVs, respectively. All of these components are explained in the following subsections. A controller is employed that decides the power flow direction among three blocks of the system depending on the change in DC bus voltage.

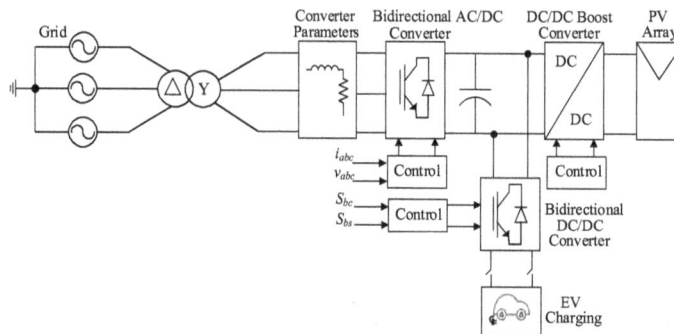

Figure 2. Charging station architecture.

2.1. Modes of Operation of the Charging Station

The direction of power flow is monitored and controlled in three different modes; parameters, including voltage magnitude at the DC link, V_{dc}, the state of charge SOC of the EV battery and the current loading on the grid I_g (this can be acquired from the data acquisition centre in a smart grid environment by using techniques given in [34]), are used to determine the required action to cause the power to flow in the required direction. In the perspective of intermittent solar power generation and the consequent change in voltage level at the DC bus, control of the system is subject to the sensing and regulation of voltage levels at the common DC bus, which is switched to operate in different modes [35]. In fact, for the purpose of simplicity, but without loss of generality, the design principles and modelling are implemented in MATLAB/Simulink in an educational way to support academic research. The presented EV charging station is formulated in an educational manner that allows its further implementation and research on the subject. In order to maintain the simplicity of the educational research problem, different values are taken for clear and better understanding of the modes of operation with corresponding voltage levels.

P_{req} is the amount of power required to recharge EVs at time t; P_{pv} is the power available from PV; EV_{soc} represents the state of charge of the EV battery; P_{tot} is the total power available at time t; $P_{pv,min}$ is the minimum available power through PV that is greater than zero; P_g is the grid power available at t; $P_{g,max}$ is the maximum power demand from the grid; and $EV_{soc,th}$ is the energy stored in EV greater than the required amount for the operation of the vehicle.

Mode I: EV charging through PV power:

$$P_{pv} \geq P_{req} \text{ and } EV_{soc} < EV_{soc,th}$$

In this mode, PV generates enough power to recharge EVs at time t, and no power is required from the grid to meet the EV charging power demand; in this case, the bidirectional DC/DC power converter acts in buck mode for recharging the EV battery. However, the charging is stopped when the SOC of the EV is maximum; then, the PV power is supplied to the grid, and grid-tied power converter inverts DC into AC. Thus, the operation of Mode I may lie in two cases: Case I is when the whole PV power is used for recharging electric vehicle, and Case II is observed when PV is redirected to the grid.

Mode II: Meeting EV charging power demand through both sources—PV power and the utility grid:

$$P_{pv} > P_{pv,min} \text{ and } P_{pv} < P_{req}$$

In this case, PV power is not sufficient to support EV charging demand, and thus, the latter is supplied by a combination of PV power and the grid. In this mode, the grid-tied power converter acts in rectification mode, and the EV interfaced bidirectional DC/DC converter serves in buck mode.

Mode III: Vehicle-to-grid mode: the EV is discharged (V2G) to contribute towards the stability of the power grid, and/or subject to availability, the PV power may also be directed to the grid.

$$P_{pv} > P_{req} \text{ or } EV_{soc} > EV_{soc,th} \text{ and } P_g < P_{g,max}$$

In the third mode of operation, the concept of V2G is performed. In this situation, the EV battery interfaced bidirectional DC/DC converter acts in boost mode and supplies energy to the grid, and PV power is also directed to the grid depending on the availability and concurrent operating conditions of all elements of the system. EV is used as an auxiliary power source and helps stabilize the power grid during peak hours. However, the availability of power from EVs depends on the availability and vehicle driver's willingness to either sell energy by discharging EV or not.

3. Electric Vehicle Battery

An EV is represented by a battery that is charged through its charger. The charger model shall be explained in the next section. An appropriate modelling of the battery is necessary for its smooth operation. Some battery models, including electrical, electro-chemical and mathematical models, are presented in [36]. For this study, an electrical model of the battery is considered accurate in terms of its characteristics. The electro-chemical behaviour of a battery has been presented in [37] while accounting for the battery's state of charge (SOC), internal resistance, discharging current, battery terminal voltage, etc. Generally, an EV battery refers to a lithium-ion-type battery with appropriate specifications of the model. Mathematically, the charge equation of the battery is expressed as given in Equation (1).

$$V_b = V_c - i_b R - K\frac{Q}{Q - bt}bt + {}^{-Bbt}$$

(1)

where V_b is the voltage of the battery, V_c represents the battery constant voltage, i_b is the battery current, K stands for the polarization constant, Q shows the battery capacity in Ah, bt is the actual charge of the battery, A stands for the zone amplitude and B is the zone time constant. The value of all of these parameters can be found using a characteristic curve given by the manufacturer with the data including the fully-charged voltage V_{full}, the capacity and value of the voltage at the nominal zone (Q_{nom}, V_{nom}) and the capacity and value of the voltage at the exponential zone (Q_{exp}, V_{exp}), where the battery parameters can also be found, assuming that R is constant, and during the charging process, it is 0.05 ohms, where A, B, V_c and K can be determined using Equation (2):

$$A = V_{full} - V_{exp}$$
$$B = \frac{3}{Q_{exp}}$$
$$V_c = V_{full} - A + i_b R i_b$$
$$K = V_c - V_{nom} - R i_b + A exp\left(-3Q_{nom}/Q_{exp}\right)$$

(2)

In EV battery applications, the required energy storage capacity and terminal voltage are acquired by combining several cells in series or in a parallel arrangement. The series arrangement of cells determines the battery stack voltage, and cells in parallel determine the current carrying capacity of the battery stacks. The capacity of the stacks can be found using Equation (3):

$$C_{tot} = C_c n_s n_p$$

(3)

C_{tot} is the capacity in Ah; C_c represents the capacity of an individual cell; n_s stands for cells arranged in series; and n_p represents the number of battery cells connected in parallel. A state-of-the-art Thevenin equivalent-based model is considered in this work; an equivalent circuit is shown in Figure 3. A Thevenin battery model uses a series resistor R_s that models the voltage and current characteristic and an repetitive control (RC) parallel circuit with R_t and C_t that helps determine the predictive response to transients at an instant SOC of the EV battery. This model was further improved with the addition of some components for enabling one to determine and predict run-time and DC response, as given in [38,39]. For instance, [39] adds a capacitor to represent a non-linear open circuit voltage and state of charge instead of open circuit voltage V_{oc}, and the model gives the relationship of V_{oc} and SOC, but does not account for transients.

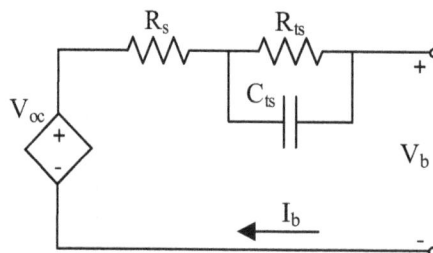

Figure 3. Thevenin battery model.

4. Components of the Smart Charging Station

The smart charging station consists of a PV array, a bidirectional grid-interfaced power converter, a DC-DC boost converter and an EV battery charger with a bidirectional DC-DC buck-boost converter for recharging and discharging of EVs, respectively. All of these components are explained in Section 4.1, Section 4.2, Section 4.2.1 and Section 4.3. A controller is employed that decides the power flow direction among the three blocks of the system depending on the change in the DC bus voltage.

4.1. Grid-Interfaced Bidirectional Power Converter

In the proposed configuration of the PV-based DC charging station, the bidirectional PWM power converter deals with power exchange between the grid and the EV battery charger. When an EV is discharged, it exchanges power from EV to the grid under V2G mode, and conversely, power is delivered to the common DC bus from the grid when required. The configuration of the power converter is shown in Figure 4. The controller design for the AC/DC power converter is based on the internal model principle to achieve a minimum tracking error and ripple-free output DC voltage V_{dc}, as described in [40,41].

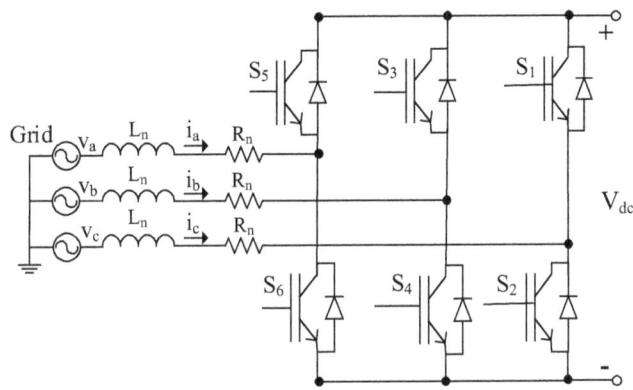

Figure 4. Configuration of the bidirectional AC/DC converter.

The control scheme of a double loop structure is shown in Figure 5 that contains an inner current loop and outer loop for voltage. The employed control scheme basically combines an outer loop that is based on a PI voltage controller and an inner loop deadbeat controller in addition with a repetitive controller for the current control [42]. The implementation of a plug-in repetitive control (RC) system is shown in Figure 6 Transfer function $C_{rc}(z)$, with control gain k_{rc}, of the RC and low pass filter $q(z)$ can be found as given in Equation (4). Generally, $q(z) = 1$ is considered for a DC signal. As per Equation (4), the integral controller is considered a special case of a repetitive controller. In the case of the DC reference, the PI controller is another particular case of the control scheme. The control scheme for the single-phase current loop is shown in Figure 7. A sample equation for each phase with standard parameters is given in Equation (5).

$$C_{rc}(z) = \frac{k_{rc}z^{-N_1}}{1 - z^{-N}}q(z) = \frac{k_{rc}z^{-N_1}}{z^N - 1}q(z) \tag{4}$$

$$i_j(k+1) = \frac{(b_1 - b_2)}{b_1}i_j(k) + \frac{1}{b_1}e_j(k) - \frac{v_d(k)}{2}\frac{1}{b_1}u_j(k) \tag{5}$$

where $b_1 = \frac{L_{nom}}{T}$, $b_2 = R_{nom}$, L_{nom} and R_{nom} are the nominal values of L and R, respectively. If T is the sampling period, then $u(k) = \frac{(2t+(k)-T)}{T}$, and in this case, the transfer function for $G_p(z)$ would be as given in Equation (6). The equation for the deadbeat controller is given in Equation (7). Obtaining $i_j(k+1) = i_{ref}(k)$, the transfer function for each control loop without the repetitive controller would become $H(z) = z^{-1}$. The deadbeat controller offers a quick response with a one-sampling period delay only.

$$G_p(z) = \frac{i(z)}{u(z)} = -\frac{v_d(z)}{2} \frac{1}{b_1 z - b_1 + b_2} \tag{6}$$

$$u_j(k) = \frac{2}{v_d(k)} \left[e(k) - b_1 i_{ref}(k) + (b_1 - b_2) i_j(k) \right] \tag{7}$$

The deadbeat controller is based on an accurate model of the AC/DC power converter. Practically, some uncertainties are found in the parameters of converter, such as: $\triangle L = L - L_{nom}$ and $\triangle R = R - R_{nom.}$. Thus, it becomes difficult to achieve a tracking phase error of zero in accordance with the references [43,44].

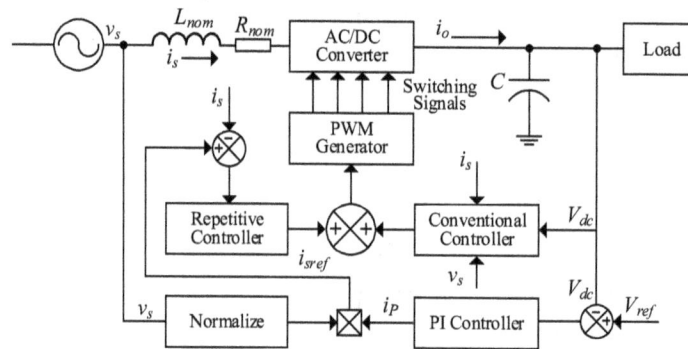

Figure 5. Control scheme for the AC/DC converter.

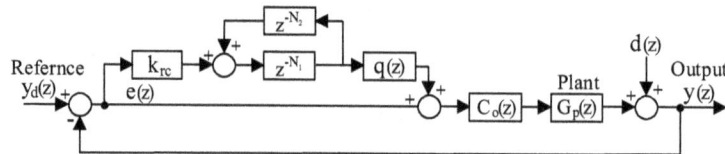

Figure 6. Implementation of the repetitive control system.

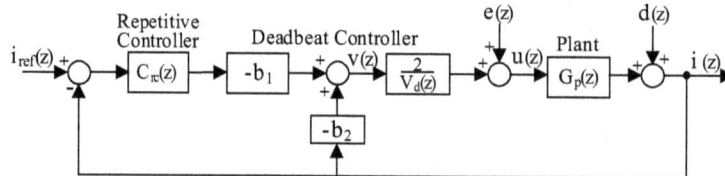

Figure 7. Control scheme for the current loop.

4.2. PV Structure

The characteristics of the PV module are described using a single-diode model, as per the I-V curve presented in [45]. Mathematically, it is expressed in Equation (8):

$$I = N_{pe} \left[I_{pg} - I_d \left\{ X_a \right\} - \frac{\frac{V}{N_s} + \frac{I}{N_c} R_{se}}{R_{pl}} \right]$$
$$X_a = exp \left(\frac{q}{AkT} \left(\frac{V}{N_{se}} + \frac{I}{N_{pe}} R_s \right) \right) - 1 \tag{8}$$

where V and I represent the voltage and current of the PV array, I_{pg} stands for the photo-generated current, I_d is the saturation current of the diode and T, A, q, k represent the operating temperature, diode quality factor, electron charge and Boltzmann constant, respectively. R_{se} and R_{pl} stand for the series and parallel resistances of the cell; N_{se} and N_{pe} are the number of cells arranged in series and parallel in cell strings. Values of these parameters can be found using an approach given in [46], and I_d can be calculated using Equation (9):

$$I_d = \frac{I_{sc}}{exp\left(\frac{qV_{oc}}{kTAN_{se}}\right) - 1} \tag{9}$$

where I_{sc}, V_{oc} represent the short circuit current and open circuit voltage, respectively. Many PV models have been presented for different purposes with a certain accuracy. Authors in [47–50] demonstrate the effect of recombined carriers using an extra diode with the inclusion of additional effects. However, a single-diode model is offers both fair accuracy and simplicity; in the literature, it has been mostly used with a simplified structure of the current source and a parallel diode. Generally, electric generators are considered as either the voltage or current source, but PV arrays show a hybrid behaviour in terms of current and voltage. Typically, series resistance substantially influences the device, when operating in the voltage source region, and the device's parallel resistance affects it while operating the current source region. The series resistance is found by summing many structural resistances, and parallel resistance is characterized by the leakage current. However, these resistances are sometimes ignored for the simplification of PV modelling [51], and the parameters are important to determine the photo-generated current. An assumption that I_{pg} photo-generated current is equal to the short circuit current is also used considering the R_s and R_p to be very low and high, respectively; I_{pg} depends on irradiation and is effected by temperature, and it can be found using Equation (10).

$$I_{pg} = (I_{pg,n} + K_1 \nabla T)\frac{G}{G_n} \tag{10}$$

where $I_{pg,n}$ represents photo-generated current at the nominal condition and ΔT is the difference of the actual and nominal temperature. G and G_n stand for irradiation in watts per square meters and nominal irradiation, respectively. I_o is the saturation current that depends on temperature and is expressed as in Equation (11).

$$I_o = I_{o,n}\left(\frac{T_n}{T}\right)^3 exp\left[\frac{qE_g}{ak}\left(\frac{1}{T_n} - \frac{1}{T}\right)\right] \tag{11}$$

where E_g and $I_{o,n}$ stand for band-gap energy (its value is 1.12 eV) and nominal saturation current, respectively, and the latter can be found using Equation (12)

$$I_{o,n} = \frac{I_{sc,n}}{exp\left(\frac{V_{oc,n}}{\alpha V_{t,n}}\right) - 1} \tag{12}$$

where $V_{t,n}$ represents the thermal voltage of the cells at T_n and α is the diode constant. I_o of PV cells is dependent on the saturation current density, J_o, of the semiconductor material, and the latter further depends on the intrinsic properties of the cell. $I_{o,n}$ is found at the $I_{sc,n}$ nominal open-circuit assuming that $V = V_{oc,n}$, $I = 0$ and $I_{pg} \approx I_{sc,n}$; where the value of α can be randomly chosen between $1\leq$ and ≤ 1.5 depending on the $I - V$ model, as it represents the degree of ideality of the diode; so, it is mostly empirical, and initially an appropriate value can be chosen; then it might be improved later for the purpose of model fitting, as its value influences the $I - V$ curve.

4.2.1. DC/DC Boost Converter

A single-phase boost stage is used that boosts the PV voltage while tracking the maximum power point of the PV array. For this, the input voltage and current are sensed and used by the algorithm.

The boost power converter regulates the input terminal voltage to the maximum power point at the PV to facilitate the maximum power point tracking (MPPT) during the normal operating condition. The control strategy of the converter is illustrated in Figure 8. An average current-mode control scheme is applied that ensures robust control as compared to the peak current-mode control strategy. A low pass filter LF is used to reduce the noise and ripples. Linearized averaged state-space equations of the DC/DC converter can be written as in Equations (13) and (14) [52]:

$$L_{pv}\frac{di_{Lpv}}{dt} = v_{pv} + V_{dc,n}m_d \tag{13}$$

$$C_{pv}\frac{dv_{pv}}{dt} = -i_{Lpv} + \frac{v_{pv}}{r_{dpv}} \tag{14}$$

where $V_{dc,n}$ represents the nominal DC-link voltage, r_{dpv} stands for the PV dynamic resistance and m_d is the averaged control input. Then, transfer functions $G_{m_d,i_{Lpv}}$ and $G_{i_{Lpv},v_{pv}}$ from m_d to i_{Lpv} and from i_{Lpv} to v_{pv}, respectively, can be found using Equations (15) and (16):

$$G_{m_d,i_{Lpv}}(s) = \frac{i_{Lpv}(s)}{m_d(s)} = \frac{\left(C_{pv}r_{dpv} - 1\right)V_{dc,n}}{L_{pv}C_{pv}r_{dpv}s^2 - L_{pv}s + r_{dpv}} \tag{15}$$

$$G_{i_{Lpv},v_{pv}}(s) = \frac{v_{pv}(s)}{i_{Lpv}(s)} = \frac{r_{dpv}}{C_{pv}r_{dpv}s - 1} \tag{16}$$

The employed PI controllers are designed by using the frequency response of the system with diverse values of r_{dpv} found at $r_{dpv,v}$ and $r_{dpv,c}$ voltage and current source regions, respectively, and the I–V characteristics, and the stability of the system is observed with $r_{dpv,v}$ and $r_{dpv,c}$.

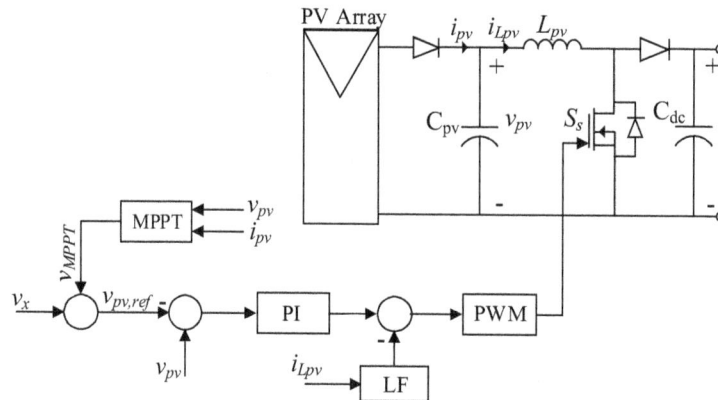

Figure 8. Control of the DC/DC boost converter.

4.3. EV Battery Charging Converter

A model of the battery charger is shown in Figure 9. It is composed of a bidirectional DC/DC power converter that enables two-way power transfer capability, with two switches (IGBT) controlled by a complimentary controller. In discharging mode of the battery with the V2G concept, switch S_{bs} operates; the converter acts in boost mode; it increases the voltage V_b and current I_b in the inductor L_b and flows towards capacitor C_{dc}. On the other hand, in the case of the battery charging process, the converter acts as a buck converter when switch S_{bc} is ON, and I_b rather flows in the opposite direction from the capacitor to the inductor for recharging the EV battery. On the other hand, the control of the battery charger depends on the state of charge SOC of the battery; where the configuration of the EV battery charger is shown in a simplified manner for educational research purposes, and it aims to demonstrate that the corresponding control methodologies can be realized depending on the required charging strategy (i.e., charging or discharging) under the constant current or constant voltage strategy. However, the model is implemented in SimPowerSystems with certain required simulation aspects, the integration method, the time step, the grid parameters, the reactive power, the inverter modulation index, the switching frequency, etc.

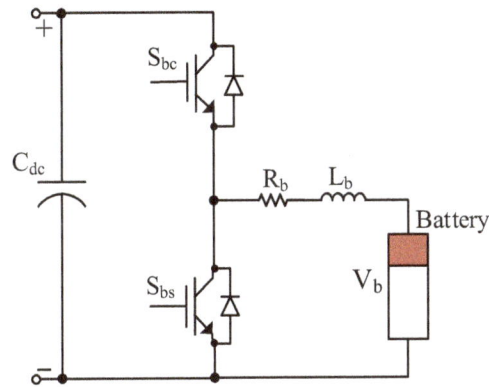

Figure 9. EV battery charger configuration.

For the battery charger, two control techniques, including constant voltage and constant current, are applied for battery charging and discharging modes, respectively. In the constant voltage strategy, the battery is operated as a voltage source, and the resulting duty ratio d_v describes the converter operation in buck mode for battery charging, as shown in Figure 10. A constant current strategy, in which the EV battery operates as a current source, is employed for controlling the charger while operating in V2G mode. In this mode, the converter serves in boost mode, and the scheme is shown in Figure 11.

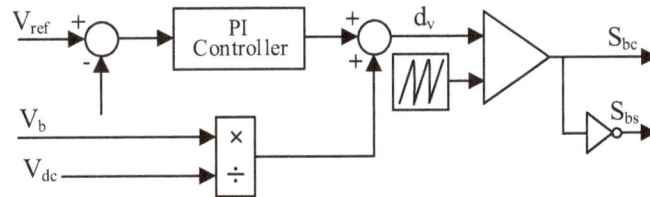

Figure 10. Control scheme for the DC/DC converter in buck mode.

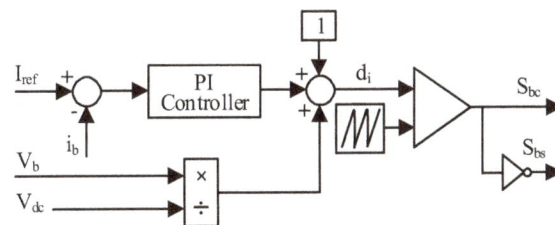

Figure 11. Control scheme for the DC/DC converter in boost mode.

5. Business Model of EV Charging

A sustainable expansion of EVs in the transportation sector requires substantial charging facilities, up-gradation of the distribution network and aggregated control strategies for market integration. Installation cost for charging facilities and equipment depends on the required charging mode and physical location. According to the information given in [53,54], let us assume a scenario of EV charging at a public lot only and estimate the total cost of the charging stations, including installation cost, equipment, electrical connections, connection with the grid, maintenance and administrative costs. The considered public lot has the following characteristics; it can charge an EV in both slow and fast modes; charging time ranges from 15 min to 12 h depending on the employed charging mode (whether the slow or fast charging method); and the charging facility is available 24 h with V2G property. Its cost is €2130 for a slow charging station of a maximum power of 3.5 kW at a public lot, and the cost will be €44,692 for a fast charging station of a power up to 10 kW. Generally, aggregators offer EV charging facilities to EV users through public charging stations. The aggregators act as retailers; they purchase energy from the spot market and sell to the ultimate customers.

It is assumed that the number of EVs to be recharged at a public lot is 48. As the charging station is equipped with V2G technology, so a public charging station needs an electronic counter for distant control, management and interaction with the market. Studio Scada is an example of such an electronic counter described in [55], and its total cost is around €240k. The EVs' ambient impact considered in this study includes the effect of emissions on human health, change in climate and noise pollution.

The impact of emission on human health is estimated using the criteria given in [56] for a small suburban area, and it is found that the mortality rate is significantly reduced with a reduction in emissions. In terms of climate change, the cost of CO_2 emissions may be estimated based on their impact on climate change. For example, in the year 2010, the average value of CO_2 per ton was €25. Additionally, EVs make no noise and ensure acoustic comfort, though the levels are different for different countries depending on the noise tolerance.

Cost Estimation

The cost for deployment of the EV charging infrastructure (EV_{DCI}) can be estimated (in €) as given in Equation (17):

$$EV_{DCI} = EV_{PL}.T_{CV}.\left(C_{pp} + C_{RU} + C_{CMS} + C_{Eq.} + C_{DNE}\right) \qquad (17)$$

where EV_{PL} is the EV penetration level, T_{CV} represents the total cost of conventional vehicles and C_{pp} and C_{RU} stand for the cost of charging stations at public parking lots and the cost of charging stations at single residential units. C_{CMS} is the cost of the control and management software; $C_{Eq.}$ stands for the cost of equipment; and C_{DNE} represents the cost of the expansion distribution network.

The cost of ambient factors in €can be calculated as expressed in Equation (18):

$$C_{AF} = AL_{EV}\left(TC_{Em.} + TC_{cc} + TC_N\right) \qquad (18)$$

TC_{AF} is the total cost of ambient factors; AL_{EV} stands for the average life of an EV; $TC_{Em.}$ is the total cost of emissions; TC_{cc} represents the cost of climate change; TC_N stands for the noise cost.

5.1. Incentivizing EVs' Participation in the Demand Response and Regulation Services

It is generally considered that most consumers would choose convenience over cost to recharge their EVs. A few of the effective techniques for providing incentives include different pricing schemes with respect to varying times, such as real-time, peak pricing and time-of-use. Time-based pricing splits a day into different chunks of time and applies different unit prices for energy use at particular instants. A pricing strategy usually involves an off-peak and on-peak price. Creating a time-of-use-based schedule for EV users will allow them to use timers to utilize energy at a certain time. The ability to contribute to the time-of-use schedule might encourage consumers to carefully utilize energy.

Two possible options for recharging of an EV can be offered to an EV driver with the perspective of revenue maximization for an aggregator: (i) flat charging (f-charging) set in which the EV recharging price is fixed by the aggregator, and EVs are reached at the maximum rate of power; (ii) f-charging in addition to service charging (s-charging) in which an aggregator bids a lower price for recharging, and in exchange, the recharging process provides regulation to the power grid.

In both options/settings, an EV driver is free to select none of the offered options by the aggregator, i.e., an alternative option of no-charging is also available. Let us consider an aggregator of numerous charging stations, which purchases energy at a wholesale price represented by WSP (wholesale price) (in $/kWh), and then offers a fixed a retail price (RP) in $/kWh for EV drivers. As a general fact, most vehicles are assumed to be parked during the daytime, and the same may be expected with EVs; thus, comparatively low recharging power may be adequate to recharge a battery of a few EVs. This fact is considered for the basic assumption that EV drivers are ready to accept reduced power at a less expensive price. In this case, an aggregator can avail itself of this opportunity to have increased revenue. As low recharging power is satisfactory for a few clients, so the aggregator can reduce it when it is cost-effective to do so. Consider that the variable EV recharging power level contributing towards auxiliary control may be termed as a regulation service. In order to perceive EV charging as a resource for regulation, there must be some margins to decrease or increase its power consumption. Regarding monetary compensation for such regulation, the aggregator pays $\Delta wspP_n$ per EV with its regulation slot. In the case of regulation, the aggregator is paid for tumbling demand to zero. The incentive is symbolized as a fraction $r_u \geq 1$ of the wholesale price, and the aggregator receives an amount, as expressed in Equation (19).

$$\Delta wspr_u \left(P_n - 0\right) = \Delta wspr_u P_n \tag{19}$$

In other words, the system operator re-purchases the energy at a unit price $r_u WSP \geq WSP$. For regulation, where EVs should consume more than scheduled, the operator bids a concession ratio of $r_d \geq 0$ on the normal price WSP, so that the aggregator purchases the spare energy at a reduced price.

Aggregating all payments and their occurrence probabilities, the anticipated revenue for the aggregator from one regulating EV is expressed in Equation (20).

$$A_r = \Delta wsp \left(\rho_u r_u P_c - \rho_d \left(1 - r_s\right) \left(P_d - P_c\right) - P_c\right) \tag{20}$$

where A_r stands for anticipated regulation, wsp is the price at which the aggregator buys energy, r_u is the ratio of the remuneration for regulation, r_s is the discount ratio, the probability of a signal when no regulation is need is denoted by ρ_u and P_c is the charging power.

5.2. Smart EV Charging and Pricing

This subsection contains a description of the potential smart operation of EVs' charging manner for the benefit of the system. It is based on a self-determined charging decision and ancillary control through variable pricing. At a certain moment, prices may typically reflect energy generation cost only. In order to use the pricing factor for altering EV charging, the prices must be able to reflect the system operating conditions. Prices that achieve this goal must be cost-reflective of generation and network cost. EVs' driving routes affect the location of the charging load. Likewise, utilization patterns influence the quantity of available energy to be supplied to the grid under V2G operation.

A controlled and smart operation of EVs can spread EV charging load over time and re-position EV demand to reduce peak power consumption and network congestion. EVs' smart operation can further increase the feasibility by reducing the driving cost. This discussion is concentrated on prices with the aim to incentivize the system charging.

Smart EV charging operation can be performed through a direct control strategy on an EV charging process or through dynamic price-based indirect control [57]. Depending on the model, a control scheme may be applied only when an EV is parked and being charged at a particular location, or a control scheme may be implemented directly by an operator or an aggregator. The right to control may be allowed by an EV user in lieu of a certain rebate payment. Indirect control is exercised through dynamic pricing. There may be several diverse pricing extents to steer EV charging operation with the aim to lessen energy peaks. These price dimensions vary from a simple tariff to greatly distinguished pricing schemes.

A probable dimension of pricing is the diversity of price according to the time of charging. Time-differentiated prices are characterized by the utilization of the network and can aid synchronized EV charging with the system operation. Such synchronization ensures a less expensive system by steering the EV charging towards off-peak hours. Real-time pricing involves different prices at different time slots in accordance with the real operating conditions. Another dimension is the difference in location, where the EV charging process may have a substantial influence on network load. Location-based steering can diminish distribution losses and avert network expansion costs, thereby reducing inclusive cost for the ultimate consumers. Differentiation in terms of location can be illustrious as per price differences. A certain benefit of location-based pricing for EV charging originates from the mobile nature of EVs and their wide-spread energy requirements. Price differentiation based on volume is another potential dimension. The amount of energy consumed at a particular time influences the capacity required for power generation, transmission and distribution. In order to direct energy consumption towards better tariffs, applying a higher price per kilowatt-hour may persuade billing less per unit for incentivizing the total turnover. Depending on the needs of a particular system, this must be addressed via price incentives. Price dimensions may greatly help produce the desired incentives for optimal EV charging operation.

6. Results

Simulations are carried out to validate the presented operation of the proposed system model. The value of the DC link bus voltage V_{dc} is taken between $V_{dc,min}$ and $V_{dc,max}$ for shifting from one mode of operation to another. DC bus voltage is variable, and thus, the power source for EV charging varies, respectively. The reference DC bus voltage value is chosen based on a mode where the load of an EV is taken as constant and the irradiation varies gradually.

The reference values at the DC bus are kept at 250 V and 350 V, and these are selected based on the change in available solar energy from morning till evening. Varying the voltage value on the DC bus causes switching of the power source (either from the grid or PV) to recharge an EV. It can be seen in Figure 12 that the variations in the solar irradiance from morning until afternoon on a sunny day cause a corresponding change in the DC bus voltage. In a public parking lot, EVs can be assumed to be parked at different times of a day; thus, an analogous varying load of EVs can also be considered. The range of operational voltage can be split into any levels, and based on that, the charging station may be subjected to different modes of operation. In this study, the DC bus voltage is chosen while taking into account the change in solar irradiance during a day. The PV panel starts yielding power when the DC bus voltage is above its threshold value. In this way, the PV system provides the required amount of power when the DC bus voltage is 250 V, and with the further increase in the DC bus voltage, it starts producing extra power, which is then directed to the grid. In other words, the DC bus voltage does correspond with the solar irradiation and the output power, and so do the criteria for switching of the operating mode from one to another with reduced complexity.

The results for all described modes of operation under steady-state conditions at the output of the various power converters employed are obtained in this work. This is aimed at showing that the supply source for recharging an EV changes when variation in the bus voltage is seen. For instance, the EV will be fully recharged through PV power while operating under Mode I if the DC bus voltage is greater than 400 V; it will serve in Mode II when the voltage at the DC bus is in the range of 250 to 350; a combination of PV power redirection and V2G operation will be observed when the load on the grid is greater than the available supply. The outputs of the system, downstream of the power converters, under Mode II, when EV charging power demand is being met through both PV and the power grid, are shown in Figures 13 and 14. Figure 13 shows that when the DC bus voltage is 250 V, this means the corresponding power from the PV system is not enough to satisfy the EV charging power demand, and thus, it has to be supplied through both sources, i.e., the PV system and the grid. In this case, the required amount of power is drawn from the grid, and the DC bus voltage is raised to 350 V, which is analogous to the sufficient amount of power for recharging EVs; where the upper part of Figure 14 is the measured grid voltage (showing that the grid is operating at off-peak hours and enough power is available for EVs), and its lower part illustrates that the DC bus voltage has reached the required level while meeting the required power demand of EV charging.

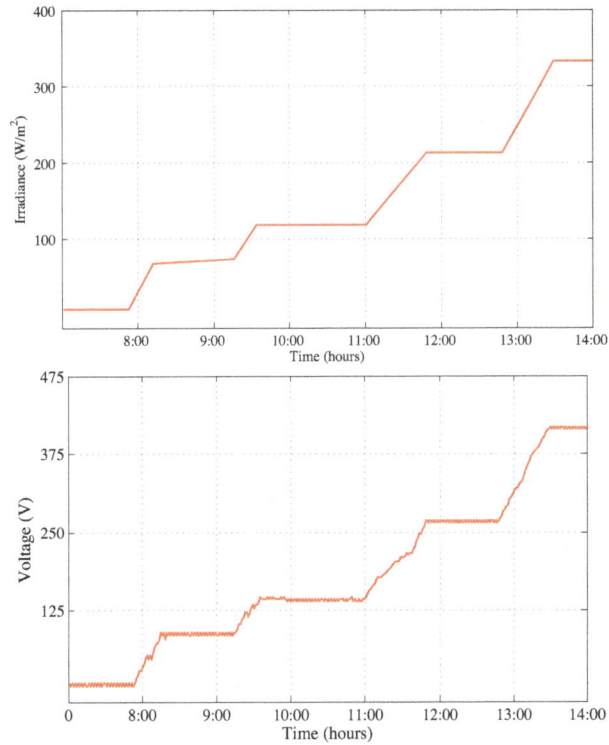

Figure 12. Changing DC bus voltage, corresponding to the PV generated power, with respect to change in solar irradiation.

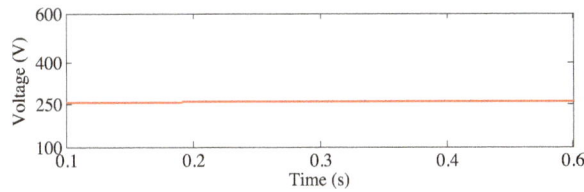

Figure 13. DC bus voltage under Mode II operation.

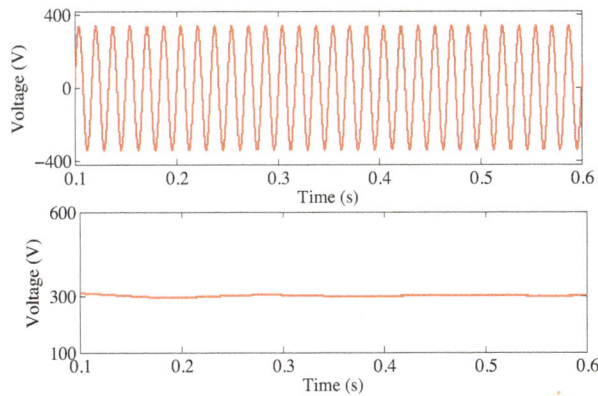

Figure 14. Grid voltage (**top**) and output voltage of the bidirectional converter in rectification mode (**bottom**).

Results including the DC bus voltage, grid supply voltage and output of the grid-connected bidirectional power converter under the system transition from Mode I to Mode II are also obtained and shown in Figures 15 and 16. It could be observed that when the bus voltage is less than 350 V, the utility grid also supplies power to charge EVs, and as soon as the DC bus voltage is sufficiently available from the PV power, then grid supply is no longer provided. Thus, in Mode I, an EV is solely recharged through PV power, and the output of the grid interfaced power converter becomes zero.

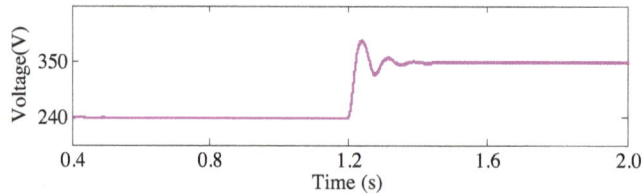

Figure 15. DC bus voltage under the transition from Mode II to Mode I.

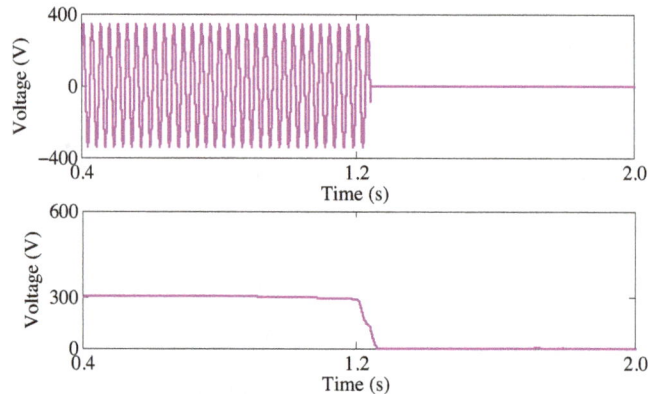

Figure 16. Grid voltage (**top**) and output voltage of the bidirectional converter in rectification mode (**bottom**) under the transition from Mode II to Mode I.

This implies that when the solar irradiation changes to the higher value of W/m^2, it causes an increase in the PV generated power, and the corresponding DC bus voltage is raised to the value of 350 V, which is termed as the transition of the mode from II to I, as shown in Figure 15. In the previous case, the charging power was being supplied by both combined sources (PV, as well as the grid), whereas, with the increase in solar irradiance, enough power becomes available through PV only. The top of Figure 16 demonstrates the grid voltage in which it can be seen that a supply is being drawn from the grid from 0.4 to 1.2 s; afterwards, no supply is required from the grid end; just to symbolize it, the supply from the grid for EV charging is reduced to zero; this means that at that moment, no power is drawn from the grid. The bottom of Figure 16 illustrates the corresponding DC bus voltage in which the rectified output of the bidirectional power converter after 1.2 s becomes zero, and it shows that no power is being rectified from the grid end; rather. the whole power demand of EVs is being met through PV power generation. For further clarification, the output of the bidirectional converter will be at some value until the power is drawn from the grid, and when no power is taken from the grid, then the output of the converter will become zero.

The grid loading and its operating condition are assessed by determining the value of the current on the secondary side of the distribution transformer. At the start, when the grid is operating at off peak hours, it delivers some amount of power required to charge EVs, as shown in the first half interval (from 0.4 till 1.2 s) of Figure 16. The charging supply is ended as soon as the load on the grid increases, and the grid interfaced side of the bidirectional power converter draws no power from the grid. In other words, zero power is taken from the grid, and it is shown in the second half interval in both the upper and lower parts of Figure 16, i.e., in this situation, zero voltage is measured at the upstream and downstream of the grid-tied bidirectional power converter.

Figure 17 shows Mode I operation under the transition from Case I to Case II when firstly all solar energy is used to recharge EVs and the DC bus voltage is close to 350 V. If it continues to increase or EV charging demand is satisfied, then the power is supplied to the grid. In this case, it can be seen that firstly, the whole EV charging demand is being supported or supplied through PV power generation only, as a sufficient amount of solar energy is available. Thus, the top of Figure 17 illustrates the DC bus voltage value, showing that from 0.4 s till 1.1 s, the DC bus voltage value is 350 V, which is sufficient (i.e., its corresponding generated power) to meet EV power demand, and no power is required to be drawn from the grid for this period. This implies that the output of the grid interfaced bidirectional converter is zero (otherwise, there would be some rectified output) from 0.4 till 1.1 s (i.e., power is being neither supplied nor drawn from the grid). Consequently, in the second half interval (from 1.1 till 2.0 s) of Figure 17, it is obvious that as soon as the value of the DC bus voltage goes above 350 V (again, here, the increase in the DC bus voltage corresponds to the increase in solar irradiation and, thus, an increase in generated power), the power management algorithm, which is based on the variation in the DC bus voltage, causes the surplus amount of power to flow towards the grid, and in this situation, the grid interfaced bidirectional converter becomes active in inversion mode and starts supplying to the grid. Therefore, the second half of the bottom of Figure 17 shows the inverted output of the respective power converter. In this way, the first and second halves of the top and bottom of Figure 17 are found to be appropriate in functioning, while validating the designed algorithm of the power flow under different operating modes.

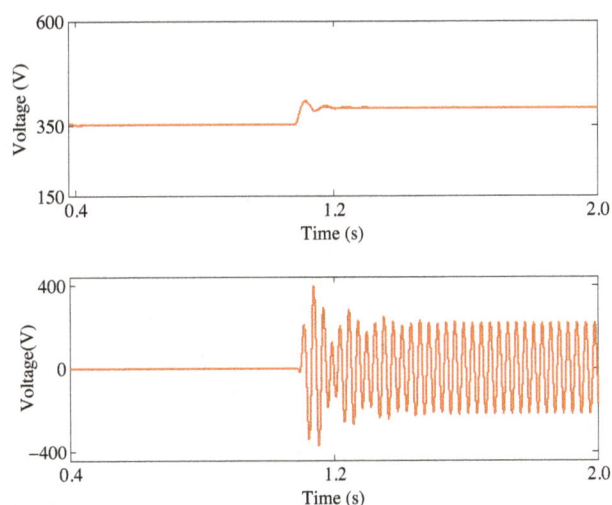

Figure 17. DC bus voltage (**top**) and supply to the grid (**bottom**).

Simulation of the charging station under V2G operation is shown in Figure 18 with reactive power compensation. It demonstrates that initially, the charging station operates at around 48 kW, and the reactive power is zero; for first quarter of a second, no reactivepower is demanded by the controller while the EV is being recharged. When time is almost 0.24 s, some reactive power injection is demanded by the controller, and the bidirectional EV charger starts providing reactive power. In V2G operation, changing the current direction affects voltage variation. It is observed that the reactive power is zero when the system is in charging mode, and the value of reactive power is also zero. At around 0.3 s, EV discharging under V2G mode is requested, and the system starts injecting power to the utility grid; the value of reactive power becomes positive. Thus, the controller ensures seamless transition among different modes of operation. The reactive power support is consider the most beneficial parameter, as it does not affect the battery performance of an EV.

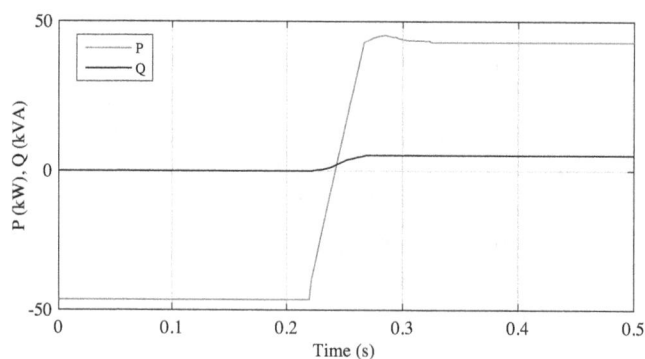

Figure 18. V2G operation for reactive power compensation: reactive power (**top**), active power (**bottom**).

7. Conclusions and Future Research Direction

A model of a PV-based fast charging station with a four-way power flow interaction is presented in this paper. The model and corresponding control strategy of each component, including the PV interfaced DC/DC boost converter, the EV interfaced DC/DC buck-boost converter and the grid-tied bidirectional AC/DC converter, are explained along with their implementation in Simulink. The employed control strategy is based on the variation in the DC link voltage that causes a change in the power flow direction, and thus, it operates in various different modes. The charging station also supports V2G operation. Simulations validate the dynamic behaviour of the DC bus voltage, battery voltage and the current. In view of the presented results, it could be safely established that the PV-based grid-connected EV charging is a promising factor to support the expected overwhelming load of EVs and the intermittent nature of renewable energy, while using EVs as an energy storage device upon availability. This paper also presents a brief model for the cost-benefit analysis of EVs' penetration in a suburban area. The theatrical analysis shows that EVs would be a profitable initiative in the transportation and energy market with significant human health and environmental benefits. Nonetheless, deployment of charging infrastructure will require substantial investments. It has been discussed that incentives and rebates can be used to encourage EV drivers' active participation in the market. A coordinated EV charging operation could increase the system's efficiency through good incentivization, as dynamic electricity pricing is a potential strategy to aid the power system. However, opportunities for price-based control of EV charging operation are still indefinite, and further research is required regarding the effects that EVs can experience in a local network. Basically, this research paper is aimed at describing the presented model and its implementation in an educational manner. For experimental verification, its further implementation is planned to be carried out using industry-oriented parameters/values. In view of the fact that the EV characteristics differentiating from the distributed resources are mobility and duality, as they can behave either as controllable loads or controllable micro-generators, the authors are working to enhance this presented model with the inclusion of the aggregation strategies for PV integrated EV charging management under the virtual power plant (VPP) concept. The authors are hoping to be able to obtain the experimental validation in the next few months and will strive to publish that work as an extended Part II of this manuscript in the same journal. As we believe that the addition of further sections in this paper will extraordinarily increase the length of this paper and, sometimes, such a lengthy paper does not appeal to potential readers, thus the extended research work with the experimental results shall be submitted for potential publication as Part II.

Acknowledgments: I, Azhar ul-haq, sincerely acknowledge that this work would have not been completed, if financial as well as technical support from University of L'Aquila under the supervision of Carlo Cecati had not been provided. Additionally, I remain grateful to Ehab El-Saadany for his outstanding supervision during my exchange studentship at ECE Department of University of Waterloo, Canada.

Author Contributions: This work has principally been done by Azhar Ul-Haq under the esteemed guidance of Carlo Cecati. During the manuscript preparation Essam A. Al-Ammar's suggestions helped improve the paper content's quality.

Conflicts of Interest: The authors declare no conflict of interest.

References

1. Deilami, S.; Masoum, A.; Moses, P.; Masoum, M. Real-time coordination of plug-in electric vehicle charging in smart grids to minimize power losses and improve voltage profile. *IEEE Trans. Smart Grid* **2011**, *2*, 456–467.
2. Kempton, W.; Tomic, J. Vehicle-to-grid power implementation: From stabilizing the grid to supporting large-scale renewable energy. *J. Power Sources* **2005**, *144*, 280–294.
3. SAE International. Charging Configurations and Ratings Terminology. Available online: http://www.sae. org/smartgrid/chargingspeeds.pdf (accessed on 3 May 2016).
4. International Electrotechnical Commission (IEC). Available online: http://webstore.iec.ch (accessed on 17 April 2016).
5. CHAdeMO Association. Available online: http://www.chademo.com (accessed on 2 March 2016).
6. Sortomme, E.; El-Sharkawi, M. Optimal charging strategies for unidirectional vehicle-to-grid. *IEEE Trans. Smart Grid* **2011**, *2*, 131–138.
7. Masoum, A.; Deilami, S.; Moses, P.; Masoum, M.; Abu-Siada, A. Smart loadmanagement of plug-in electric vehicles in distribution and residential networks with charging stations for peak shaving and loss minimisation considering voltage regulation. *IET Gener. Transm. Distrib.* **2011**, *5*, 877–888.
8. Han, S.; Han, S.; Sezaki, K. Development of an optimal vehicle-togrid aggregator for frequency regulation. *IEEE Trans. Smart Grid* **2010**, *1*, 65–72.
9. Wu, C.; Mohsenian-Rad, H.; Huang, J. Vehicle-To-Aggregator Interaction Game. *IEEE Trans. Smart Grid* **2012**, *3*, 434–442.
10. Sbordone, D.; Bertini, I.; Di Pietra, B.; Falvo, M.C.; Genovese, A.; Martirano, L. EV fast charging stations and energy storage technologies: A real implementation in the smart micro grid paradigm. *Electr. Power Syst. Res.* **2015**, *120*, 96–108.
11. Mouli, G.C.; Bauer, P.; Zeman, M. System design for a solar powered electric vehicle charging station for workplaces. *Appl. Energy* **2016**, *168*, 434–443.
12. Bianchi, F.D.; Domínguez-García, J.L.; Gomis-Bellmunt, O. Control of multi-terminal HVDC networks towards wind power integration: A review. *Renew. Sustain. Energy Rev.* **2016**, *55*, 1055–1068.
13. Arancibia, A.; Strunz, K. Modeling of an Electric Vehicle Charging Station for Fast DC Charging. In Proceedings of the 2012 IEEE International Electric Vehicle Conference (IEVC), Greenville, SC, USA, 4–8 March 2012; pp. 1–6.
14. Aggeler, D.; Canales, F.; Zelaya, H.; de la Parra; Coccia, A.; Butcher, N.; Apeldoorn, O. Ultra-Fast DC-Charge Infrastructures for EV-Mobility and Future Smart Grids. In Proceedings of the 2010 IEEE PES Innovative Smart Grid Technologies Conference Europe (ISGT Europe), Gothenberg, Sweden, 11–13 October 2010; pp. 1–8.
15. Berthold, F.; Ravey, A.; Blunier, B.; Bouquain, D.; Williamson, S.; Miraoui, A. Design and development of a smart control strategy for plug-in hybrid vehicles including vehicle-to-home functionality. *IEEE Trans. Transp. Electr.* **2015**, *1*, 168–177.
16. Zhou, X.; Wang, G.; Lukic, S.; Bhattacharya, S.; Huang, A. Multifunction Bi-Directional Battery Charger for Plug-in Hybrid Electric Vehicle Application. In Proceedings of the 2009 IEEE Energy Conversion Congress and Exposition, San Jose, CA, USA, 20–24 September 2009; pp. 3930–3936.
17. Kramer, B.; Chakraborty, S.; Kroposki, B. A Review of Plug-in Vehicles and Vehicle-to-Grid Capability. In Proceedings of the 34th Annual Conference of the IEEE Industrial Electronics Society (IECON), Orlando, FL, USA, 10–13 November 2008; pp. 2278–2283.
18. Kisacikoglu, M.; Ozpineci, B.; Tolbert, L. Effects of V2G Reactive Power Compensation on the Component Selection in an EV or PHEV Bidirectional Charger. In Proceedings of the IEEE Energy Conversion Congress and Exposition (ECCE), Atlanta, GA, USA, 12–16 September 2010; pp. 870–876.

19. Kisacikoglu, M.; Ozpineci, B.; Tolbert, L. Reactive Power Operation Analysis of a Single-Phase EV/PHEV Bidirectional Battery Charger. In Proceedings of the IEEE 8th International Conference on Power Electronics (ECCE Asia), Jeju, Korea, 30 May–3 June 2011; pp. 585–592.

20. Neumann, H.; Schar, D.; Baumgartner, F. The Potential of Photovoltaic Carports to Cover the Energy Demand of Road Passenger Transport. *Prog. Photovolt. Res. Appl.* **2012**, *20*, 639–649.

21. Locment, F.; Sechilariu, M.; Forgez, C. Electric Vehicle Charging System with PV Grid-connected Configuration. In Proceedings of the IEEE Vehicle Power and Propulsion Conference (VPPC), Lille, France, 1–3 September 2010.

22. Erickson, L.E.; Robinson, J.; Brase, G.; Cutsor, J. *Solar Powered Charging Infrastructure for Electric Vehicles: A Sustainable Development*; CRC Press: Boca Raton, FL, USA, 2016.

23. Tulpule, P.J.; Marano, V.; Yurkovich, S.; Rizzoni, G. Economic and Environmental Impacts of a PV Powered Workplace Parking Garage Charging Station. *J. Appl. Energy* **2013**, *108*, 323–332.

24. Birnie, D.P., III. Solar-to-Vehicle (S2V) Systems for Powering Commuters of the Future. *J. Power Sources* **2009**, *186*, 539–542.

25. Carlsson, F.; Johansson-Stenman, O. Costs and Benefits of Electric Vehicles a 2010 Perspective. *J. Transp. Policy* **2003**, *37*, 1–28.

26. Markel, T.; Simpson, A. Cost-Benefit Analysis of Plug-in Hybrid Electric Vehicle Technology. In Proceedings of the 22nd International Battery, Hybrid and Fuel Cell Electric Vehicle Symposium and Exhibition (EVS-22), Yokohama, Japan, 25–28 October 2006.

27. Anair, D.; Mahmassani, A. *Electric Vehicles' Global Warming Emissions and Fuel-Cost Savings Across the United States*; UCS Publications: Cambridge, MA, USA, 2012.

28. Leterme, W.; Ruelens, F.; Claessens, B.; Belmans, R. A Flexible Stochastic Optimization Method for Wind Power Balancing with PHEVs. *IEEE Trans. Smart Grid* **2014**, *5*, 1238–1245.

29. Kockar, I.I.; Papadaskalopoulos, D.; Strbac, G.; Pudjianto, D.; Galloway, S.; Burt, G. Dynamic Pricing in Highly Distributed Power Systems of the Future. In Proceedings of the 2011 IEEE Power and Energy Society General Meeting, Detroit, MI, USA, 24–28 July 2011; pp. 1–4.

30. O'Connell, N.; Wu, Q.; Ostergaard, J.; Nielsen, A.H.; Cha, S.T.; Ding, Y. Electric Vehicle (EV) Charging Management with Dynamic Distribution System Tariff. In Proceedings of the 2011 IEEE Innovative Smart Grid Technologies, Manchester, UK, 5–7 December 2011.

31. Faruqui, A.; Harris, D.; Hledik, R. Unlocking the V53 Billion Savings from Smart Meters in the EU: How Increasing the Adoption of Dynamic Tariffs Could Make or Break the Eu's Smart Grid Investment. *Energy Policy* **2009**, *38*, 6222–6231.

32. Jaske, M.; Rosenfeld, A. *Dynamic Pricing, Advanced Metering, and Demand Response in Electricity Markets*; Center for the Study of Energy Markets Working Paper Series, No. 105; University of California Energy Institute: Berkeley, CA, USA, 2002.

33. Brandstätt, C.; Brunekreeft, G.; Friedrichsen, N. Locational Signals to Reduce Network Investments in Smart Distribution Grids: What Works and What Not? *Util. Policy* **2011**, *19*, 244–254.

34. Kirkham, H. Current Measurement Methods for the Smart Grid. In Proceedings of the Power & Energy Society General Meeting, Calgary, AB, Canada, 26–30 July 2009.

35. Kouro, S.; Leon, I.; Vinnikov, D.; Franquelo, L.G. Grid-Connected Photovoltaic Systems: An Overview of Recent Research and Emerging PV Converter Technology. *IEEE Ind. Electron. Mag.* **2015**, *9*, 47–61.

36. Gholizadeh, M.; Salmasi, F.R. Estimation of state of charge, unknown nonlinearities, and state of health OFA lithium-ion battery based on a comprehensive unobservable model. *IEEE Trans. Ind. Electron.* **2014**, *61*, 1335–1344.

37. Baba, A.; Kinnosuke, I.; Teranishi, N.; Edamoto, Y.; Osamura, K.; Maruta, I.; Adachi, S. *Simultaneous Estimation of the Soc and Parameters of Batteries for HEV/EV*; SAE Technical Paper, No. 2016-01-1195; The Society of Automotive Engineers (SAE): Warrendale, PA, USA, 2016.

38. Layadi, T.M.; Champenois, G.; Mostefai, M.; Abbes, D. Lifetime estimation tool of lead acid Batteries for hybrid power sources design. *Simul. Model. Pract. Theory* **2015**, *54*, 36–48.

39. Carter, R.; Cruden, A.; Hall, P.J.; Zaher, A.S. An improved lead-acid battery pack model for use in power simulations of electric vehicles. *IEEE Trans. Energy Convers.* **2012**, *27*, 21–28.

40. Buccella, C.; Cecati, C.; Khalid, H.A.; Ul-Haq, A. On flatness-based control for series-connected VSC for voltage dip mitigation. In Proceedings of the 2014 IEEE 23rd International Symposium on Industrial Electronics (ISIE), Istanbul, Turkey, 1–4 June 2014; pp. 2637–2642.

41. Zhou, K.; Wang, D.; Zhang, B.; Wang, Y. Plug-in dual-mode-structure repetitive controller for CVCF PWM inverters. *IEEE Trans. Ind. Electron.* **2009**, *56*, 784–791.

42. Escobar, G.; Valdez, A.; Leyva-Ramos, J.; Mattavelli, P. Repetitive-based controller for a UPS inverter to compensate unbalance and harmonic distortion. *IEEE Trans. Ind. Electron.* **2007**, *54*, 504–510.

43. Zhou, K.; Low, K.S.; Wang, Y.; Luo, F.-L.; Zhang, B. Zero-phase odd-harmonic repetitive controller for a single-phase PWM inverter. *IEEE Trans. Power Electron.* **2006**, *21*, 193–201.

44. Lu, W.; Zhou, K.; Wang, D.; Cheng, M. A general parallel structure repetitive control scheme for multiphase DC–AC PWM converters. *IEEE Trans. Power Electron.* **2013**, *28*, 3980–3987.

45. Rauschenbach, H.S. *Solar Cell Array Design Handbook*; Van Nostrand Reinhold: New York, NY, USA, 1980.

46. Xiao, W.; Edwin, F.; Spagnuolo, G.; Jatskevich, J. Efficient approaches for modelling and simulating photovoltaic power systems. *IEEE J. Photovolt.* **2013**, *3*, 500–508.

47. Khalid, M.S.; Abido, M.A. A novel and accurate photovoltaic simulator based on seven-parameter model. *Electr. Power Syst. Res.* **2014**, *116*, 243–251.

48. Vimalarani, C.; Kamaraj, N. Modeling and performance analysis of the solar photovoltaic cell model using embedded matlab. *Simulation* **2015**, *91*, 217–232.

49. Nishioka, K.; Sakitani, N.; Uraoka, Y.; Fuyuki, T. Analysis of multicrystalline silicon solar cells by modified 3-Diode equivalent circuit model taking leakage current through periphery into consideration. *Sol. Energy Mater. Sol. Cells* **2007**, *91*, 1222–1227.

50. Verma, D.; Savita, N.; Shandilya, A.M.; Soubhagya, K.D. Maximum power point tracking (MPPT) techniques: Recapitulation in solar photovoltaic systems. *Renew. Sustain. Energy Rev.* **2016**, *54*, 1018–1034.

51. Eftekharnejad, S.; Vijay, V.; Gerald, T.H.; Keel, B.; Jeffrey, L. Small signal stability assessment of power systems with increased penetration of photovoltaic generation: A case study. *IEEE Trans. Sustain. Energy* **2013**, *4*, 960–967.

52. Xiao, W.; Dunford, W.G.; Palmer, P.R.; Capel, A. Regulation of photovoltaic voltage. *IEEE Trans. Ind. Electron.* **2007**, *54*, 1365–1374.

53. Nemry, F.; Leduc, G.; Munoz, A. *Plug-in Hybrid and Battery Electric Vehicles*; Institute for Prospective Technological Studies: Seville, Spain, 2010.

54. Avila, F.; Gonzalez, F. *Conexión de Vehículos a la Red Eléctrica (V2G)*; Pontificia Universidad Católica de Chile: Región Metropolitana, Chile, 2009.

55. Lizarraga de Miguel, A. *Implementation of Power Studio Scada Software*; Universidad Politecnica de Cataluna: Barcelona, Spain, 2011.

56. Perez, L.; Sunyer, L.; Kunzli, N. Estimating the health and economic benefits associated with reducing air pollution in the barcelona metropolitan area. *Gac. Sanit.* **2009**, *23*, 287–294.

57. Dallinger, D.; Wietschel, M. *Grid Integration of Intermittent Renewable Energy Sources Using Price-Responsive Plug-in Electric Vehicles*; Working Paper Sustainability and Innovation No. S 7/2011, FhG-ISI; Fraunhofer Institute for Systems and Innovation Research: Karlsruhe, Germany, 2011.

PERMISSIONS

LIST OF CONTRIBUTORS

Li Zhai, Liwen Lin and Chao Song
National Engineering Laboratory for Electric Vehicle,
Beijing Institute of Technology, Beijing 100081, China
Co-Innovation Center of Electric Vehicles in Beijing,
Beijing Institute of Technology, Beijing 100081, China

Xinyu Zhang
Beijing Institute of Radio Metrology and Measurement,
Beijing 100854, China

Hae Jin Kang
SAMOO Architects and Engineers, Seoul 138-240,
Korea

Johannes Schalk
MTU Friedrichshafen GmbH, Maybachplatz 1, 88045
Friedrichshafen, Germany

Harald Aschemann
Chair of Mechatronics, Rostock University, Justus-
von-Liebig Weg 6, 18059 Rostock, Germany

**Tariq Abdulsalam Khamlaj and Markus Peer
Rumpfkeil**
300 College Park Kettering Labs, University of Dayton,
Dayton, OH 45469-0238, USA

Dong Cheng, Datong Qin and Qingbo Xie
State Key Laboratory of Mechanical Transmissions
& School of Automotive Engineering, Chongqing
University, Chongqing 400044, China

Yi Zhang
Department of Mechanical Engineering, University of
Michigan-Dearborn, Dearborn, MI 48128, USA

Zhenzhen Lei and Yonggang Liu
State Key Laboratory of Mechanical Transmissions
& School of Automotive Engineering, Chongqing
University, Chongqing 400044, China
Key Laboratory of Advanced Manufacture Technology
for Automobile Parts, Ministry of Education, Chongqing
University of Technology, Chongqing 400054, China

Wenxin Huang
Department of Electrical Engineering, Nanjing
University of Aeronautics and Astronautics, Nanjing
211106, China

Chris Mi
Department of Electrical and Computer Engineering,
San Diego State University, San Diego, CA 92182, USA

Bing Xia
Department of Electrical and Computer Engineering,
San Diego State University, San Diego, CA 92182, USA
Department of Electrical and Computer Engineering,
University of California San Diego, San Diego, CA
92093, USA

Yunlong Shang
Department of Electrical and Computer Engineering,
San Diego State University, San Diego, CA 92182, USA
School of Control Science and Engineering, Shandong
University, Jinan 250061, China

Jufeng Yang
Department of Electrical Engineering, Nanjing
University of Aeronautics and Astronautics, Nanjing
211106, China
Department of Electrical and Computer Engineering,
San Diego State University, San Diego, CA 92182, USA

**Xiancheng Zheng, Husan Ali, Xiaohua Wu, Haider
Zaman and Shahbaz Khan**
Department of Electrical Engineering, School of
Automation, Northwestern Polytechnical University,
Xi'an 710000, China

Cihan Turhan
Mechanical Engineering, Izmir Institute of Technology,
Gulbahce Campus, Urla, 35430 Izmir, Turkey

Silvio Simani
Dipartimento di Ingegneria, Università degli Studi di
Ferrara. Via Saragat 1E, 44122 Ferrara (FE), Italy

Ivan Zajic
Control Theory and Applications Centre, Coventry
University, Coventry CV1 5FB, UK;

Gulden Gokcen Akkurt
Energy Engineering Program, Izmir Institute of
Technology, Gulbahce Campus, Urla, 35430 Izmir,
Turkey

**Thomas Levermore, M. Necip Sahinkaya and Yahya
Zweiri**
Faculty of Science, Engineering and Computing,
Kingston University London, London SW15 3DW, UK

Ben Neaves
Jaguar Land Rover Limited, Gaydon CV35 0RR, UK

Azhar Ul-Haq
DSIM, University of L'Aquila, 67100 L'Aquila, Italy
College of E&ME, National University of Science and
Technology (NUST), H-12 Islamabad, Pakistan

Essam A. Al-Ammar
Department of Electrical Engineering, King Saud
University, Riyadh 12372, Saudi Arabia

Carlo Cecati
DSIM, University of L'Aquila, 67100 L'Aquila, Italy

Index